# Urban Expansion, Land Cover and Soil Ecosystem Services

T0187969

More than half of the world population now lives in cities, and urban expansion continues as rural people move to cities. This results in the loss of land for other purposes, particularly soil for agriculture and drainage. This book presents a review of current knowledge of the extension and projected expansion of urban areas at a global scale.

Focusing on the impact of the process of 'land take' on soil resources and the ecosystem services that they provide, it describes approaches and methodologies for detecting and measuring urban areas, based mainly on remote sensing, together with a review of models and projected data on urban expansion. The most innovative aspect includes an analysis of the drivers and especially the impacts of soil sealing and land take on ecosystem services, including agriculture and food security, biodiversity, hydrology, climate and landscape.

Case studies of cities from Europe, China and Latin America are included. The aim is not only to present and analyse this important environmental challenge, but also to propose and discuss solutions for the limitation, mitigation and compensation of this process.

**Ciro Gardi** works in the Animal and Plant Health Unit of the European Food Safety Authority, Parma, Italy. Previously, he was a Senior Scientist at the Land Resource Management Unit of the Joint Research Center of the European Commission and Professor of Soil Science at the University of Parma. He has served as an independent expert and consultant for the European Commission, World Bank, OECD and several NGOs and is currently a member of the Scientific Advisory Committee of the Global Soil Biodiversity Initiative, representing it in the Global Soil Partnership (FAO).

# Urban Expansion, Land Cover and Soil Ecosystem Services

Edited by Ciro Gardi

Routledge
Taylor & Francis Group
LONDON AND NEW YORK

from Routledge

First published 2017 by Routledge

2 Park Square, Milton Park, Abingdon, Oxfordshire OX14 4RN
52 Vanderbilt Avenue, New York, NY 10017

*Routledge is an imprint of the Taylor & Francis Group, an informa business*

First issued in paperback 2018

*British Library Cataloguing-in-Publication Data*
A catalogue record for this book is available from the British Library

*Library of Congress Cataloging in Publication Data*
Names: Gardi, Ciro, editor.
Title: Urban expansion, land cover and soil ecosystem services /
    edited by Ciro Gardi.
Description: London ; Boston : Routledge, 2017. | Includes
    bibliographical references and index.
Identifiers: LCCN 2016042613| ISBN 9781138885097 (hbk) |
    ISBN 9781315715674 (ebk)
Subjects: LCSH: Urban ecology (Biology) | Urbanization—
    Environmental aspects. | Urbanization—Case studies. |
    Soil ecology.
Classification: LCC QH541.5.C6 U72 2017 | DDC 577.5/6—dc23
LC record available at https://lccn.loc.gov/2016042613

ISBN: 978-1-138-88509-7 (hbk)
ISBN: 978-0-367-17279-4 (pbk)

Typeset in Bembo
by Swales & Willis Ltd, Exeter, Devon, UK

To Sofia and Maria Isabella, to stimulate them struggling for a better world.

# Contents

# Figures

# Tables

# Contributors

**Adrian Guillermo Aguilar** is a Senior Researcher at the Institute of Geography and Professor for Urban Geography in the graduate studies of the National Autonomous University of Mexico (UNAM). He conducts research on the areas of urban-regional analysis, medium-sized cities and metropolitan development, globalization and megacities, and urban sustainability. He has co-authored 11 books and numerous articles and chapters in different books. He received a PhD from the University of London, UK.

**Ece Aksoy** is an interdisciplinary researcher with a BSc in Urban and Regional Planning, an MSc on GIS and a PhD in Agricultural and Soil Science. She worked as a scientific officer in the European Commission, Joint Research Centre (JRC), Institute for Environment and Sustainability (IES) from 2011 to 2014, where she strengthened her experiences on EU agricultural and environmental policies and agricultural technologies in the EU. She has worked on numerous applied research projects in Turkey on the environment, land management, agricultural information systems and soil studies at different scales (farm, watershed, urban–rural). She also worked as an international consultant of FAO-UN and developed a National Geospatial Soil Fertility and Soil Organic Carbon Information System for Turkey. She has worked in the European Topic Centre on Urban, Land and Soil Systems (ETC-ULS of EEA) as a senior expert since 2014, dealing with digital soil mapping, urban sustainability studies, quantitative assessment of land degradation, potentials of soil biodiversity functions, impact analysis on land resource productivity and providing support to biodiversity strategy targets – CAP greening actions by assessing multi-mono functionality of the soils.

**Jean-Philippe Aurambout** holds an agronomy engineering degree from ENSAIA, France (2000), an MSc (2002) and PhD (2005) in Natural Resources and Environmental Sciences from the University of Illinois at Urbana-Champaign, USA. Between 2005 and 2013 he worked as a Senior Research Scientist with the Spatial Information Sciences group of the Victorian Department of Primary Industries in Melbourne, Australia. He is currently working as s Scientific Officer at the European Commission, Joint Research Centre in Italy. His research interests focus on the modelling of complex dynamic systems at the spatial scale.

**Claudia Baranzelli** is from Milan, Italy, where she graduated in Environmental Engineering in 2005, with particular focus on Environmental Modelling and Territorial Planning, and completed a PhD in Urban, Regional and Environmental Planning in 2010. From 2006 to 2010 she worked at the Politecnico di Milano as researcher and teaching assistant, with particular focus on urban and regional planning, and assessment methods applied to buildings and urban areas. In 2010 she joined the Joint Research Centre and since then she has been working on the development and application of the LUISA Territorial Modelling Platform in several fields, mainly related to energy, agriculture and the urban environment.

**Ana Luisa Barbosa** graduated in Geography (2005) and has a Master of Geographic Information System and Science (2009). With over eight years of experience working for European Institutions (European Topic Centre – European Environmental Agency and the Joint Research Centre at the European Commission), she participated in several initiatives to support the impact assessments studies of EU regional and environmental policies. As a scientific officer at the JRC (2011–2015) she has been involved in the development of the LUISA Territorial Modelling Platform, a platform used for the ex-ante evaluation of EC policies that have a direct territorial impact. She specializes in scenarios development, land use modelling and the development of indicators for impact assessment.

**Ricardo Barranco** graduated in 2008 in Environmental Engineering at University of Algarve, Portugal, doing his final Master's thesis at Wageningen University and Research Centre, the Netherlands, focusing on the use of Geographical Information Systems (GIS) on urban areas. From 2008 to 2012 he worked in Genova, Italy at IREN's Environmental and Technological Services department. From 2012 to the present he has been part of the LUISA Territorial Modelling Platform project. Working mainly on urban indicators development, more recently he has been collaborating on energy-related projects and data visualization techniques.

**Filipe Batista e Silva** graduated in 2006 in Geography, and was later awarded a Master's in Geographic Information Systems and Spatial Planning. He worked in the Department of Geography at the University of Porto in Portugal until 2010, lecturing and contributing to urban and regional planning projects. He then joined the Joint Research Centre of the European Commission and has since been involved in the development and application of the LUISA Territorial Modelling Platform. He is particularly interested in population and activity mapping, demography–economy–land use interactions, regional modelling, spatial analysis and downscaling methods. He is finalizing his PhD at the VU University Amsterdam, the Netherlands.

**Carolina Perpiña Castillo** received her PhD in Cartography and Geodesy Engineering in 2012 from the Universitat Politecnica de Valencia, Spain, as applied to biomass logistics and transport optimization using spatial techniques. As a researcher, she worked at the Institute of Energy Engineering from 2006 to 2012 (funded by the Spanish Ministry of Science and Innovation), specifically in the area of renewable energies. Additionally, she earned an MSc in Land Use Planning and Transport at the same university. Since 2012, Carolina has been working at EC-JRC (Ispra, Italy), focused on the development of the LUISA Modelling Platform, especially on the impact assessment of food, feed and energy production in Europe. Other contributions to the modelling platform have included various technical improvements and the integration of an economic evaluation of agricultural land.

**Andrea Colantoni** is a Researcher at the Department of Agriculture and Forestry Science, Tuscia University (Viterbo, Italy). He has a PhD in Agriculture Mechanization. He is a member of: the scientific committee of PhD Engineering for Energy and Environment at Tuscia University; the Italian Association of Agricultural Engineers, which is part of the European Society of Agricultural Engineers (EurAgEng); the International Commission of Agricultural Engineering (CIGR); and the Italian Association of Scientific Agricultural Societies (AISSA). His scientific activity is mainly focused on: the mechanization of harvesting of tree crops (hazelnut), industrial crops (tomato) and fibre plants; safety and quality in the agro-food and agro-industrial chains; ergonomics and analysis of working loads in agro-forestry; safety and risk assessment in agro-industrial workplaces; and the use of renewable energy sources in agriculture.

**René R. Colditz** has been a Remote Sensing Specialist at the National Commission for the Knowledge and Use of Biodiversity (CONABIO), Mexico, since 2007. His main fields of research are land cover classification and change detection, time series generation and analysis, and multi-resolution analysis using medium to coarse (30–1000 m) spatial resolution satellite data. In 2008 he received his PhD in Geography and Remote Sensing from the University of Würzburg, Germany, in cooperation with the German Aerospace Center (DLR).

**Luca Congedo** has a Master's degree in Environmental Engineering. His major fields of specialization are remote sensing and GIS analysis. He has conducted several studies related to the semi-automatic classification of remote sensing images such as the land cover classification of Dar es Salaam (Tanzania) in the context of the ACC Dar Project, and the mapping of asbestos-cement roofing in the Latium Region (Italy) using MIVIS hyper-spectral images. He has been involved in the verification and enhancement of the Copernicus High Resolution Layers 2012, and cooperated with the Italian National Institute for Environmental Protection and Research (ISPRA) for the report on soil consumption monitoring.

**María Isabel Cruz López** works as Head of the Remote Sensing Division at the National Commission for the Knowledge and Use of Biodiversity (CONABIO), Mexico. Since 1998 she has coordinated and developed projects on ecosystem monitoring and early warning systems using remote sensing data. Her main research fields are forest fires, land cover changes and vegetation monitoring with remote sensing and geographic information system (GIS). She holds a Master's degree in Geography from the National Autonomous University of Mexico (UNAM).

**Marie Cugny-Seguin** is Project Manager for 'Urban and Territorial Issues' at the European Environment Agency, Copenhagen, Denmark. She has professional experience in environmental reporting at city, regional and national level. Her interests are focused on the way to develop more resource-efficient cities, the analysis of green infrastructure inside and around cities, and the integrative analysis of cities taking into account the complexity of urban systems and the diversity of cities.

**José Manuel Dávila Rosas** worked for seven years in the private sector before joining the National Commission for the Knowledge and Use of Biodiversity (CONABIO), Mexico, as Head of Geographical Information Systems in 2012. His main interests are development of GIS software, automation of geo-data processing, digital cartography and geographic analysis for biodiversity in urban areas. He graduated as Geographer from Autonomous University of the State of Mexico.

**Daniele Ehrlich** holds a PhD from the University California at Santa Barbara, USA. He is a senior staff member of the Joint Research Centre (JRC), European Commission, Ispra, Italy. He has over 25 years of experience in remote sensing and GIS applied to a variety of disciplines including crop area estimation, tropical forest mapping, crisis management with a focus on damage assessment and humanitarian assistance. His current research focuses on quantifying the extent and the dynamics of settlements using high-resolution satellite imagery. He uses derived settlement information for a systematic analysis of the global built-up environment, for population estimations and for generating physical exposure databases for global disaster risk assessments.

**Stefan Fina**, PhD in Geography, is a land use planning researcher at the Institute of Spatial and Regional Planning at the University of Stuttgart, Germany. He has an international research focus on issues of land use change and urban sprawl, both in academia and planning practice. His dissertation and a range of research articles are about indicators and quantification methods, as well as topics of demographic change, infrastructure planning, environmental impacts, and segregation studies. As a planner he worked for a number of years in transport and as an integrated planning analyst for a council in Auckland, New Zealand, and as a private consultant in the development of planning support systems and land use modelling. His teaching is mainly about planning tools and assessment methods in national and international Master's programmes at the University of Stuttgart. Stefan currently extends his research interests into areas of health-related urban development and participatory planning approaches.

**Aneta Jadwiga Florczyk** received a PhD in Computer Science from the University of Zaragoza, Zaragoza, Spain, in 2012. In November 2007, she started to collaborate with the Advanced Information Systems Laboratory (IAAA), University of Zaragoza, Spain. Currently, she works as a Scientific Project Officer (Post-doctoral Grant Holder) with the Joint Research Centre (JRC), European Commission, Ispra, Italy. Her research interests include remote sensing, GIScience, image processing, machine learning, statistics, high-performance computing and big data analytics.

**Ciro Gardi** is currently working in the Plant Health team of the European Food Safety Authority, Parma, Italy. Agronomist, soil scientist/ecologist, with a PhD in Crop Science, he has a deep knowledge of agricultural systems and of the interactions between land management, soil quality and ecosystem service provision. He is actively involved in all aspects related to soil degradation, from research to policy support and awareness raising. He taught soil science at the University of Parma, Italy, and in international Master's and other courses. His main research activities are on the relationships between land use, agronomic management and soil quality, with particular emphasis on soil degradation processes and their relationships with soil organic matter and soil biodiversity. He is experienced in GIS, remote sensing and soil survey research, which he has carried out in Italy and abroad. He has been a consultant and served as an independent expert for the European Commission, World Bank, OECD and several NGOs and he is currently a member of the Global Soil Partnership (FAO).

**Mirko Gregor** holds a Diploma degree in Applied Physical Geography (2003) from the University of Trier, Germany, with a particular focus on geomorphology, remote sensing and climatology. After working as a GIS and remote sensing expert for the Luxembourg-based private company GIM (2003–2008), he became a project manager at space4environment (formerly GeoVille Environmental Services), Luxembourg. Since then, he has been actively involved and leading space4environment's activities on land resource efficiency and urban sustainability assessments in the framework of the European Topic Centre on Urban, Land and Soil Systems (ETC-ULS). He is also working on urban biodiversity topics in the context of the European Topic Centre on Biodiversity (ETC-BD) and as Technical Project Manager for the ESA project Earth Observation in support of the City Biodiversity Index.

**Fernanda Guerrieri** holds an MSc in Agronomy with honours from the University of Bologna, Italy, and a Diploma in Watershed Management from the International Institute for Aerial Survey and Earth Science (ITC), Enschede, the Netherlands, with distinctions. Ms Guerrieri started her career in 1982 at the University of Bologna, Italy, working on soil conservation and land evaluation/watershed management. She joined the Food and Agriculture Organization of the United Nations (FAO) in 1988 as Associate

Project Operations Officer on the Latin American and Caribbean Desk, Forestry Department, FAO Headquarters. From 1990 to 1995, she served as Programme Officer/Deputy FAO Representative in Equatorial Guinea, and then, from 1992 to 1995, in Côte d'Ivoire and Mozambique. In 1995, she returned to FAO Headquarters as Project Analyst, Technical Cooperation Programme (TCP). In 1998, she was appointed FAO Representative in Vietnam. She returned to FAO Headquarters in 2002 as Chief, Emergency and Rehabilitation Operations Service. In August 2008, she was appointed Deputy Regional Representative, Subregional Coordinator for Central and Eastern Europe in Budapest, Hungary. From May 2010 until January 2013, she served as Assistant Director-General and Regional Representative for Europe and Central Asia, in Budapest, Hungary. From January 2013 until July 2015, she served as Assistant Director-General/Directeur de Cabinet, Office of the Director-General in Rome.

**Chris Jacobs-Crisioni** is from Amsterdam, the Netherlands, where he graduated in Urban Planning at the University of Amsterdam and is currently finalizing a PhD in Spatial Economics at VU University Amsterdam. His main research interest is in the expansion processes of transport networks and the subsequent land use and sustainability effects of these processes. He has published in reputed international scientific journals such as *Environment and Planning A* and the *Journal of Geographical Systems*. Between 1999 and 2013, he worked for various municipalities, transport consultancy agency Goudappel Coffeng and VU University Amsterdam. In those years he learned many facets of geographic data gathering, geographical information systems and spatial decision support systems. In 2013, he joined the Joint Research Centre's LUISA team, where he is occupied with the further development of the LUISA model.

**Andreea Julea** received a PhD in Electronics, Telecommunication and Computer Science from the Polytechnic University of Bucharest, Romania, and Grenoble University (Savoy University, Annecy), France, in 2011. Between 2005 and 2013, she was a scientific researcher with the Institute of Space Science, Magurele-Bucharest, Romania. Since 2013, she has been Scientific Project Officer (Post-doctoral Grant Holder) with the Joint Research Centre (JRC), European Commission, Ispra, Italy. Her main research interests are in the areas of knowledge discovery in databases, data mining, image processing and remote sensing applications.

**Thomas Kemper** received a PhD in Geosciences from the University of Trier, Germany, in 2003. He is a Scientific Officer with the Joint Research Centre (JRC), European Commission, Ispra, Italy. From 2004 to 2007, he worked with the German Aerospace Center (DLR), Cologne, Germany, where he helped in setting up the Center for Satellite-Based Crisis Information (ZKI), which provides rapid mapping information after natural disasters. Since 2007, he has been working on the analysis of human settlements, in particular informal settlements such as slums and IDP/refugee dwellings.

**Mert Kompil** is Scientific/Technical Project Officer in the LUISA Modelling Platform group at the Joint Research Centre, Italy. He worked as postdoctoral researcher on transport policy analysis and modelling at the Institute for Prospective Studies of the Joint Research Centre based in Spain. Before that, he worked as a research and teaching assistant at the Department of City and Regional Planning under the Izmir Institute of Technology in Turkey. He completed his PhD in the same department with a specialization on travel demand analysis. His main areas of research include spatial interaction models, trip distribution models, travel demand analysis, accessibility, land use modelling and territorial impact assessment.

**Rattan Lal**, PhD, is a Distinguished University Professor of Soil Science at the Ohio State University, USA, and was Senior Research Fellow with the University of Sydney, Australia (1968–1969), and Soil Physicist at IITA, Ibadan, Nigeria (1969–1987). His research focus is on climate-resilient agriculture, soil carbon sequestration, sustainable intensification, enhancing the use efficiency of agroecosystems, and the sustainable management of soils of the tropics. He was included in Thomson Reuters 2014 list of World's Most Influential Scientific Minds (2002–2013). He was President the Soil Science Society of America (2005–2007), is President Elect of the International Union of Soil Sciences, Vienna, Austria (2014–), and is Chair of the Advisory Committee to UNU-FLORES, Dresden, Germany (2014–). He has mentored 105 graduate students, 54 postdoctoral researchers and 145 visiting scholars. He has authored/co-authored more than 1940 research publications including 761 refereed journal articles and 421 book chapters, and has written 16 and edited/co-edited 58 books.

**Carlo Lavalle**, has over 25 years of experience in applied geophysics and integrated modelling. Since 1990 he has been with the Joint Research Centre of the European Commission, with involvement in dossiers related to environment, energy, urban and regional development. He is coordinating the development of the LUISA Territorial Modelling Platform.

**Manuel Löhnertz** holds a Diploma degree in Applied Environmental Science (2006) from the University of Trier, Germany, with a special focus on soil science, physical geography and remote sensing. From 2006 to 2008 he worked as GIS and Remote Sensing expert for the private company GIM in Luxembourg. Since 2009 he has been working for space4environment, Luxembourg (formerly GeoVille Environmental Services) as a GIS and remote sensing expert. Having a strong background in remote sensing applications, application development and database management, his current focus is spatial data processing and control in the framework of space4environment's activities for the European Topic Centre on Urban, Land and Soil Systems (ETC-ULS) and Biodiversity (ETC-BD).

**Klaus Lorenz**, PhD, is Assistant Director/Research Scientist at the Carbon Management and Sequestration Center, Ohio State University, USA. From 2011 to 2013, he was Chief Soil Scientist/Research Fellow at the IASS Institute for Advanced Sustainability Studies in Potsdam, Germany.

He studied biology at University of Freiburg, Germany, and obtained his PhD in Agricultural Sciences from University of Hohenheim, Germany. His research focuses on agricultural, forest and urban soil use and management to enhance soil organic carbon sequestration for climate change adaptation and mitigation. He has written the book *Carbon Sequestration in Forests Ecosystems* (co-author Rattan Lal), authored/co-authored 33 peer-reviewed journal articles and 18 book chapters.

**Geertrui Louwagie** graduated as Bio-Engineer in Land and Forest Management, Soil and Water Management at Ghent University, Belgium (1995), and has a PhD in Science (earth sciences) from Ghent University (2004). She started her career as an academic assistant (in soil science and land evaluation) at Ghent University, where she also did research on land evaluation and palaeo-environment reconstruction and provided policy support in the area of archaeological site management. In 2005 she joined University College Dublin as a postdoctoral researcher in an EU research consortium, developing methods for the evaluation of agri-environmental schemes in different EU contexts. From 2008 to 2011, she was a postdoctoral researcher at the European Commission's Joint Research Centre (Institute for Prospective Technological Studies, Spain) and coordinated policy support projects on sustainable agriculture and soil conservation in the EU, as well as on rural development in the Western Balkans. Since 2012, she has been the Project Manager 'Soil Assessments and Reporting' at the European Environment Agency, Denmark.

**Luca Montanarella** has been working as a scientific officer in the European Commission since 1992. He has been leading the Soil Data and Information Systems (SOIL Action) activities of the Joint Research Centre in support of the EU Thematic Strategy for Soil Protection and numerous other soil-related policies, like the Common Agricultural Policy (CAP), the UNCCD, UNFCCC, CBD, among others. He is responsible for the European Soil Data Centre (ESDAC), the European Soil Information System (EUSIS) and the European Soil Bureau Network (ESBN). More recently he has been in charge of supporting the establishment of the Global Soil Partnership at FAO. He is currently the Chair of the Intergovernmental Technical Panel on Soils (ITPS) of the GSP and the Co-chair of the Intergovernmental Platform for Biodiversity and Ecosystem Services (IPBES) Land Degradation and Restoration Assessment (LDRA). He has more than 300 publications, books and reports, and has received numerous awards and memberships.

**Michele Munafò** has been at ISPRA since 2000. He is currently head of the Environmental Monitoring and Pressures Database Unit and of the Italian Land Take monitoring group. He has a PhD in Urban Planning. He is temporary Professor in Regional and Urban Planning, Strategic Environmental Assessment and Geographical Information Systems at the University of Rome Sapienza, Italy. He is a member of the scientific committee for PhDs in Landscape and Environment. He is responsible for enhancement and validation activities for the national Copernicus Land Monitoring High Resolution

Layers, National Reference Centre of the European Environment Information and Observation Network (European Environment Agency), involved in the production of Italian Corine Land Cover, and project manager for several projects regarding land monitoring and environmental information.

**Erika Orlitova** graduated in Automatic Control Systems at the Czech Technical University in Prague and postgraduate study in GIS Application at Technical University of Ostrava. She is a senior data manager experienced in spatial analysis, thematic mapping, satellite data processing and interpretation and geo-information assessment, taking part in various domestic and international projects related to spatial data processing. She was involved in the MARSOP3 project supporting the JRC-AGRI4CAST and JRC-FOOD-SEC and has experience working with EEA since 2004 within the ETC-LUSI framework on various data related tasks.

**Mitchell Pavao-Zuckerman** is an Assistant Professor in the Department of Environmental Science and Technology and Institute for Sustainability in the Built Environment at the University of Maryland, College Park, MD, USA. Mitch is an ecosystem ecologist focusing on the responses of ecosystems to urbanization and land-use change. He has research focuses on: (1) the ecosystem services and biogeochemical cycling of green infrastructure; (2) the influence of urbanization on the ecohydrology and physiological ecology of soils and plants; (3) the resilience of urban socio-ecological systems to climate and land-use change, and (4) interdisciplinary approaches to studying urban ecosystems, including connections to environmental design, planning, policy, and governance.

**Luigi Perini**, at CREA (the Italian Council for Research in Agriculture), Rome, Italy, since 2004, is currently Director of the Research Institute of Climatology and Meteorology applied to Agriculture (CREA-CMA). Graduating in agronomy, he obtained subsequent specializations in meteorology and agrometeorological modelling. He was involved in various scientific projects and institutional collaborations. He participated to found the National Agrometeorological Service for the Italian Ministry of Agriculture and contributed professionally to constitute several regional agrometeorological services. He collaborated with the National Commission to combat drought and desertification (CNLSD) in order to identify the vulnerabilities of the Italian territory. He is a member of the technical group on long-term forecast for the Italian Department of Civil Protection and he is the Italian component of the Agricultural Commission for the World Meteorological Organization (ONU-WMO). Currently he is involved in experimentation in land monitoring techniques by unmanned aerial vehicles (UAV).

**Martino Pesaresi** graduated in Town and Regional Planning, University of Venice, Italy, in 1992. He pursued research activities on remote sensing, spatial statistics and urban analysis with the Centre d'Analyse et de Mathématique Sociales of the Ecole des Haute Etudes en Sciences Sociales (EHESS), Paris, France, in 1991–1992, with the Laboratoire d'Informatique Appliquée,

ORSTOM, Paris, France, in 1992–1993 and for the Department of Urban and Regional Planning, University of Venice until 1995. From 1997 to 2000, he was on a postdoctoral research contract on urban analysis using satellite data with the EC Joint Research Centre, Space Applications Institute. Since 2004, he has been working with the Joint Research Centre (JRC), European Commission, Ispra, Italy, contributing to several programmes dealing with the use of space technologies for automatic image information retrieval and decision support systems. Since 2014, he has been scientifically responsible for the GLOB-HS and E-URBAN projects exploiting remote sensing technologies for fine-scale systematic analysis of human settlements, respectively, at the global and European levels.

**Alberto Pistocchi**, PhD, is a chartered environmental engineer and land planner and an Associate Professor of Land and Urban Planning at the Joint Research Center of the European Commission, Ispra, Italy. He is a scientific officer and project leader at the European Commission DG JRC since 2013. Since 1997 he has been working as a professional hydraulic engineer and land planner, focusing on river basin management, flood risk management, groundwater resources, water supply and distribution networks, environmental assessment and project appraisal. His scientific interests concern hydrological and water quality modelling, and spatial decision support systems. He has authored/co-authored several scientific publications and the book *GIS Based Chemical Fate Modeling: Principles and Applications* (2014).

**Richard V. Pouyat** received his PhD in Ecology from Rutgers University, USA, in 1992 and an MSc in Forest Soils and BSc in Forest Biology at the College of Environmental Science and Forestry, State University of New York, USA, in 1983 and 1980, respectively. He is the National Program Lead for Air and Soil Quality Research for Research and Development at the Washington DC headquarters of the United States Forest Service. He is currently on a detail to the White House Office of Science, Technology, and Policy (OSTP). He is an original co-principal investigator of the Baltimore Ecosystem Study – a Long Term Ecological Research site funded by the National Science Foundation.

**Gundula Prokop** (MSc, MBA) is a senior expert and project manager at the Austrian Environment Agency (Umweltbundesamt). Since 1996 she has been with the Austrian Federal Environment Agency, mainly providing consultancy to the European Environment Agency and working as co-ordinator or project partner in EU Research and Territorial Co-operation Projects. Furthermore, she has been working under contracts for the European Commission, EUROSTAT, the World Bank and the Joint Research Centre. She has been in charge of developing an indicator for contaminated sites for the EEA (CSI015, LSI003), indicators for land take and soil sealing for the European Commission and land use indicators for regional surveys on behalf of EUROSTAT. She is an active networker and member of several national and international working groups, including the

National Reference Centre for Land Use (EEA), member of the UNECE-ITU Technical Advisory Board for the Sustainable Development Goal 11 'sustainable cities and communities'. Gundula is the author and editor of national and international publications related to contaminated land management, brownfield recycling and land take.

**Gianluca Renzi**, having graduated in Agricultural Science and Technology, obtained a subsequent specialization PhD in Sciences and Technologies for Forest and Environmental Management. He has worked at CREA (the Italian Council for Research in Agriculture), Rome, Italy, since 2009.

**Rainer A. Ressl** is the General Director of Geomatics at the National Commission for the Knowledge and Use of Biodiversity (CONABIO), Mexico. His main research interests include satellite remote sensing for ecosystem monitoring, development of operational monitoring systems based on remote sensing and GIS data, e.g. for forest fire, land cover, and ocean products and GIS applications related to biodiversity. He graduated with a PhD in Geography and Remote Sensing from the Ludwig Maximilians University of Munich, Germany in cooperation with the German Aerospace Center (DLR) in 1999.

**Stefano Salata** has a PhD in Territorial Government and Urban Design XXVI Cycle at the Department of Architecture and Urban Studies (DAStU), Politecnico di Milano. He is a contract Professor of Urban and Territorial Analysis at Politecnico di Milano and works on the project LIFE SAM4CP with the Interuniversity Department of Regional and Urban Studies and Planning, Politecnico di Torino. Graduated in Urban Planning and Territorial Policies he is involved in research activities on land take analysis and its environmental effects. He is a member of Ecosystem Service Partnership (ESP) and a member of the National Research Centre on Land Take – ITALY (CRCS). He is the co-editor of CRCS' Italian National Reports on land take (2010, 2012, 2014 and 2016). He is also the co-author of the book *L'insostenibile consumo di suolo* (2013) and author of many international publications on land take and its related effect on ecosystem services.

**Luca Salvati** has two degrees (Ecology: 2000; Demography and Social Sciences: 2004), a Master's degree in Economic Statistics, a specialization degree in Geography and Environment and a PhD in Economic Geography. He is adjunct Professor of Cartography and GIS, Multivariate Statistics, and Strategic Environmental Assessment at Third University of Rome. He collaborates with University of Rome 'La Sapienza' in the field of urban and rural geography. He is currently permanent researcher at the Italian Council of Agricultural Research and Economics (CREA). He has also collaborated, as a research fellow, with various research institutions in the framework of both national and European projects on the following themes: desertification, sustainable agriculture, land use, climate change, urban sprawl and polycentric development at the regional scale, with reference to the

Mediterranean basin. He is the author of more than 300 scientific articles in English, 20 books, essays and cartographical atlases.

**Christoph Schröder** graduated in Geography and is a GIS specialist and environmental researcher at the University of Malaga, Spain, where he works for the European Topic Centre on Urban, Land and Soil Systems to support land and soil related activities of the European Environment Agency. His main field of research is the application of Geographic Information Systems to land use/cover change analysis from local to global scale with particular interest in the Mediterranean. Over the past couple of years he has been involved in analysing the impact of land cover changes on soil functions at a European scale.

**Song Xiaoqing**, PhD, is a Lecturer on Geographical Sciences, Guangzhou University, China, and was a visiting scholar of the Cluster of Excellence on Integrated Climate System Analysis and Prediction (CliSAP) at Hamburg University, Germany. His research focus is on urbanization, land use transition and multifunctional land management. He has authored more than 15 refereed journal articles.

**Vasileios Syrris** received a PhD in Computational Intelligence from Aristotle University of Thessaloniki, Thessaloniki, Greece, in 2010. He has worked as an Assistant Lecturer/Tutor with the Automation and Robotics Laboratory, Aristotle University of Thessaloniki, and the Department of Informatics and Electronics, Technological Educational Institute of Thessaloniki, Thessaloniki. Currently, he works as a researcher with the Global Security and Crisis Management Unit, Institute for the Protection and Security of the Citizen, Joint Research Centre, European Commission, Ispra, Italy. His research interests include high-performance computing, machine learning, robotics, automation, computer vision, remote sensing, control engineering, statistics and big data analytics.

**Ine Vandecasteele** has a Master's degree in Hydrogeology (2007) and Conflict and Development (2008), both from Ghent University, Belgium. She has been working at the JRC linking land use modelling with hydrological management since 2011. She completed her PhD on this subject area in 2014 with the Vrije Universiteit Brussel.

**Pilar Vizcaino** received her Bachelor and MSc degrees as a Forest Engineer from the Polytechnic University of Madrid, Spain, in 2001. She worked for several years for the Hydrology Department of her university, and is co-founder of a spin-off company that provides consultancy on the management of ecosystems. She has worked in the Joint Research Centre of the European Commission between 2006 and 2009 and since 2013, where she has mainly worked on the development and application of models integrated in geographical information systems. Her fields of interest are spatial analysis of growth and development indicators and application of quantitative methods (machine learning techniques, statistics, and optimization) to spatial problems.

**Zhifeng Wu**, PhD, is Professor of Geographical Sciences, Guangzhou University, China. He plays important roles in landscape ecology, urban studies, remote sensing and GIS in China. His research focuses on Anthropocene landform processes, urban remote sensing, the urbanization spatial process and its environmental-ecological effects. He also has organized many international co-researches. He is the committee member of the Sino-EU Panel on Land and Soil (SEPLS) and Vice-Director of the International Association for Landscape Ecology, China. He has supervised 25 graduate students and nine postdoctoral researchers. He has authored/co-authored more than 1940 research publications including 267 refereed journal articles and five book chapters (in Chinese).

# Foreword

An anonymous saying states 'whatever mankind does, we depend on 10 centimetres of soil and a few drops of water'.

Soil is a finite resource, meaning its loss and degradation is not recoverable within a human lifespan. As a core component of land resources, agricultural development and ecological sustainability, the soil is the base for food, feed, fuel and fibre production and for many critical ecosystem services. Current demographic trends and predicted growth in global population (to exceed 9 billion by 2050) are estimated to result in a global 60 per cent increase in demand for food, feed and fibre by 2050 (FAO, IFAD and WFP, 2015). By 2050, 70 per cent of the world population will live in urban areas. Cities will expand dramatically and mainly in Africa and other low- and middle-income areas. Demographic changes create a series of challenges that differ from one country to another, such as ageing populations, shrinking or rapidly expanding cities or intense processes of suburbanization. The population in some areas of Europe has increased significantly in recent years while other areas have depopulated (Piirto et al., 2010), and as life expectancy increases, the average age of the population will rise. Overall, this means more people to house, with higher expectations of the size and appropriate location of their homes. Despite a notable decrease in the average number of people in a household, in high-income countries the recent changes in the society such as lifestyles and consumption patterns demand more land and often good agricultural land. The pressure on soils for urbanization and the subsequent soil sealing phenomenon have never been higher.

In the late Anthropocene (Crutzen and Stoermer, 2000) there is often a general lack of appreciation as to the value of soil (and landscape), which is not recognized as a limited and non-renewable resource. This is a cause of serious concern, because soil formation is a very slow process, often taking centuries to build up even a centimetre.

The objective of this book is to create awareness on soil sealing due to the expansion of urbanization. The book proposes best practices to limit, mitigate or compensate the phenomenon. These best practice examples may be of interest to competent national authorities, professionals dealing with land planning and soil management, and stakeholders in general, but individual citizens may also find them useful.

Fernanda Guerrieri
FAO, Rome, Italy

## References

Crutzen, P. J., and Stoermer, E. F. (2000). *The Anthropocene IGBP Newsletter*, 41. Stockholm: Royal Swedish Academy of Sciences.

FAO, IFAD and WFP (2015) *The State of Food Insecurity in the World 2015. Meeting the 2015 International Hunger Targets: Taking Stock of Uneven Progress.* Rome: FAO.

Piirto, J., Johansson, A., and Lang, V. (2010). *Europe in Figures: Eurostat Yearbook 2010.* Luxembourg: Publications Office of the European Union.

# Acknowledgements

I express my gratitude to Tandra Fraser, James Cottrell and Martha Dunbar for the revision of texts. A special thanks go to Arwyn Jones for inspiring me with the 'Apple' concept of soil as limited resource. And of course I acknowledge the authors, for trusting me since the proposal for this book was just a fuzzy idea in my mind.

Ciro Gardi

# Part I

# Introducing and understanding the process

# 1 Is urban expansion a problem?

*Ciro Gardi*

## Introduction

Every mark traced on the territory by a new road or highway assumes the meaning of the ditch ploughed by Romulus, imposing the boundaries of the rising ancient Rome: another piece of land to be filled by buildings, generally low-quality buildings lacking aesthetics. Our approach to the use of land is much the same as the American pioneer, even 200 years later, even in crowded continents like Europe or China.

The evident environmental failure of liberal, and also communist, economic systems is caused essentially by the limits of the monetary aspects involved in the production of goods and services, ignoring the externalities (or, in the best case, under evaluating them). The globalization process that was announced as the *panacea* for these problems resulted in an unlimited amplification of environmental issues.

It is evident that if we consider only the direct costs (production and transport for instance) associated with the production of a good in China, for example (with labour and social costs one tenth that in Europe or North America), the market competition will be very unfair. To perform an environmentally correct evaluation, in addition to the evident, direct costs, we should add the impact associated to the extraction, production and use of fuel needed for the transport of the goods. This disproportional competition, and incorrect evaluation of environmental costs, has resulted in unsustainable development processes and severe environmental impacts.

The uncontrolled, and often unmotivated urban sprawl is an example of this inaccurate evaluation of the environmental consequences of our actions and decisions. The 'flooding' of concrete and asphalt is progressing, with little consideration of the irreversible consequences of these practices. Degradation of the landscape, increase in traffic and air pollution, flooding events, the loss of agricultural and natural areas, have been ineffective in raising awareness and stimulating action to protect one of our most precious resources and our collective identity: our land. In addition to the local impacts, we then have to consider the cumulative consequences of our local actions at the global scale.

Soil is becoming, more and more, a limited and strategic resource; increases in population and food demand and the production of biofuels, are driving an increase in biomass demand, and consequently demands on agricultural lands. At the same time, urban expansion and the intensification of agricultural practices, are causing the degradation of agricultural soils and pushing agriculture onto marginal lands or into natural habitats.

What is happening to the soil is a representation of the frequent 'tragedies of commons' (Hardin, 1968), that are replicated every day, in every corner of the planet, for water, biodiversity, climate, etc.

There is, however, increasing attention to these global environmental issues in the public opinion (still too limited), driving also changes in the international and local political agenda.

New concepts, such as 'climate justice' (Robinson and Miller, 2009), or 'soil security' (McBratney et al., 2014), have been introduced, and a new deal and renovated environmental commitment by the Catholic Church, started by Pope Francesco, are marked by the recent encyclical Laudato si (Francesco, Pope, 2015), in which the Pope is waking humanity up to care for the planet.

Analysis of global land cover maps, e.g. Global Land Cover 2000, indicates that urban areas were covering 0.2 per cent of the Earth's land surface at the end of the previous millennium. This number can be considered relatively small and, therefore, not necessarily a major threat to our planet's ecosystem services. Let us focus, however, on one of the most pertinent ecosystem services, at least from an anthropocentric point of view: food production. If we consider the potential threat represented by urban expansion on food productivity, it becomes clear that the topic merits a more in-depth assessment.

With regard to estimating the extent of urban areas/artificial surfaces, it is possible to ascertain values from other global land cover maps, most of them derived from satellite images. Schneider et al. (2009) produced a land cover map, derived from Modis images, and focused on urban areas. Based to this map, they provided a very accurate estimate that the extent of urban areas in 2000 was 657,000 km². Values obtained from other data sources, reviewed in the same paper, ranged between 276,000 km² up to 3,524,000 km². These values highlight the importance of reliable data in understanding urban growth processes.

## Urban expansion dynamics in the world

Over the past few decades, human activities have reached such intensity that they represent the most significant factor modifying our planet. Among human activities, the processes related to urbanization most certainly play a major role.

Urbanization can be determined by several factors: population growth, a positive balance between immigration and emigration, the economic growth of a given area, and speculation processes where the expansion of built-up areas is not related to the needs of increasing residential, commercial or industrial infrastructures.

At the global level, especially with regard to developing countries, demographic pressures and the migration of rural populations towards urban centres are the main drivers of urban expansion.

The global urban population increases by 200,000 every day, amounting to 70 million people every year that become part of the Earth's urban centres.

Currently, the urban population represents approximately 53 per cent of the total human population, and it is expected that by 2050 this proportion will reach 70 per cent. It is considered that 60 per cent of urban growth is determined by the demographic growth of urban populations, while 40 per cent is determined by processes of migration and reclassification of land uses (with the expansion of urban areas).

We know that the first urban centres were associated with the introduction of agricultural practices, in the fertile crescent,[1] and with our ability to produce surplus food that would allow the maintenance of a certain proportion of the population not directly engaged in agriculture. This is how the first cities in the Middle East, the Mediterranean, Asia and South America were established. The earliest signs of human aggregation, in the form of rural villages, dates back to 8500 BC, and the first city that we know of is Jericho, whose construction is attributed to 8000 BC.

Following the first signs of urbanization we enter the modern era, which is commonly divided into three phases:

- The first phase of urban growth in the modern era coincides with the important innovations in energy production technologies and, therefore, with the industrial era. From 1750 to 1950 Europe, North America and some areas of Asia were the centres of attention. Since then we have witnessed the birth of a new urban and industrial society involving significant population growth. In 1950 there were two megacities in the world, with more than 10 million inhabitants (New York–Newark, USA, and Tokyo, Japan).
- The second phase is represented by the rapid growth of urban areas in developing nations, where population growth and urbanization are usually accompanied by economic growth. This second phase, which is currently underway, is developing at a much faster rate than the previous one.
- The third phase is characterized by extremely rapid growth of urban areas, occurring in countries with fast growing economies, where rapid growth of the Gross Domestic Product (GDP) is associated with urbanization processes and/or demographic growth.

As of 2014, there were 488 cities in the world with a population of more than 1 million inhabitants, and 28 urban areas classified as megacities, as characterized by a population of over 10 million inhabitants (Table 1.1). It is expected that within 15 years, 13 additional cities will be added to the list of megacities: Ahmadabad, Bangalore, Chennai, Hyderabad (India), Bangkok (Thailand), Bogota (Columbia), Chengdu (China), Dar es Salaam (United Republic of

Table 1.1 Past, actual (2014) and predicted population of the world's 28 megacities

| Country | City | Population (× 1,000) | | | | Average annual rate of change (%) | | |
|---|---|---|---|---|---|---|---|---|
| | | 1970 | 1990 | 2014 | 2030 | 1970–1990 | 1990–2014 | 2014–2030 |
| Japan | Tokyo | 23,298 | 32,530 | 37,833 | 37,190 | 1.67 | 0.63 | −0.11 |
| India | Delhi | 3,531 | 9,726 | 24,953 | 36,060 | 5.07 | 3.93 | 2.30 |
| China | Shanghai | 6,036 | 7,823 | 22,991 | 30,751 | 1.30 | 4.49 | 1.82 |
| Mexico | Mexico City | 8,831 | 15,642 | 20,843 | 23,865 | 2.86 | 1.20 | 0.85 |
| Brazil | Sao Paolo | 7,620 | 14,776 | 20,831 | 23,444 | 3.31 | 1.43 | 0.74 |
| India | Mumbai (Bombay) | 5,811 | 12,436 | 20,741 | 27,797 | 3.80 | 2.13 | 1.83 |
| Japan | Osaka | 15,272 | 18,389 | 20,123 | 19,976 | 0.93 | 0.38 | −0.05 |
| China | Beijing | 4,426 | 6,788 | 19,520 | 27,706 | 2.14 | 4.40 | 2.19 |
| United States | New York–Newark | 16,191 | 16,086 | 18,591 | 19,885 | −0.03 | 0.60 | 0.42 |
| Egypt | Cairo | 5,585 | 9,892 | 18,419 | 24,502 | 2.86 | 2.59 | 1.78 |
| Bangladesh | Dhaka | 1,374 | 6,621 | 16,982 | 27,374 | 7.86 | 3.92 | 2.98 |
| Pakistan | Karachi | 3,119 | 7,147 | 16,126 | 24,838 | 4.15 | 3.39 | 2.70 |
| Argentina | Buenos Aires | 8,105 | 10,513 | 15,024 | 16,956 | 1.30 | 1.49 | 0.76 |
| India | Kolkata (Calcutta) | 6,926 | 10,890 | 14,766 | 19,092 | 2.26 | 1.27 | 1.61 |
| Turkey | Istanbul | 2,772 | 6,552 | 13,954 | 16,694 | 4.30 | 3.15 | 1.12 |
| China | Chongqing | 2,237 | 4,011 | 12,916 | 17,380 | 2.92 | 4.87 | 1.86 |
| Brazil | Rio de Janeiro | 6,791 | 9,697 | 12,825 | 14,174 | 1.78 | 1.16 | 0.62 |
| Philippines | Manila | 3,534 | 7,973 | 12,764 | 16,756 | 4.07 | 1.96 | 1.70 |
| Nigeria | Lagos | 1,414 | 4,764 | 12,614 | 24,239 | 6.08 | 4.06 | 4.08 |
| United States | Los Angeles | 8,378 | 10,883 | 12,308 | 13,257 | 1.31 | 0.51 | 0.46 |
| Russia | Moscow | 7,106 | 8,987 | 12,063 | 12,200 | 1.17 | 1.23 | 0.07 |
| China | Guangdong | 1,542 | 3,072 | 11,843 | 17,574 | 3.45 | 5.62 | 2.47 |
| Democratic Republic of Congo | Kinshasa | 1,070 | 3,683 | 11,116 | 19,996 | 6.18 | 4.60 | 3.67 |
| China | Tianjin | 3,318 | 4,558 | 10,860 | 14,655 | 1.59 | 3.62 | 1.87 |
| France | Paris | 8,208 | 9,330 | 10,764 | 11,803 | 0.64 | 0.60 | 0.58 |
| China | Shenzhen | 22 | 875 | 10,680 | 12,673 | 18.44 | 10.42 | 1.07 |
| United Kingdom | London | 7,509 | 8,054 | 10,189 | 11,467 | 0.35 | 0.98 | 0.74 |
| Indonesia | Jakarta | 3,915 | 8,175 | 10,176 | 13,812 | 3.68 | 0.91 | 1.91 |

Source: UN (2015).

Note: Cities are ranked according to the actual population.

Tanzania), Johannesburg (South Africa), Lahore (Pakistan), Lima (Peru), Luanda (Angola), and Thành Pho Ho Chí Minh (Vietnam).

The process of growth of urban agglomerations, also known as urban sprawl, can be analysed from many points of view: social, economic, environmental; however, the aim of this discussion is limited to the assessment of urbanization with regard to its direct impacts on a limited and non-renewable resource, such as soil.

## The global picture

In 2008, for the first time in history, the urban population reached 50 per cent of the Earth's total human population (Figure 1.1). It could be argued that a limit, which is not only psychological, has been exceeded.

From the graphic in Figure 1.1 we can see that the projections indicate that, while from 2020 onwards the rural population will begin to decline, the urban population will continue to grow and by 2050 the rural population will represent just one-third of the total (United Nations Department of Economic and Social Affairs, 2008). It is clear that the environmental impacts resulting from such a radical change will be enormous, albeit difficult to assess and predict with accuracy. Considering only the flows of matter and energy necessary to sustain an urban ecosystem, and the consequent production of waste and disposal thereof, we have an indication of the dimensions and the type of problems that will need to be addressed in the future.

If we consider the organic matter cycle, it is easy to see how much more balanced and sustainable a widespread system (rural area) would be, in which

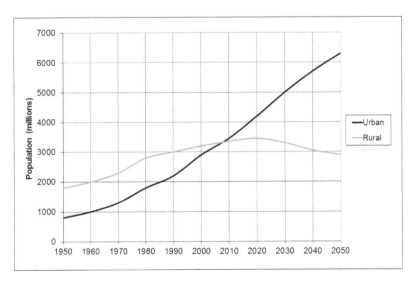

*Figure 1.1* Dynamics of urban and rural populations of the world

Table 1.2 Population dynamics in some Asian urban agglomerations

| City | Country | Population (thousands) | | | Urbanized area (km²) | | | Population density (pop./ha) | | |
| --- | --- | --- | --- | --- | --- | --- | --- | --- | --- | --- |
| | | 1990 | 2000 | Var. (%) | 1990 | 2000 | Var. (%) | 1990 | 2000 | Var. (%) |
| Anging | China | 1,003 | 1,055 | 0.5 | 54 | 78 | 3.6 | 186 | 135 | -3.0 |
| Bacolod | Philippines | 462 | 510 | 1.3 | 13 | 33 | 12.3 | 343 | 155 | -9.8 |
| Bandung | Indonesia | 2,942 | 3,628 | 2.2 | 109 | 182 | 5.4 | 271 | 199 | -3.1 |
| Cebu | Philippines | 1,118 | 1,524 | 3.0 | 54 | 66 | 1.9 | 206 | 231 | 1.1 |
| Changzhi | China | 1,160 | 1,254 | 1.2 | 104 | 156 | 6.4 | 111 | 104 | -1.1 |
| Coimbatore | India | 552 | 613 | 1.1 | 99 | 156 | 4.7 | 56 | 39 | -3.4 |
| Guangzhou | China | 7,712 | 13,156 | 5.5 | 452 | 979 | 8.1 | 171 | 134 | -2.4 |
| Hyderabad | India | 4,888 | 5,708 | 1.3 | 167 | 302 | 5.1 | 293 | 189 | -3.6 |
| Jaipur | India | 2,116 | 2,779 | 2.5 | 59 | 141 | 8.3 | 360 | 197 | -5.4 |
| Jalna | India | 445 | 556 | 2.1 | 11 | 25 | 7.5 | 395 | 223 | -5.0 |
| Kanpur | India | 1,124 | 1,442 | 2.3 | 34 | 60 | 5.4 | 110 | 79 | -2.9 |
| Kolkata | India | 6,646 | 7,834 | 1.7 | 288 | 484 | 5.3 | 231 | 162 | -3.5 |
| Kuala Lumpur | Malaysia | 2,733 | 4,959 | 5.0 | 383 | 805 | 6.2 | 71 | 62 | -1.2 |
| Leshan | China | 608 | 670 | 0.9 | 75 | 146 | 6.4 | 81 | 46 | -5.1 |
| Manila | Philippines | 14,044 | 17,335 | 2.4 | 444 | 660 | 4.5 | 316 | 263 | -2.0 |
| Mumbai | India | 14,224 | 17,070 | 2.1 | 344 | 451 | 3.1 | 413 | 378 | -1.0 |
| Pune | India | 3,510 | 4,042 | 2.1 | 93 | 191 | 11.0 | 379 | 211 | -8.1 |
| Rajshahi | Bangladesh | 491 | 600 | 1.8 | 11 | 20 | 5.8 | 452 | 296 | -3.8 |
| Saidpur | India | 503 | 596 | 1.4 | 9 | 16 | 5.5 | 564 | 366 | -3.9 |
| Songkhla | Thailand | 220 | 244 | 1.0 | 14 | 19 | 3.0 | 159 | 129 | -1.9 |
| Vijayawada | India | 981 | 1,117 | 1.3 | 40 | 62 | 4.5 | 244 | 179 | -3.0 |
| Yigang | China | 1,108 | 1,135 | 0.5 | 49 | 100 | 14.7 | 227 | 114 | -12.4 |

the production of biomass (food, fibre, biomass for energy production) and its decomposition take place in a distributed way throughout the territory. In contrast, in a megalopolis, fluxes of thousands of tons of organic matter entering the urban ecosystem are then disposed through artificial processes (landfills, incineration, anaerobic digestion) in a concentrated way.

From the point of view of land use, processes of urban growth and the urbanization of rural populations involve a net increase in impervious surfaces (sealed areas), increasing the consumption of land through a phenomenon known as 'land take'.

If we consider, for example, areas with strong economic growth in some Asian cities, we can observe a much higher expansion of urban areas, which is not correlated to the increase in the resident population (Table 1.2). The growth of city populations considered increased at an average of 23 per cent in the period 1990–2000, while the average growth of the urban areas during the same period amounted to 80 per cent.

It seems clear then that the growth of urbanized areas is not only determined by a demand for more living space, which would lead to a linear correlation between the two variables. Trigger processes, that are only partly related to the hypothetical improvement of living conditions, but are largely linked to speculation, are occurring. The space requirements of urban growth are not only determined by the housing spaces for the new inhabitants, which in some cases can legitimately aspire to having a larger home compared to their previous dwellings, but are determined mainly by the need for new transport infrastructure, manufacturing, commercial and entertainment facilities, logistical infrastructures and areas for the treatment and disposal of waste. To all this, which is a functional requirement of the urban ecosystem, we should add the speculative drivers that result in the proliferation of business and shopping centres, hotels, residential settlements, not always necessary, but rather seen as a potential investment. In an article published in *Urban Studies*, Julian Marshall (2007) analysed the relationship between population growth and the growth of urban areas; the author noted that this relationship can be expressed by the following equation:

$$\propto = P^n$$

The urban areas (A) are growing proportionally to the number of inhabitants (P), to the power (n); n assumes, in several of the cases analysed by Marshall, a value close to 2, indicating that on average recently established citizens tend to use a larger amount of land, compared to their predecessors (i.e. a 3 per cent population increase will determine approximately a 9 per cent increase in area).

If these processes have been, somehow, managed and controlled in Western countries, thanks to the tradition in land and urban planning, the same processes can be devastating when they occur in territories or countries without any process of land use planning, management and control.

Among the 30 cities characterized by the highest demographic growth, there is only one Western city (Table 1.3). In Africa for instance, at the beginning

*Table 1.3* The top 30 fastest growing urban agglomerations

| Country | Urban area | Annual growth % |
| --- | --- | --- |
| China | Beihai | 10.58 |
| India | Ghaziabad | 5.20 |
| Yemen | Sana'a | 5.00 |
| India | Surat | 4.99 |
| Afghanistan | Kabul | 4.74 |
| Mali | Bamako | 4.45 |
| Nigeria | Lagos | 4.44 |
| India | Faridabad | 4.44 |
| Tanzania | Dar es Salaam | 4.39 |
| Bangladesh | Chittagong | 4.29 |
| Mexico | Toluca | 4.25 |
| Congo | Lubumbashi | 4.10 |
| Uganda | Kampala | 4.03 |
| Bolivia | Santa Cruz | 3.98 |
| Angola | Luanda | 3.96 |
| India | Nashik | 3.90 |
| Congo | Kinshasa | 3.89 |
| Kenya | Nairobi | 3.87 |
| Bangladesh | Dhaka | 3.79 |
| Madagascar | Antananarivo | 3.73 |
| India | Patna | 3.72 |
| India | Rajkot | 3.63 |
| Guinea | Conakry | 3.61 |
| India | Jaipur | 3.60 |
| Mozambique | Maputo | 3.54 |
| Somalia | Mogadishu | 3.52 |
| Pakistan | Gujranwala | 3.49 |
| India | Delhi | 3.48 |
| India | Pune | 3.46 |
| United States | Las Vegas | 3.45 |

Source: City of Mayors, www.citymayors.com/.

of the previous century 95 per cent of the population was living in rural areas and only 5 per cent was settled in urban areas. In 1960 the percentage of city dwellers reached 20 per cent, in 2010 it was 43 per cent and it is estimated to reach 50 per cent by 2030. The growth of several African cities has been tumultuous and chaotic: Nairobi, Dar es Salaam, Lagos, Kinshasa grew seven-fold in the period between 1950 and 1980 (Brundtland, 1987). The pressures on the environment, also in terms of land take, made by 'citizens' are much larger than the pressures of the rural populations. The problems of water supply, solid waste and waste water production and management increased dramatically, without having adequate infrastructures for water provision and plants for waste treatment.

Around the world there are already more than 17 urban agglomerations with areas larger than 1,000 km$^2$ (Table 1.4). Often without any system or process of urban planning, urban growth occurs at the expense of the most fertile and productive agricultural lands. Among the possible examples to be mentioned, one of the provinces of the People's Republic of China, the Shandong province, where the expansion of urban areas and the growth of transport infrastructures occurred between 1996 and 2003, was built on agricultural lands (Hong, 2007). According to other data sources, however (Demographia, 2015), the number of urban agglomerations larger than 1,000 km² were more than 100 in 2014. These discrepancies depend on the definition of urban agglomeration.

In Europe the situation is quite heterogeneous: we have countries like Spain, where the urban area is lower than 2 per cent, while small, densely populated countries like Belgium or the Netherlands have urbanized areas that are close to 20 per cent of the entire national territory.

*Table 1.4* Extension, population and population density of the 30 largest urban agglomerations

| Country | City | Area (km²) | Population | Population density (pop./km²) |
|---|---|---|---|---|
| United States | New York | 11,642 | 20,955,600 | 1,800 |
| Japan | Tokyo–Yokohama | 8,547 | 37,606,800 | 4,400 |
| United States | Los Angeles | 6,299 | 15,117,600 | 2,400 |
| Russia | Moscow | 4,403 | 15,850,800 | 3,600 |
| China | Beijing | 3,497 | 18,184,400 | 5,200 |
| United States | Phoenix | 3,276 | 3,931,200 | 1,200 |
| Brazil | Sao Paolo | 3,173 | 20,624,500 | 6,500 |
| France | Paris | 2,845 | 10,811,000 | 3,800 |
| Indonesia | Jakarta | 2,784 | 26,726,400 | 9,600 |
| Thailand | Bangkok | 2,331 | 14,452,200 | 6,200 |
| Malaysia | Kuala Lumpur | 1,943 | 6,606,200 | 3,400 |
| Egypt | Cairo | 1,658 | 15,087,800 | 9,100 |
| United Kingdom | London | 1,623 | 9,575,700 | 5,900 |
| Philippines | Manila | 1,437 | 21,267,600 | 14,800 |
| India | Kolkata | 1,204 | 14,568,400 | 12,100 |
| Russia | St Petersburg | 1,191 | 4,883,100 | 4,100 |
| Spain | Barcelona | 1,075 | 4,622,500 | 4,300 |
| Ghana | Accra | 945 | 3,969,000 | 4,200 |
| Nigeria | Lagos | 907 | 12,063,100 | 13,300 |
| Belgium | Brussels | 751 | 1,952,600 | 2,600 |
| Mexico | Guadalajara | 699 | 4,543,500 | 6,500 |
| Brazil | Brasilia | 673 | 2,422,800 | 3,600 |
| Greece | Athens | 583 | 3,498,000 | 6,000 |
| Tanzania | Dar es Salaam | 570 | 3,705,000 | 6,500 |
| Kenia | Nairobi | 557 | 4,456,000 | 8,000 |
| India | Mumbai | 546 | 17,308,200 | 31,700 |
| New Zealand | Auckland | 544 | 1,305,600 | 2,400 |
| Singapore | Singapore | 518 | 5,283,600 | 10,200 |
| Finland | Helsinki | 492 | 1,180,800 | 2,400 |
| Denmark | Copenhagen | 453 | 1,223,100 | 2,700 |

Source: UN-Habitat, www.unhabitat.org.

During the last two decades (1990–2008), however, the less urbanized countries in Europe have also experienced a boom in the real estate sector, driven mainly by speculative processes. Especially in countries such as Spain and Ireland during that period, there was a proliferation of new houses, buildings and commercial centres that largely exceeded the real demand. In fact, in these countries the rapid economic growth was the driving factor for the explosion of the building sector during that period.

If we consider the case of Ireland, where thanks to a tempting fiscal policy during the last few decades it was possible to attract capital, and to boost the economy, creating jobs and also promoting an increase in prices. This positive economic situation prompted demographic growth, causing a increase in demand for residential and productive infrastructures on one side, but on the other side the rising prices of real estate were the driving factor behind speculative processes. In Dublin, for instance, it is foreseen that there will be an increase of 250,000 inhabitants during next years; part of this population increase will be determined by immigration from eastern EU countries (Poland, the Baltic states), and will largely determine changes to the structure of the city: new houses, schools, hospitals, roads, supermarkets, etc. The problem is that, during these large immigration processes, for instance, for 1,000 ha of new buildings in Dublin, it will not be possible to dismantle or de-seal the same amount of land in the Netherlands or the Baltic states. This vicious cycle, associated with unlimited economic growth, will cause a progressive and continuous depletion of soil resources.

Let us consider the case of Germany: during the last few decades it has reached the level of 100–120 ha day$^{-1}$ of land take for urban expansion (60 ha day$^{-1}$ was the average in the period 1990–2000 according to Corine Land Cover). Germany has decided to reduce the rate of this process, with a target value of 30 ha day$^{-1}$ for the near future. This would be a great achievement, but in any case would represent a continuous dissipation of a finite resource, such as the land.

In addition to the urban growth associated with the needs of growing populations, growing economies that require new industrial or commercial districts, we have luxury and unessential goods: we need only read on-board magazines to be fatally attracted by a luxurious, small villa within a golf resort in the Algarve, at the same price that we would pay for a car garage or a car-box in one of the largest European towns. In Italy for instance, until a few years ago, it was common at the entrance to highways to see advertisements for very nice little houses or villas in (probably not in the most renowned) touristic areas at very cheap prices.

## Soil: a strategic and limited resource

The occupation and destruction of soil caused by urban expansion should be considered an irreversible event in the human time scale. Several thousands of years are necessary to form a deep, fertile and mature soil, such as the soils of the US corn belt, Argentinian pampas or Russian steppe region. The development and the economic models that we are adopting often represent, from a soil perspective, a non-return option: an area of urban expansion, occupied by a commercial centre or a logistic facility, in the case of economic crisis or displacement of economic activities, can be converted to different uses, but could never be re-used as an agricultural area (unless after extremely expensive operations of de-sealing). Every time a new shed, parking area or road is built, we are wasting a non-renewable, finite natural capital, cared for and valued by our predecessors, to constitute heritage for the future generations. To seal a piece of land with asphalt or concrete means not only preventing for our selves and for future generations the provision of ecosystem services, but also to lose the rent associated with the economic productivity that soil can ensure over the long term.

The increasing rate of agricultural land take, the lack of care taken when disposing of land, and the absence of critical voices and opposition to the irreversible transformations of land, make evident the progressive loss of competitiveness by agriculture with respect to other economic activities more profitable in the short term, but also the limited strategic value that society attributes to this natural capital. In practice we are misusing and wasting agricultural soils, as if they were extremely abundant and indefinitely replaceable resources. A quick look at land use and food consumption at the global scale, however, demonstrates the weakness of this assumption.

At a global scale, agricultural soils, and in particular fertile and resilient soils, are already limited resources, destined to become even more limited in the future.

According to the UN's Millennium Ecosystem Assessment (2005), agricultural systems cover a quarter of continental lands, which is already the vast majority of soils suitable for agriculture (UN, 2005a). From these agro-ecosystems we obtain more than 90 per cent of the carbohydrates and proteins needed by the human population. After the Second World War, agricultural mechanization and the 'green revolution' made it possible to increase agricultural productivity by several folds, at rates much higher than the increase in human population (Alés and Solbrig, 2001; UN, 2005a). If we consider the average values at the global scale, the amount of available per capita calories has steadily increased, but despite this benefit it was not evenly distributed among the human population. According to current UN estimates, about 793 million people are undernourished globally (FAO, 2015).

Recent global data on agricultural production and food consumption, however, indicate a possible trend reversal, shifting from a global market of food commodities characterized by surplus, to one characterized by scarcity (Brown, 2005).

According to the Millennium Ecosystem Assessment scenarios, by 2050 the demand for food will increase between 70 and 85 per cent, compared to current values – this takes into account the combined effect of demographic and economic growth. In the next 35 years, an additional 2.5 billion people will live on the planet, in addition to the current 6.5 billion, with increments of 70 million per year; more than 9 billion people by 2050 (UN, 2005b). The increased food demand, generated solely by the increased number of people living on the Earth, will be added to the relevant increase in food demand generated by the improved economic conditions reached by hundreds of millions people as a result of economic growth. The effect of improved economic conditions is generally associated with a shift from purely subsistence diets, based on direct consumption of cereals, to more diverse diets, richer in fat and animal proteins. This transition towards higher trophic levels in the food pyramid will determine a relevant increase in the per capita needs of agricultural, productive land. Compared to a pure subsistence diet, the increased consumption of meat, cheese and other milk derivatives, and alcoholic beverages, will require two to three times more agricultural land (Gerbens-Leenes and Nonhebel, 2002).

The same factors determining economic growth and, consequently, the rise in food demand will be the driving factors of urban expansion and land take, in order to create new residential, industrial and commercial structures, as well as roads, railways, waste disposal sites, mining sites, etc., thereby threatening the agricultural production capacity. It is estimated that every day approximately 4,000 ha of land, most of which is agricultural, is covered by asphalt for roads, highways and parking lots. If we consider that two of the most rapidly growing economies, China and India, are primarily adopting transport on wheels for transporting goods, it becomes clear how this process can directly affect food security at a local and global scale. In order to understand the potential impact associated with road transport, it is sufficient to consider that in China and India the motorization rate (number of vehicles per 1,000 people) was 6.7 and 6.0, respectively, in 2002, and already 91 and 20 in 2013 (OICA, 2013). Considering that every car implies an asphalt coverage ranging between 0.02 and 0.07 $m^2$, if these growing countries reach the motorization rate of Europe (565) or the USA (790), we can easily calculate the impact on agricultural land. According to Lester Brown of the Earth Policy Institute, considering only the population growth, more than 3 million ha, most of them agricultural, will be taken. Furthermore the urban growth, a consequence of population and economic growth, will take place over the most fertile soils of the planet, where towns are typically settled. In addition to the competition for land, agriculture and cities will also compete in the future for water, a fundamental resource for irrigated crops and for ensuring high productivity in many areas of the world.

It is very unlikely that in the future we will be able to tackle the increased demand for food, and the reduction of available agricultural land, simply by increasing the productivity of the remaining agricultural areas. It is true that during the Green Revolution it was possible within 40 years to triple (even quadruple in some cases) crop yields, but this type of increase seems very difficult to achieve in the future (Hawken *et al.*, 1999). Despite some very ambitious programmes that aim to further increase the productivity of some crops (e.g. 'wheat 20:20'), it is more realistic to think that, at least in Western agriculture, the selection of new crop varieties, the use of mechanization, fertilizers, growth regulators and pesticides has already exploited most of its potential. Furthermore, genetically modified organism (GMO) crops, independent of the issues concerning consumer and ecosystem safety, cannot be exempted from the limits imposed by crop physiology and by the efficiency of photosynthetic processes. Biotechnology has mainly contributed, so far, to developing GMO crops characterized by resistance to pests or to specific herbicides, or better adapted to grow and produce in unfavourable pedoclimatic conditions (Brown, 2005).

We must also consider that the significant increase in crop productivity achieved during the Green Revolution was obtained using large quantities of water and energy, often with negative impacts on the environment. Among the impacts associated with intensive, conventional agriculture, we have eutrophication of surface waters (especially where there are excessive phosphorus and nitrogen loads), and the dispersion in the environment of toxic and persistent chemicals that have accumulated throughout trophic levels. Often, increases in crop yield, in the short term, have been obtained by compromising natural capital and concomitant ecosystem services.

In the most vulnerable agricultural areas, after an initial promise to increase crop productivity, severe and often irreversible declines in yields have occurred as a consequence of soil degradation processes (soil erosion, compaction, organic matter decline, salinization, etc.) or of the depletion of water resources. In some extreme cases this type of intensive exploitation has caused desertification processes, resulting in the complete loss of soil fertility, thus preventing any possibility of agricultural or pastoral activities. Examples of these processes are common in the Sahel or in north-western China, where the desert expands over hundreds of thousands of hectares every year, promoting the formation of dust storms or dust bowls, that tarnish the sun over Beijing and other areas of the country.

The solution to the increasing food demand will not be found in the oceans. The quantity of fish caught that increased fivefold between 1950 and 1990, has been declining since the end of the 1990s, and the possibility of reversing this decline is very unlikely due to the overexploitation of fish stock of several species in several areas. However, the reduction in fish catch has been compensated by the coastal and off-shore pisciculture. This type of activity, albeit very efficient for protein production, does not represent a realistic alternative to agricultural food production because it is essentially based on the use of feed obtained in terrestrial agricultural systems

In order to satisfy the future food demand, in absence of significant increases in crop yields and without any possible alternative to food derived from agriculture, it would be necessary to extend, where the pedoclimatic conditions are suitable, the arable and pastoral lands. This extension of agricultural land would allow us to compensate for the land taken by the urban and infrastructure expansion, but would cause loss of natural or semi-natural ecosystems, and the associated impacts on biodiversity and ecosystem services. The process will eventually progress, including marginal lands, where risks of desertification and soil degradation are higher, and where the inputs (water, fertilizers, energy) required for sustaining productivity will be higher. In the future, the conversion of natural habitats into agricultural systems will represent one of the most important processes, resulting in environmental degradation and biodiversity losses on Earth (Vitousek, 1994; Tilman *et al.*, 2001; Foley *et al.*, 2005; UN, 2005a). The Millennium Ecosystem Assessment scenarios show that a percentage ranging between 10 and 20 per cent of current grasslands and forested areas will be converted into agricultural land. This land use change will occur mainly in tropical and sub-tropical areas, among the most biodiverse ecosystems on the planet. At the same time the intensification of existing agricultural systems, and the increased impacts associated with the use of chemicals, will cause changes and deterioration of the provision of ecosystem services. Despite all these changes and environmental impacts, there is no certainty that by 2050 we can achieve global food security and the eradication of malnutrition.

The environmental changes and crises and the food security crisis described by the Millennium Ecosystem Assessment are, unfortunately, realistic and very much occurring. From the 1980s onwards, global grain production (a good indicator of food security), started to increase at a lower rate compared to population growth. Between 2000 and 2007, grain production was lower than consumption in seven out of eight years, causing the progressive depletion of global grain reserves. In 2008 the lowest level of the last 35 years was reached. In 2008 the quantity of grain stocks was sufficient to cover only 54 days of global consumption. These changes have been caused by several combined processes: some of the traditional grain exporters became importers, such as China which in 2004 was the largest importer with more than 8 million tonnes of grain. Between 1998 and 2003 China lost 18 per cent of its cereal production, mainly as a consequence of the decline in cultivated areas. In order to tackle the rapidly growing demand for protein, directly connected with its economic growth and dietary changes, China also became the largest soybean importer (22 million tonnes), while up until 1997 it was self-sufficient.

Soybeans represent the most important protein source for animal feed. Soybean production has been booming since the 1970s, driven by the constant increase in meat consumption at the global scale, and by the progressive reduction of alternative sources of proteins, such as fishmeal. In Brazil, the worlds leading soybean exporter, the production of this commodity increased from 1 to 66 million tonnes in 35 years. This astonishing result was obtained by extending the cultivation area by 23 million hectares. Half of this expansion took place after 1996, by converting the Cerrado (a mixed bush–low forest ecosystem) into arable systems at the rate of 1.5 million ha year$^{-1}$, but also part of the Amazon rainforest. The environmental impacts of this large-scale land use change are huge, and at the same time very difficult to assess: from the biodiversity losses, to severe modification of the hydrological cycle, from the soil degradation to the release in the atmosphere of enormous quantities of carbon dioxide. The theoretical possibility to go further, and convert 75 million ha of Cerrado into agricultural systems may allow for tripling soybean production, but it would also represent an environmental catastrophe without precedent, with a high probability of severe degradation of the soils of the area.

The soybean represents a compensation for the loss of productive lands in other areas of the planet. Not only for China, but also for the extremely intensive dairy and meat production systems in Europe. Imported soybeans represent the classical case of indirect land use change: the loss of agricultural and forage production areas in Europe (due also to urban expansion), is compensated for by the import of proteins from elsewhere (Brazil, Argentina, etc.), where cultivated lands are expanding, pushing (as is the case of Argentina) the pasture land towards marginal and less productive areas.

In other words, we can consider that the effect of urban expansion occurring in China, Europe, and North America, is directly responsible for the deforestation of at least part of the Cerrado or the Amazon basin and for the loss of natural grasslands in Argentina.

## Note

1 Middle East–Northern African area, also known as the Cradle of Civilization.

## References

Alés, R.F. and Solbrig, O.T. (2001) 'Are Famine and Malnutrition a Question of Supply or Demand: Implications for Environmental Rural Sustainability', in O.T. Solbrig, R. Paarlberg and F. Di Castri (eds) *Globalization and the Rural Environment*, Harvard University Press, Cambridge, MA, 49–71.

Brown, L.R. (2005) *Outgrowing the Earth: The Food Security Challenge in an Age of Falling Water Tables and Rising Temperatures*, Norton & Company, New York and London.

Brundtland, G.H. (1987) *Our Common Future*, Report of the World Commission on Environment and Development. UN.

Demographia (2015) *Demographia World Urban Areas: 11th Annual Edition*, www. demographia.com, accessed 10 September 2015.

FAO (2015) 'The State of Food Insecurity in the World 2015', www.fao.org/hunger/ key-messages/en/, accessed 14 September 2015.

Foley, J.A., DeFries, R., Asner, G.P., Barford, C., Bonan, G., Carpenter, S.R., Chapin, F.S., Coe, M.T., Daily, G.C., Gibbs, H.K., Helkowski, J.H., Holloway, T., Howard, E.A., Kucharik, C.J., Monfreda, C., Patz, J.A., Prentice, I.C., Ramankutty, N. and Snyder, P.K. (2005) 'Global Consequences of Land Use', *Science*, 309, 570–574.

Francesco, Pope (2015) *Laudato si*. Libreria Editrice Vaticana, Rome.

Gerbens-Leenes, P.W. and Nonhebel S. (2002) 'Consumption Patterns and Their Effects on Land Required for Food', *Ecological Economics*, 42, 185–199.

Hardin, G. (1968) 'The Tragedy of the Commons', *Science*, 162(3859), 1243–1248.

Hawken, P., Lovins, A. and Lovins, L.H. (1999) *Natural Capitalism: Creating the Next Industrial Revolution*, Little, Brown, Boston, MA.

Hong, X. (2007) 'Mutual Conversion of Land Use between Urban and Rural Area in the Process of Urbanization: A Case Study of Shandong Province', *Chinese Journal of Population, Resources and Environment*, 5(2), 93–96.

McBratney, A., Field, D.J. and Koch, A. (2014) 'The Dimensions of Soil Security', *Geoderma*, 213, 203–213.

Marshall, J.D. (2007) 'Urban Land Area and Population Growth: A New Scaling Relationship for Metropolitan Expansion', *Urban Studies*, 44(10), 1889–1904.

OICA (2013) 'Motorization Rate 2013 – Worldwide', www.oica.net, accessed 10 July 2015.

Robinson, M. and Miller, A. (2009) 'Expanding Global Cooperation on Climate Justice'. Bretton Woods Project, London, 1 December.

Schneider, A., Friedl, M.A. and Potere, D. (2009) 'A New Map of Global Urban Extent from MODIS Satellite Data', *Environmental Research Letters*, 4(4), 044003.

Tilman, D., Fargione, J., Wolff, B., D'Antonio, C., Dobson, A., Howarth, R., Schindler, D., Schlesinger, W.H., Simberloff, D. and Swackhamer, D. (2001) 'Forecasting Agriculturally Driven Global Environmental Change', *Science*, 292, 281–284.

United Nations (UN) (2005a) 'Millenium Ecosystem Assessment – Synthesis Report', United Nations, New York.

United Nations (UN) (2005b) 'Population Challenges and Development Goals', United Nations, New York.

United Nations Department of Economic and Social Affairs: Population Division (2008) 'World Urbanization Prospects: The 2007 Revision', United Nations, New York.

Vitousek, P.M. (1994) 'Beyond Global Warming: Ecology and Global Change', *Ecology*, 75(7), 1861–1876.

# 2 Measuring and monitoring land cover

## Methodologies and data available

*Michele Munafò and Luca Congedo*

## Introduction

Land Cover Change (LCC) and its environmental consequences are global challenges, as pointed out by IPCC (2001): climate processes are indirectly affected by land surfaces and the materials on the ground, and soil has a major role in carbon fluxes and greenhouse gas emissions. Moreover, soil provides ecosystem services that are fundamental for humanity and environmental sustainability, such as food and timber production, biodiversity and habitat support, carbon sequestration and climate regulation (IPCC, 2001; Lal, 2005; TEEB, 2010; Munafò *et al.*, 2015). Furthermore, soil has a major role in the mitigation of and adaptation to floods or droughts and extreme events in general (European Commission, 2014).

One of the main drivers of LCC is urban development, especially in the form of soil consumption that is the conversion from natural to artificial land cover. In Europe, soil consumption is a major issue related to the demand for residential, industrial, commercial infrastructures, and transportation, without a direct correlation to demographic growth (Indovina, 2006; European Commission, 2006).

During the last decade, soil consumption has been addressed by various institutions, especially in Europe, with the major objective of ensuring soil protection (European Environmental Agency, 2006). The European Thematic Strategy for Soil Protection (European Commission, 2006) defined good practices for reducing negative impacts of urban development. In 2012, the European Commission described the implementation of the Soil Thematic Strategy, highlighting the importance of raising awareness about soil, supporting research projects and monitoring soil at regular intervals (European Commission, 2012a).

Several studies have demonstrated the utility of remote sensing and Geographic Information Systems (GIS) for monitoring the built-up expansion and for mapping land cover in general (Brook and Davila, 2000). The results of land cover monitoring are fundamental for developing effective policies for sustainability and adaptation to environmental change (Cardona *et al.*, 2012).

## Basic definitions about land and soil monitoring

It is worth pointing out the main definitions related to land and soil monitoring.

Soil is the top layer of the earth's crust that has a crucial role in ecosystems, especially considering that it is a non-renewable resource with an extremely slow process of formation (European Commission, 2006).

Soil sealing is the process of permanently covering the soil surface with impervious and artificial materials, separating soil from other ecosystem compartments; however, there are alternative definitions, depending on the study approach, that highlight: the loss of functions associated to soil sealing; or the change of soil natural characteristics causing soil to behave as an impermeable medium (Burghardt et al., 2004). A wider concept is soil consumption, which is the increase of artificial land cover, defined as the physical material at the ground, which for instance is vegetation, bare soil, water, asphalt, etc. (Fisher and Unwin, 2005). Artificial land cover includes soil sealing (in terms of impervious surfaces) and other artificial surfaces that may be permeable but alter the natural soil, such as dumps, quarries and railways (Munafò et al., 2015).

The land consumption (land take) phenomenon is the increase of artificial land use at the expense of natural and semi-natural land use, therefore it includes sealed and unsealed areas such as urban green areas (European Environmental Agency, 1997). It usually does not include impervious surfaces in natural, semi-natural and agricultural areas, such as greenhouses.

It is worth highlighting that land take and land consumption refer to the use of soil, while soil consumption refers to land cover. However, soil sealing, soil consumption and land consumption are highly interrelated (Huber et al., 2008).

## Land cover monitoring in the land system science

Considering the complexity of interactions between land cover and environmental issues, it is useful to identify a framework that helps discern these relations, bearing in mind the aim of sustainability; in this sense, the focus of this chapter is on land cover.

During the last few years, land system science has attempted to understand the relationship between urban development and ecosystems (at the local and global scale) by integrating several disciplines, from ecology and social science, to remote sensing (Verburg et al., 2013).

Within this framework, land cover change is considered both cause and effect of climate change, highlighting the importance of land cover monitoring for soil protection, and improving sustainable policies and planning processes (Verburg et al., 2013).

Land cover monitoring can provide the spatial data required for assessing ecosystem services and supporting decision making (Maes et al., 2012). The use of remote sensing is a valuable approach for the efficient monitoring of land cover, especially for the built environment, and the estimation of impacts on ecosystem services (Chen et al., 2013).

In this context, Europe has developed several initiatives aimed at land cover monitoring; Copernicus (previously GMES), is a European programme of earth observation with the purpose of environmental protection and civil security. This is a complex system of data acquisition from multiple satellites and data integration with field surveys that aims also at ensuring the independence of Europe in environmental monitoring.

In the Copernicus framework, services are provided for several thematic areas (e.g. atmosphere, security, emergency) and in particular for the land theme; these data and services can foster environmental protection (e.g. management of urban areas, regional and local planning, agriculture, forest management, etc.) and increase awareness about soil consumption and environmental change (European Commission, 2014).

The sources of information are fundamental in land cover monitoring: the techniques and tools used for monitoring land cover change affect the results in terms of resolution and accuracy. In fact, the incorrect understanding of data acquisition, or interpretation of the classification system can lead to the misinterpretation of results.

Environmental analyses that are based on the study of land use and land cover change allow for the assessment of environmental evolution in time; nevertheless, data are characterized by technical and semantic features (such as the reference system, classification system and legend) that must be considered for the accurate estimation of soil consumption (Munafò *et al.*, 2010; CRCS, 2012). In particular, different systems of classifications can lead to substantial dissimilarities of results, for instance because of various definitions of homogeneous area, minimum mapping unit, or the mix of land use and land cover characteristics that define classes.

Moreover, the majority of databases is created for specific purposes (e.g. agricultural controls, land planning, environmental assessment, statistical report), having classification systems that are not suitable for soil consumption monitoring (ISPRA, 2013).

Land cover monitoring can be based on two main approaches: mapping and sampling. The former is particularly useful because the output is a spatial product that can be used as input for spatial modelling and the assessment of ecosystem services; the latter is more reliable from the statistical point of view, allowing for greater flexibility of use, ease and speed of data update.

It is fundamental that land cover monitoring is based on the integration of mapping and sampling approaches.

## An overview of Earth observation satellites

Remote sensing is the science and technology that allows for the identification or measure of object characteristics without direct contact (JARS, 1993). In general, the energy that emanates from the Earth's surface is measured by sensors, distinguished in: passive remote sensing, if the source of the measured energy is the sun; active remote sensing, if the source of the measured energy is emitted from the sensor platform (Richards and Jia, 2006).

Sensors are placed on board airplanes or satellites, measuring the electromagnetic radiation at specific ranges (i.e. spectral bands). In particular, sensors measure the radiance that corresponds to the brightness in a given direction toward the sensor; it is worth mentioning the definition of reflectance that is the ratio of reflected versus total power energy.

Measured energy is quantized and converted into a digital image, where each picture element (i.e. pixel) has a discrete value in units of Digital Number (i.e. DN) (NASA, 2013).

The characteristics of sensors (i.e. resolutions) are defined as:

- Spatial resolution 'is the resolving power of an instrument needed for the discrimination of features and is based on detector size, focal length, and sensor altitude' (NASA, 2013); spatial resolution is also referred to as geometric resolution or IFOV (Instantaneous Field Of View), and it is usually measured in pixel size.

- Spectral resolution is the number and location in the electromagnetic spectrum (defined by two wavelengths) of the spectral bands (NASA, 2013) in multispectral sensors, for each band corresponds to an image.

- Radiometric resolution, usually measured in bits (binary digits), is the range of available brightness values, which in the image correspond to the maximum range of DNs; for example an image with 8 bit resolution has 256 levels of brightness (Richards and Jia, 2006).

- The temporal resolution, related to satellites' sensors, is the time required for revisiting the same area of the Earth (NASA, 2013).

Satellite resolutions deeply influence applications in environmental studies, therefore the evaluation of satellite capabilities on which the objectives of the studies are based is fundamental. The following section examines the characteristics of the main high-resolution satellites.

### High-resolution satellites

#### Landsat

Landsat is a set of multispectral satellites developed by NASA (National Aeronautics and Space Administration) since the early 1970s, which are widely used in environmental research on land cover and soil consumption (Vogelmann *et al.*, 1998; Lu *et al.*, 2011).

The resolutions of Landsat 4 and 5, Landsat 7 and Landsat 8 are reported in Tables 2.1, 2.2 and 2.3 respectively (http://landsat.usgs.gov/band_designations_landsat_satellites.php, accessed 27 January 2016).

Landsat temporal resolution is 16 days, which allows for frequent image acquisitions and land cover change analyses (NASA, 2013). At the moment, only Landsat 7 and Landsat 8 are operational, but a vast archive of Landsat images

*Table 2.1* Band characteristics of Landsat 4 and 5

| Landsat 4, Landsat 5 bands | Wavelength (micrometres) | Resolution (metres) |
|---|---|---|
| Band 1 – Blue | 0.45–0.52 | 30 |
| Band 2 – Green | 0.52–0.60 | 30 |
| Band 3 – Red | 0.63–0.69 | 30 |
| Band 4 – Near Infrared (NIR) | 0.76–0.90 | 30 |
| Band 5 – Shortwave Infrared (SWIR) | 1.55–1.75 | 30 |
| Band 6 – Thermal Infrared | 10.40–12.50 | 120 (resampled to 30) |
| Band 7 – Shortwave Infrared (SWIR) | 2.08–2.35 | 30 |

*Table 2.2* Band characteristics of Landsat 7

| Landsat 7 bands | Wavelength (micrometres) | Resolution (metres) |
|---|---|---|
| Band 1 – Blue | 0.45–0.52 | 30 |
| Band 2 – Green | 0.52–0.60 | 30 |
| Band 3 – Red | 0.63–0.69 | 30 |
| Band 4 – Near Infrared (NIR) | 0.77–0.90 | 30 |
| Band 5 – Shortwave Infrared (SWIR) | 1.57–1.75 | 30 |
| Band 6 – Thermal Infrared | 10.40–12.50 | 60 (resampled to 30) |
| Band 7 – Shortwave Infrared (SWIR) | 2.09–2.35 | 30 |
| Band 8 – Panchromatic | 0.52–0.90 | 15 |

*Table 2.3* Band characteristics of Landsat 8

| Landsat 8 bands | Wavelength (micrometres) | Resolution (metres) |
|---|---|---|
| Band 1 – Coastal aerosol | 0.43–0.45 | 30 |
| Band 2 – Blue | 0.45–0.51 | 30 |
| Band 3 – Green | 0.53–0.59 | 30 |
| Band 4 – Red | 0.64–0.67 | 30 |
| Band 5 – Near Infrared (NIR) | 0.85–0.88 | 30 |
| Band 6 – Shortwave Infrared 1 (SWIR 1) | 1.57–1.65 | 30 |
| Band 7 – Shortwave Infrared 2 (SWIR 2) | 2.11–2.29 | 30 |
| Band 8 – Panchromatic | 0.50–0.68 | 15 |
| Band 9 – Cirrus | 1.36–1.38 | 30 |
| Band 10 – Thermal Infrared (TIRS) 1 | 10.60–11.19 | 100 (resampled to 30) |
| Band 11 – Thermal Infrared (TIRS) 2 | 11.50–12.51 | 100 (resampled to 30) |

is freely available for the past decades at the USGS EROS (http://earthexplorer.usgs.gov/, accessed 27 January 2016).

The numerous Landsat bands allow for environmental analyses such as land cover and urban areas (Bagan and Yamagata, 2012), vegetation and ecosystem monitoring (Yang *et al.*, 2012) and land surface temperature using the thermal infrared (Sobrino *et al.*, 2004). In particular, the visible bands (blue, green and red) are useful for the visualization of urban features, and the near infrared bands allows for the identification of healthy vegetation and the calculation of vegetation indices (Rouse *et al.*, 1973).

Moreover, the new Landsat 8 characteristics allow for new and enhanced applications in agriculture, coastal water and change detection (Roy *et al.*, 2014).

Spatial resolution of multispectral bands (i.e. 30 m) is a constraint because the detection of small objects (e.g. isolated buildings) is difficult, therefore Landsat is mainly used for studies at the regional scale (Patino and Duque, 2013).

### Sentinel-2

The European initiative Copernicus includes the development of earth observation satellites. In particular, the Sentinel-2 satellite (launched in June 2015) is designed to provide high-resolution images for several spectral bands (Drusch *et al.*, 2012). It is worth noting that Sentinel-2 images are provided for free by the European Space Agency (ESA).

Table 2.4 (https://sentinel.esa.int/web/sentinel/user-guides/sentinel-2-msi/resolutions/radiometric, accessed 27 January 2016) summarizes the characteristics of the Sentinel-2 sensor, which is comparable to Landsat sensors.

Sentinel-2 bands have different spatial resolutions depending on the spectral range; however, the visible and near infrared bands have a resolution of 10 m, which is remarkable if compared to Landsat pixel size (i.e. 30 m). Therefore, several applications are possible using the numerous spectral bands of Sentinel-2 that allow for the accurate identification of land cover classes, especially for vegetation; in fact, the vegetation red edge bands are very useful for deriving vegetation indices and assessing the state of crops (Clevers and Gitelson, 2013).

*Table 2.4* Band characteristics of Sentinel-2

| Sentinel-2 bands | Central wavelength (micrometres) | Resolution (metres) |
|---|---|---|
| Band 1 – Coastal aerosol | 0.443 | 60 |
| Band 2 – Blue | 0.490 | 10 |
| Band 3 – Green | 0.560 | 10 |
| Band 4 – Red | 0.665 | 10 |
| Band 5 – Vegetation Red Edge | 0.705 | 20 |
| Band 6 – Vegetation Red Edge | 0.740 | 20 |
| Band 7 – Vegetation Red Edge | 0.783 | 20 |
| Band 8 – NIR | 0.842 | 10 |
| Band 8b – Vegetation Red Edge | 0.865 | 20 |
| Band 9 – Water vapour | 0.945 | 60 |
| Band 10 – SWIR – Cirrus | 1.375 | 60 |
| Band 11 – SWIR | 1.610 | 20 |
| Band 12 – SWIR | 2.190 | 20 |

*Other satellites*

Several satellites (especially the commercial ones) are useful for land cover monitoring, offering high- and very high-resolution images. It is worth illustrating the main characteristics of the available satellites, in particular for:

- SPOT
- RapidEye
- QuickBird
- IKONOS
- WorldView.

SPOT is a system of satellites designed and developed by the French space agency Centre National d'Études Spatiales. SPOT 4 and 5 are multispectral satellites, acquiring four spectral bands with a 10 m spatial resolution: green, red, near infrared (NIR) and short-wave infrared (SWIR). The ESA provides most SPOT images for free to research studies, and considering their spatial resolution SPOT data are valuable for land cover and urban studies (Kong *et al.*, 2012).

RapidEye is a commercial satellite providing five bands (blue, green, red, NIR and red edge) with 5 m pixel size. These bands have proved to be useful for urban studies (Munafò *et al.*, 2015), especially for the identification of vegetation (Tigges *et al.*, 2013).

QuickBird is a satellite providing very high-resolution images (panchromatic at 61 cm) and 2.44 m resolution multispectral bands (i.e. blue, green, red and NIR) (www.digitalglobe.com/sites/default/files/QuickBird-DS-QB-Prod.pdf, accessed 21 November 2013). IKONOS is similar to QuickBird, having 82 cm resolution for panchromatic and 3.2 m resolution for multispectral bands (i.e. blue, green, red and NIR) (www.digitalglobe.com/sites/default/files/DG_IKONOS_DS.pdf, accessed 21 November 2013). These two commercial satellites allow for urban studies at the local scale, detecting very small objects on the Earth's surface (Myint *et al.*, 2011).

WorldView is a family of multispectral satellites with very high spatial resolution (panchromatic less than 50 cm and multispectral less than 2 m) with several bands ranging from visible to infrared (www.digitalglobe.com/sites/default/files/DG_WorldView3_DS_forWeb_0.pdf, accessed 21 November 2013); the very high resolution allows for the identification of small features, which is very useful in urban areas (Belgiu *et al.*, 2014).

## Land cover classification using remote sensing images

The use of remote sensing for land cover classifications has proved to be reliable and affordable for the detection of impervious surfaces (Brook and Davila, 2000; Fan *et al.*, 2007).

Classification methodologies rely on image resolutions producing different results in terms of land cover classes and accuracy (Richards and Jia, 2006); consequently, several approaches have been developed for land cover classifications depending on the scale of the study.

## Methodologies of land cover classification

According the image processing approach, methodologies of classification can be defined as:

- unsupervised classification
- supervised classification
- Object Based Image Analysis
- photo-interpretation.

Unsupervised classification is a per-pixel methodology (i.e. based on spectral characteristics of single pixels, often using clustering methods) where spectral classes are assigned without foreknowledge of the existence of classes. Therefore the definition of classes is performed after the classification (Richards and Jia, 2006). This method is relatively quick to execute although the a posteriori definition of classes can be difficult.

Supervised classifications (also semi-automatic classifications) are per-pixel processing techniques that use class foreknowledge for the identification of materials, based on the spectral properties (i.e. spectral signatures) of the materials on the ground (Richards and Jia, 2006). There are several algorithms in this category of classification that have been widely used with remote sensing images, such as: the Maximum Likelihood algorithm, which calculates the probability distributions (assumed in the form of multivariate normal models) for the classes, related to Bayes' theorem, estimating if a pixel belongs to a land cover class (Strahler, 1980; Richards and Jia, 2006); the Spectral Angle Mapping algorithm calculates the spectral angle between spectral signatures of image pixels and class spectral signatures defined a priori (Kruse *et al.*, 1993; Fiumi *et al.*, 2014).

With the increasing availability of very high-resolution images, the Object Based Image Analysis (OBIA) has been developed, based on image segmentation, for exploiting the spatial properties of objects on the ground (Blaschke *et al.*, 2014). This kind of classification is particularly useful for the classification of urban land cover (Myint *et al.*, 2011).

Finally, photointerpretation is the visual inspection of images that allows for very high accuracy levels (Richards and Jia, 2006). For instance, photointerpretation is used for classification of samples of the Italian monitoring network of soil consumption (ISPRA, 2013).

## Classification accuracy

The accuracy assessment of a land cover classification is a fundamental step of the monitoring process, in order to identify and measure map errors and at the same time evaluate the coherence between the classification and reality.

*Table 2.5* Schematization of an error matrix

|  | Ground truth 1 | Ground truth 2 | . . . | Ground truth k | Total |
|---|---|---|---|---|---|
| Class 1 | $a_{11}$ (correct) | $a_{12}$ (error) | . . . | $a_{1k}$ (error) | $a_{1+}$ |
| Class 2 | $a_{21}$ (error) | $a_{22}$ (correct) | . . . | $a_{2k}$ (error) | $a_{2+}$ |
| . . . | . . . | . . . | . . . | . . . | . . . |
| Class k | $a_{k1}$ (error) | $a_{k2}$ (error) | . . . | $a_{kk}$ (correct) | $a_{k+}$ |
| Total | $a_{+1}$ | $a_{+2}$ | . . . | $a_{+k}$ | n |

Note: k is the number of classes identified in the land cover classification; n is the total number of collected sample units.

In general, classification accuracy is calculated with an error matrix (also referred to as confusion matrix), which is a table that compares map information with reference data (i.e. ground truth data) for a number of sample units (Congalton and Green, 2009).

The structure of an error matrix has ground truth classes (i.e. reference data) in columns and thematic map classes (i.e. classification data) in rows (see Table 2.5). Therefore, correctly classified samples are located in the major diagonal of the matrix, while errors are in the other elements of the matrix (Richards and Jia, 2006).

The selection of sample units should be random and, depending on classification spatial resolution, sample units can be a single pixel, a cluster of pixels or a polygon (Congalton and Green, 2009). The reference data are produced by field survey, or by the photo interpretation of images having higher spatial resolution than classification.

It is possible to calculate several accuracy statistics using the error matrix. In particular, the Overall Accuracy is the ratio between the number of samples that are correctly classified and the total number of sample units (Congalton and Green, 2009):

Overall Accuracy = $a_{sum}$ / n

Where:

- $a_{sum}$ = the sum of the major diagonal
- n = total number of sample units

In addition, it is possible to calculate the User's Accuracy and the Producer's Accuracy for each class, defined as (Congalton and Green, 2009):

User's Accuracy = $a_{ii}/R_i$

Where:

- $a_{ii}$ = samples classified correctly for class i
- $R_i$ = sum of row i, which is the number of samples belonging to class i in the classification

Producer's Accuracy = $a_{ii}/Ci$

Where:

- $a_{ii}$ = samples classified correctly for class i
- $C_i$ = sum of column i, which is the number of samples belonging to class i in reality

Generally, a classification is considered good if class accuracy is at least 85 per cent (European Environmental Agency, 2012).

## Case studies about land cover monitoring

During the last few decades, Europe has been developing several initiatives in order to provide land cover information to users in the field of environmental and other terrestrial applications. In particular, the GMES/Copernicus programme developed a series of projects, starting from the GMES Fast Tracking Services (European Commission, 2005), through the INSPIRE Directive (2007/2/EC) that aims to create a European spatial data infrastructure.

The European Environmental Agency has been particularly active in monitoring land cover since the 1990s, in particular with the Land Cover project of the CORINE programme (i.e. Coordination of Information on the Environment).

EUROSTAT has developed a monitoring network (LUCAS, Land Use and Cover Area Frame Survey) that monitors land use and cover change since 2006 in the European Union, based on an *in situ* survey of samples; this survey allows for the production of homogeneous statistics for European Countries.

Now, at the European level, the activities of land cover monitoring rely on Copernicus products and services such as the High Resolution Layers (HRLs).

At the regional and local level, land cover monitoring requires higher resolution data, acquired systematically according to the pace of growth of urban areas. For these reasons, Copernicus products still require additional information and parallel studies in order to correctly assess soil consumption.

Moreover, for non-European countries where Copernicus data are not available, land cover monitoring must rely only on affordable and efficient methodologies, mainly based on remote sensing. For instance at the global level, the ESA has developed a global land cover map having 300 m spatial resolution.

In Chapters 13 and 16 two case studies are described: the first one illustrates land cover monitoring in Italy, addressed by ISPRA (the Italian National Institute for Environmental Protection and Research), which in 2015 published the second National Report on Soil Consumption (Munafò *et al.*, 2015); the second case study is the assessment of land cover change in Dar es Salaam (Tanzania) using free Landsat data, in the frame of the European project ACC Dar (Adapting to Climate Change in Coastal Dar es Salaam).

## Conclusions

At the European level, soil consumption is a major issue that in the past decades has become a priority for environmental policies; the European Commission (2012b) published the guidelines on best practice to limit, mitigate or compensate soil sealing which are the key strategies for reducing soil consumption and land cover change. The pace of land cover change is important in light of the European objectives for 2020 (European Commission, 2011).

Therefore, land cover monitoring is becoming a crucial activity in order to assess soil consumption and consequently adapt current policies at the various administrative levels. The Copernicus initiative, which aims at monitoring the earth surface and environmental changes, is becoming the main point of reference at the European and national levels, especially with the production of HRLs.

However, considering the complexity of environmental effects caused by soil consumption at the local level, it is fundamental that land cover monitoring could assess even little changes. In Italy, ISPRA has developed several methodologies and products at very high resolution with the purpose of assessing soil consumption with a high level of precision.

The VHRL will be useful for local administrations in order to assist decision making and keep the database of land cover up to date. This valuable information, homogenous and complete for the whole country, could improve urban planning and policy making from the local to the national level.

## References

Bagan, H. and Yamagata, Y. (2012) 'Landsat analysis of urban growth: How Tokyo became the world's largest megacity during the last 40 years', *Remote Sensing of Environment*, 127, 210–222.

Belgiu, M., Drăguţ, L. and Strobl, J. (2014) 'Quantitative evaluation of variations in rule-based classifications of land cover in urban neighbourhoods using WorldView-2 imagery', *ISPRS Journal of Photogrammetry and Remote Sensing*, 87, 205–215.

Blaschke, T., Hay, G.J., Kelly, M., Lang, S., Hofmann, P., Addink, E., Queiroz Feitosa, R., van der Meer, F., van der Werff, H., van Coillie, F. and Tiede, D. (2014) 'Geographic Object-Based Image Analysis: Towards a new paradigm', *ISPRS Journal of Photogrammetry and Remote Sensing*, 87, 180–191.

Brook, R.M. and Davila, J. (2000) *The Peri-urban Interface: A Tale of Two Cities*, School of Agricultural and Forest Sciences, University of Wales and Development Planning Unit, University College London, Gwynedd, Wales.

Burghardt, W., Banko, G., Hoeke, S., Hursthouse, A., de L'Escaille, T., Ledin, S., Ajmone Marsan, F., Sauer, D., Stahr, K, Amann, E., Quast, J., Nerger, M., Schneider, J. and Kuehn. K. (2004) 'Taskgroup 5: Sealing soils, soils in urban areas, land use and land use planning', in L. Van-Camp, B. Bujarrabal, A.R. Gentile, R.J.A. Jones, L. Montanarella, C. Olazabal and S.-K. Selvaradjou (eds) *Reports of the Technical Working Groups Established Under the Thematic Strategy for Soil Protection*, Volume VI: *Research, Sealing and Cross-cutting Issues*. Office for Official Publications of the European Communities, Luxembourg.

Cardona, O.D., van Aalst, M.K., Birkmann, J., Fordham, M., McGregor, G., Perez, R., Pulwarty, R.S., Schipper, E.L.F. and Sinh, B.T. (2012) 'Determinants of risk: exposure and vulnerability', in *Managing the Risks of Extreme Events and Disasters to Advance Climate Change Adaptation*, A Special Report of Working Groups I and II of the Intergovernmental Panel on Climate Change (IPCC), Cambridge University Press, Cambridge and New York, 65–108.

Chen, X., Bai, J., Li, X., Luo, G., Li, J. and Li, B.L. (2013) 'Changes in land use/land cover and ecosystem services in Central Asia during 1990–2009', *Current Opinion in Environmental Sustainability*, 5, 116–127.

Clevers, J. and Gitelson, A. (2013) 'Remote estimation of crop and grass chlorophyll and nitrogen content using red-edge bands on Sentinel-2 and -3', *International Journal of Applied Earth Observation and Geoinformation*, 23, 344–351.

Congalton, R. and Green, K. (2009) *Assessing the Accuracy of Remotely Sensed Data: Principles and Practices*, CRC Press, Boca Raton, FL.

CRCS (2012) *Rapporto 2012. Centro di Ricerca sui Consumi di Suolo*, INU Edizioni, Milano.

Drusch, M., Del Bello, U., Carlier, S., Colin, O., Fernandez, V., Gascon, F., Hoersch, B., Isola, C., Laberinti, P., Martimort, P., Meygret, A., Spoto, F., Sy, O., Marchese, F. and Bargellini, P. (2012) 'Sentinel-2: ESA's optical high-resolution mission for GMES Operational Services', *Remote Sensing of Environment*, 120, 25–36.

European Commission (2005) 'GMES: from concept to reality', COM(2005) 565. European Commission, Brussels.

European Commission (2006) 'Thematic strategy for soil protection', COM(2006) 231. European Commission, Brussels.

European Commission (2011) 'Our life insurance, our natural capital: an EU biodiversity strategy to 2020', COM (2011) 244 final. European Commission, Brussels.

European Commission (2012a) 'The implementation of the Soil Thematic Strategy and ongoing activities', COM (2012) 46. European Commission, Brussels.

European Commission (2012b) 'Guidelines on best practice to limit, mitigate or compensate soil sealing', SWD (2012) 101. European Commission, Brussels.

European Commission (2014) 'Mapping and assessment of ecosystems and their services: indicators for ecosystem assessments under Action 5 of the EU Biodiversity Strategy to 2020', 2nd Report. European Commission, Brussels.

European Environmental Agency (1997) 'The concept of environmental space: implications for policies', Environmental Reporting and Assessments. EEA, Copenhagen.

European Environmental Agency (2006) 'Urban sprawl in Europe: the ignored challenge'. Report. EEA/OPOCE, Copenhagen.

European Environmental Agency (2012) 'Guidelines for verification of high-resolution layers produced under GMES/Copernicus initial operations', (Gio) Land Monitoring 2011–2013 version 4.

Fan, F., Weng, Q. and Wang, Y. (2007) 'Land use and land cover change in Guangzhou, China, from 1998 to 2003, based on Landsat TM /ETM+ Imagery', *Sensors*, 7, 1323–1342.

Fisher, P. and Unwin, D. (2005) *Re-Presenting GIS*, Chichester, England: John Wiley and Sons.

Fiumi, L., Congedo, L. and Meoni, C. (2014) 'Developing expeditious methodology for mapping asbestos-cement roof coverings over the territory of Lazio Region', *Applied Geomatics*, 6, 37–48.

Huber, S., Prokop, G., Arrouays, D., Banko, G., Bispo, A., Jones, R.J.A., Kibblewhite, M., Lexer, W., Moller, A., Rickson, R.J., Shishkov, T., Stephens, M., Toth, G., van den Akker, J., Varallyay, G. and Verheijen, F. (2008) *Environmental Assessment of Soil for Monitoring*, Volume I: *Indicators and Criteria*, JRC, Office for the Official Publications of the European Communities, Luxembourg.

Indovina, F. (2006) *Governare la città con l'urbanistica. Guida agli strumenti di pianificazione urbana e del territorio*, Maggioli, Rimini.

IPCC (2001) *Climate Change 2001: Impacts, Adaptation, and Vulnerability: Contribution of Working Group II to the Third Assessment Report of the IPCC*, Cambridge University Press, Cambridge, UK.

ISPRA (2013) 'Il monitoraggio del consumo di suolo in Italia', *Ideambiente*, 62, 20–31, ISPRA, www.isprambiente.gov.it/files/ideambiente/ideambiente_62.pdf, accessed 23 March 2015.

JARS (1993) 'Remote sensing note: Japan Association on Remote Sensing', www.jars1974.net/pdf/rsnote_e.html, accessed 22 October 2014.

Kong, F., Yin, H., Nakagoshi, N. and James, P. (2012) 'Simulating urban growth processes incorporating a potential model with spatial metrics', *Ecological Indicators*, 20, 82–91.

Kruse, F.A., Lefkoff, A.B., Boardman, J.W., Heidebrecht, K.B., Shapiro, A.T., Barloon, P.J. and Goetz, A.F.H. (1993) 'The Spectral Image Processing System (SIPS): interactive visualization and analysis of imaging spectrometer data', *Remote Sensing of Environment*, 44, 145–163.

Lal, R. (2005) *Encyclopedia of Soil Science*, CRC Press, Boca Raton, FL.

Lu, D., Moran, E. and Hetrick, S. (2011) 'Detection of impervious surface change with multitemporal Landsat images in an urban–rural frontier', *ISPRS Journal of Photogrammetry and Remote Sensing*, 66(3), 298–306.

Maes, J., Egoh, B., Willemen, L., Liquete, C., Vihervaara, P., Schägner, J.P., Grizzetti, B., Drakou, E.G., La Notte, A., Zulian, G., Bouraoui, F., Paracchini, M.L., Braat, L. and Bidoglio, G. (2012) 'Mapping ecosystem services for policy support and decision making in the European Union', *Ecosystem Services*, 1, 31–39.

Munafò, M., Norero, C., Sabbi, A. and Salvati, L. (2010) 'Soil sealing in the growing city: a survey in Rome, Italy', *Scottish Geographical Journal*, 126(3), 153–161.

Munafò, M., Assennato, F., Congedo, L., Luti, T., Marinosci, I., Monti, G., Riitano, N., Sallustio, L., Strollo, A., Tombolini, I. and Marchetti, M. (2015) 'Il consumo di suolo in Italia: Edizione 2015'. Rapporti 218/2015, ISPRA, Roma.

Myint, S.W., Gober, P., Brazel, A., Grossman-Clarke, S. and Weng, Q. (2011) 'Per-pixel vs. object-based classification of urban land cover extraction using high spatial resolution imagery', *Remote Sensing of Environment*, 115, 1145–1161.

NASA (2013) 'Landsat 7 science data user's handbook', http://landsathandbook.gsfc.nasa.gov, accessed 23 May 2014.

Patino, J.E. and Duque, J.C. (2013) 'A review of regional science applications of satellite remote sensing in urban settings computers', *Environment and Urban Systems*, 37, 1–17.

Richards, J.A. and Jia, X. (2006) *Remote Sensing Digital Image Analysis: An Introduction*, Springer, Berlin.

Rouse, J.W., Haas, R.H., Schell, J.A. and Deering, D.W. (1973) 'Monitoring vegetation systems in the Great Plains with ERTS NASA', Goddard Space Flight Center 3d ERTS-1 Symp., 1-A, 309–317.

Roy, D., Wulder, M.A., Loveland, T.R., Woodcock, C.E., Allen, R.G., Anderson, M.C., Helder, D., Irons, J.R., Johnson, D.M., Kennedy, R., Scambos, T.A., Schaaf, C.B., Schott, J.R., Sheng, Y., Vermote, E.F., Belward, A.S., Bindschadler, R., Cohen, W.B., Gao, F., Hipple, J.D., Hostert, P., Huntington, J., Justice, C.O., Kilic, A., Kovalskyy, V., Lee, Z.P., Lymburner, L., Masek, J.G., McCorkel, J., Shuai, Y., Trezza, R., Vogelmann, J., Wynne, R.H., Zhu, Z. (2014) 'Landsat-8: science and product vision for terrestrial global change research', *Remote Sensing of Environment*, 145, 154–172.

Sobrino, J., Jiménez-Muñoz, J.C. and Paolini, L. (2004) 'Land surface temperature retrieval from LANDSAT TM 5', *Remote Sensing of Environment*, 90, 434–440.

Strahler, A.H. (1980) 'The use of prior probabilities in maximum likelihood classification of remotely sensed data', *Remote Sensing of Environment*, 10, 135–163.

TEEB (2010) *Mainstreaming the Economics of Nature: A Synthesis of the Approach, Conclusions and Recommendations of TEEB*.

Tigges, J., Lakes, T. and Hostert, P. (2013) 'Urban vegetation classification: benefits of multitemporal RapidEye satellite data', *Remote Sensing of Environment*, 136, 66–75.

Verburg, P.H., Erb, K.-H., Mertz, O. and Espindola, G. (2013) 'Land system science: between global challenges and local realities', *Current Opinion in Environmental Sustainability*, 5, 433–437.

Vogelmann, J., Sohl, T., Campbell, P. and Shaw, D. (1998) 'Regional land cover characterization using Landsat thematic mapper data and ancillary data sources', *Environmental Monitoring and Assessment*, 51, 415–428.

Yang, J., Weisberg, P.J. and Bristow, N.A. (2012) 'Landsat remote sensing approaches for monitoring long-term tree cover dynamics in semi-arid woodlands: comparison of vegetation indices and spectral mixture analysis', *Remote Sensing of Environment*, 119, 62–71.

# 3 Measuring and monitoring the extent of human settlements

From the local to the global scale

*Daniele Ehrlich, Aneta J. Florczyk, Andreea Julea,
Thomas Kemper, Martino Pesaresi and
Vasileios Syrris*

## Introduction

Population increase and urbanisation are fuelling the growth of cities and human settlements. This growth is often at the expense of valuable agricultural land from which societies draw their food base. It is also at the expense of forest and other natural land that provides timber or other ecosystem services such as clean water and fresh air. This process of growth modifies the land uses and the land cover and most importantly seals soil with built material. The extent of the growth of the built environment is much talked about but rarely quantified. This is often because of a lack of semantics and measurement technologies issues that are also briefly addressed in this chapter.

This chapter addresses the measurement of the spatial extent of human settlements and their changes in time. This measure can be used as a proxy value for the loss of soils. Settlements are part of the landscape that includes buildings, roads and transport networks that are also referred to as built-up environment. In its simpler term, we can define a settlement as any form of human habitation, which ranges from a single dwelling to a large city. Settlements' building blocks are three dimensional constructions typically referred to as buildings used for residential or other societal activities. Settlements differ in aspect and function from other land cover types. While vegetation is still found as parks and lawns, the cover is by and large dominated by concrete, asphalt and other man-made covers. It is thus completely different from other (semi-)natural land cover types.

Quantifying changes in human settlement is not trivial. Measuring settlement requires an unambiguous definition of a built-up area and changes in a built-up area, and assumes standardisation of measurement (measurement scale) and standardisation in processing or modelling information on built-up areas. This work uses the building in its different uses (i.e. as residential, commercial, industrial) as the characterising element of the built-up environment (Pesaresi *et al.*, 2008). Other constructions, roads or parking lots can be included in the built-up. However, the building is the only characterising element with the density of built-up as a measure. Density is defined as the area identified by the building footprint over a given spatial reporting unit (Pesaresi *et al.*, 2013), typically the grid cell. The changes can thus be measured as changes of density within that reporting unit (Gueguen *et al.*, 2011). The changes can be coded as *no-change*, when both images show either non-built-up or the same amount of built-up, *positive change* and *negative change* when the percentage of built-up increases or decreases respectively between the two dates.

Human settlements are studied using aerial photography and satellite imagery, also referred to as remote sensing. The most valuable characteristic of satellite remote sensing is its ability to provide a synoptic overview that allows us to outline the extent of a settlement, its size, its form and the complexity of the urban fabric. In addition, when analysed over time, imagery allows measuring the change in size and form of settlements.

Remotely sensed data are available globally at different resolutions, and offer a multi-temporal representation of the Earth. Each sensor provides unique opportunities, either spatial precision, or temporal coverage, or spectral characteristics to be used in the detection of the built environment. Each can provide information that, when combined, can provide a useful measure of the increase in the built environment.

The following sections provide an overview of remote sensing technology and its use for measuring changes in the built environment. First, we list the type of satellite imagery that has been used and that potentially can be used to derive information on the built environment. Second, we provide examples of analysis of urban growth from different sensors and using different procedures. We then show two examples of change at the city level, assuming only the city's change of interest. We then provide examples of global and regional processing that are conducted in an automatic way. Finally, we discuss the challenges in combining imagery and image processing products at different resolution.

## Satellite imagery

This section summarises the types of remotely sensed satellite images used in civilian applications and provides an outlook on future missions. We consider both the open source imagery and the commercial imagery used for the analysis of the built environment. The unique characteristic of satellite remote sensing is its ability to collect imagery globally. The data acquired by a satellite is stored in large imagery archives, which allow temporal comparison, and thus urban change analysis, even at global scale. Remote sensing has been widely recognised as the most economic and feasible approach to derive land cover information over large areas (Cihlar, 2000). Today, the continuous remotely sensed observations of the Earth's land surface offer unique opportunities to perform multi-temporal analysis of global phenomena. Satellite programmes continue to proliferate. Civilian, military/intelligence and commercial communities enjoy the imaging capabilities of polar-orbiting satellites. Since the first Earth Observation satellite was launched by the USA in 1972, almost 200 satellites have been launched with a global land cover mission; and at the end of 2013, 50 per cent of them were still operating (Belward and Skien, 2015).

The most relevant long-term missions that offer a major data source for developing continental to global scale land cover and change products at spatial resolutions necessary for many surface phenomena are the Landsat,[1] MODIS,[2] SPOT Vegetation[3] and Sentinel[4] missions. The Landsat mission is an ideal source of data because of its 40-year acquisition legacy (Markham and Helder, 2012), which provides long-term inventory of global land cover change at a sub-hectare resolution (30–80 m). Currently, the Global Land Survey (GLS) datasets (Gutman et al., 2013), i.e. collections of orthorectified, cloud-minimised Landsat-type satellite images, aim at providing mosaics of near complete coverage of the global land area and are available per decade since the early 1970s, centred on 1975, 1990, 2000, 2005, and 2010. The Landsat record will continue to grow with the currently operational Landsat-8 and the planned Landsat-9 in 2023.

Another source of long-term land cover observations is the MODerate resolution Imaging Spectrometer (MODIS) on board the Earth Observing System (EOS) Terra and Aqua satellites (Ardanuy et al., 1991). In particular, Terra's MODIS, in operational mode from 2002, is specifically designed to monitor land properties at global scales, and acquires multispectral data with medium resolution but high temporal frequency (almost daily).

In Europe, the SPOT Vegetation programme offers long-term imagery that is relevant for observing and analysing the evolution of land surfaces and understanding land changes over large areas (Henry et al., 1996; Mucher and de Badts, 2002). The 1 km SPOT Vegetation (VGT) data have been acquired by SPOT 4–5 satellites from 1998 till 2015 on a daily basis. Recently, the European Space Agency (ESA) has launched the first satellite of the Sentinel 2 constellation, which is designed to provide systematic global acquisition of high-resolution, multispectral images allied to a high revisit frequency (Berger et al., 2012; Malenovský et al., 2012). These observation data will be the base for the next generation of operational products, such as land cover maps, land change detection maps and geophysical variables. Sentinel 2 is one of the Sentinel missions developed by ESA within the European programme Copernicus for the establishment of a European capacity for Earth observation beyond 2025 (Aschbacher and Milagro-Pérez, 2012).

Additional sources for mapping human presence from space at global scale are the long-term time series of night light imagery. There is, for example, the imagery produced by the Defense Meteorological Satellite Program (DMSP) Operational Linescan System (OLS) (Croft and Colvocoresses, 1979; Imhoff et al., 1997) and the Suomi-NPP satellite that carries a panchromatic Day/Night Band (DNB) radiometer, namely VIIRS (Miller et al., 2013). However, the night light imagery includes also temporary light sources such as wild fires and volcanic eruptions that have to be taken into account. In addition, they indicate human economic activities. Hence, settlements without illumination will be neglected.

Table 3.1 presents a comparison of Landsat, MODIS and Sentinel with other selected sources. We can differentiate between panchromatic (PAN), multispectral and synthetic aperture radar (SAR) sensors. A panchromatic sensor is sensitive to all visible colours and usually has the highest spatial resolution. Some sensors, such as MODIS on EOS-Terra, are able to provide data at two or more spatial resolutions. Also, some missions carry more than one sensor, which capture data at different resolutions. SAR missions, such as ESA Sentinel 1, offer variable resolution image acquisitions through different operational modes.

Other examples of a SAR mission can be ESA Envisat (ASAR instrument) (ESA, 1993), ESA ALOS (PALSAR instrument) (Henderson and Lewis, 1998)

*Table 3.1* Examples of global observing missions and selected technical specifications. Repeat cycle in days/minutes depends on latitude, cloud conditions (in case of optical sensors) and constellation configuration (e.g. Sentinel 2 using the full two-satellite constellation configuration)

| OWNER Platform (Sensor) | Operational period (platform) | Spatial resolution (m) (total number of bands) | Revisiting time in days (d) or minutes (min) | Main objective (operational purpose) |
|---|---|---|---|---|
| NASA Terra/Aqua (MODIS) | 1999–... (T) 2002–... (A) | 250m (2) 500m (5) 1km (28) | 1–2 d | Multiple |
| NASA/USGS Landsat 8 (OLI/TIRS) | 2012–2017 | 15m (PAN) 30m (9) 100m (2) | 16 d | Land cover (continuity) |
| NASA/USGS Landsat 7 (ETM+) | 1999–2003 | 15m (PAN) 30m (6) 60m (1) | 16 d | Land cover (continuity) |
| NASA Landsat 4–5 (MSS/TM) | 1982–2001 (4) 1984–2013 (5) | 80/30m (4/6) 120m (1) | 16 d | Land cover (continuity) |
| NASA Landsat 1–3 (MSS) | 1972–1978 (1) 1975–1982 (2) 1978–1983 (3) | 80m (4) | 18 d | Land cover |
| CNES SPOT 5 | 2002–2015 | 2.5 or 5m (PAN) 10m (4) 20m (1) | 2–3 d | Land cover (continuity) |
| Spot Image SPOT 6–7 | 2012–2023 (6) 2014–2023 (7) | 1.5m (PAN) 6m (4) | 1–3 d | Land cover (continuity) |
| ESA Sentinel 3 A/B/C | 2015–2025 (A) 2017–2025 (B) 2022–2025 (C) | 300m–1km | 27 d | Global ocean and land monitoring (continuity ENVISAT, SPOT VEG) |
| ESA Sentinel 2 A/B/C | 2014–2020 (A) 2016–2022 (B) 2021–2025 (C) | 10m (3) 20m (6) 60m (3) | 2–5 d (with 2 satellites) | High-resolution and optical imaging for land services (continuity SPOT and Landsat MT) |

| ESA Sentinel 1 A/B/C (SAR) | 2014–2020 (A) 2015–2021 (B) 2021–2025 (C) | 5m (Strip Map and wave modes) | <1–3 d | All-weather, day and night radar-imaging for land and ocean services |
|---|---|---|---|---|
| NASA/NOAA Suomi-NPP (VIIRS) | 2011–. . . | 750m (PAN DNB) 750m (21) | 102 min | continuity with MODIS, DMSP-OLS and NOAA AVHRR |
| US DoD DMSP (OLS) | 1972–2013 | 2.7km (daily) | 101 min | Meteorology |

or the German Aerospace Center (DLR) TerraSAR-X (Eineder and Runge, 2002). Since a SAR sensor scans the Earth's surface with the help of micro-waves, it has advantages over optical instruments, i.e. it works also at night and despite cloud cover, and it can be used to obtain reliable geophysical measurements (e.g. backscatter constants, distances).

Global Earth Observation missions aim at improving spatial and radiometric (i.e. quantisation) resolution of the provided imagery (see Table 3.2). In general, we can observe an increment in spatial and spectral resolution. The third dimension of improvement focuses on the spectral resolution, which is referred to as hyperspectral images. So far, hyperspectral sensors are operated mainly on airborne remote sensing platforms such as the Airborne Visible InfraRed Imaging Spectrometer (AVIRIS). There are also space borne missions (Bioucas-Dias *et al.*, 2013), such as the Italian Space Agency (ASI) Hyperspectral Precursor and application mission (PRISMA) or the DLR EnMap (Environmental Mapping and Analysis Program) (Table 3.3).

*Table 3.2* Examples of optical space-borne missions (in orbit, approved and planned) grouped based on resolution nomenclature commonly used in the Copernicus programme (i.e. Low Resolution (LR), Medium Resolution (MR), High Resolution (HR), Very High Resolution (VHR)) according to the highest resolution on board.

| LR (>300m) | MR1/MR2 (30–300m) | HR2 (10–30m) | HR1 (4–10m) | VHR2 (4–1m) | VHR1 (<1m) |
|---|---|---|---|---|---|
| DMSP 1972–2013; Envisat AATSR/ MERIS 2002–2013; Sentinel 3 2015 | Terra MODIS 1999; Aqua MODIS 2002; PROBA-V 2013–2015 | Sentinel 2 2014 | *RapidEye 5, Follow-on 2008;* Sentinel 1 2014 | *SPOT 4–7 1998; IKONOS 1999;* Seosat/ Ingenio 2 2014 | *WorldView 1–2 2007; GeoEye 1 2008; Pleiades 1–2 2011; Deimos 2 2014; DMC3 2015* |

Note: Italic text indicates commercial missions.

*Table 3.3* Examples of hyperspectral platforms

| Platform (Sensor) | Launch year | Spatial resolution (total number of bands) | Spectral resolution of hyperspectral bands | Comments |
|---|---|---|---|---|
| Space-borne DLR EnMap | 2018 | 30m (228) | 6.5nm in a range of 420–1000nm (VNIR); 10nm in a range of 900–2450nm (SWIR) | German mission to support ecosystem applications Status: planned Revisit: 4–27 days |
| Space-borne ASI PRISMA | 2015 | 20–30m (238); 2.5–5m (PAN) | 10nm in a range of 400–2500nm (VNIR and SWIR regions) | Italian mission of demonstrative/ technological and pre-operational nature supporting multiple applications Status: planned |
| Airborne NASA/JPL ERS (AVIRIS) | 1987 | 20m (224) | 10nm in a range of 380–2500nm | Climate change (not limited to) Status: operational |

## Change analysis at city level

This section provides two examples on change detection at the city level using two different sensors. The first example illustrates automatic built-up mapping and change detection by combining recent VHR SPOT imagery and older HR SPOT imagery over Alger. SPOT data are commercial, thus with limited access. The second case relies on open access Landsat imagery from archives that date back to the mid-1970s. An analysis using Landsat images over Bangalore is briefly presented. The two case studies use concepts and procedures that can be applied to detect changes at continental scale.

### Alger case study

The Alger case study uses SPOT 1 images from 8 July 1986 and SPOT 5 from 9 February 2009, which are available as panchromatic bands at 10 and 2.5 m resolution, respectively. SPOT imagery is used to detect changes in the built environment due to its fine spatial resolution that captures, by and large, most of the built-up structures. Some authors have also attempted to measure the changes only for the built environment by comparing built-up maps produced at different moments in time (Tiede *et al.*, 2012). This work follows a new research trend where changes are measured by directly comparing features computed from the imagery (Ehrlich and Bielski, 2011; Gueguen *et al.*, 2013). The conceptual and methodological issues on change detection and this specific case study are fully reported in Ehrlich *et al.* (2015).

This change detection method uses imagery from two different sensors and two different spatial resolutions. The processing procedure consists in six main steps: (1) pre-processing imagery for geometric correction; (2) calculation of built-up presence index (BUPI) features (Pesaresi *et al.*, 2008) for each image; (3) combining (stacking and resampling) the BUPI features into a single two-band image in order to perform principal component (PC) analysis; (4) processing the PC2 as a change feature; (5) thresholding the change feature into a built-up change map; (6) modelling the change feature according to a regular grid. This last step associates the BUPI features with the desired information 'built-up' or 'not built-up' and 'built-up change' or 'not built-up change' as described in Ehrlich *et al.* (2015).

For the sake of simplicity, we provide below a descriptive summary of processing steps and the results. The main processing relates to the computation of a texture-related feature (Pantex) as described in Pesaresi *et al.* (2008). These derived texture measurements have been shown to be highly correlated with the presence of built-up land. In fact, the 2009 texture image is used to generate a binary built-up map, which is produced by simply thresholding the texture values.

The change analysis is based on identifying changes in textures, and thus changes in built-up. The changes are quantified using principal component (PC) analysis between texture measures computed for the images collected in 1986 and 2009. PC 1 captures the region of the image with similar texture and is not used in the analysis. PC 2 captures the changes in texture between the 1986 and 2009 images and thus the changes in built-up and is thus our change information layer. We analysed only the *positive changes* as reported in PC2, that is a change from low texture (i.e. agricultural land) into landscapes with high texture (i.e. built-up).

In order to make the changes in built-up information more explicit, we have applied a threshold on the PC2 obtaining the built-up change map of interest. Finally, the change information was aggregated at a spatial unit of 100 × 100 m to provide the gridded change map. This aggregation into grids of 100 m allows the fine tuning of changes by removing unwanted artefacts corresponding to small patches or to those whose change signal is of low magnitude. The artefacts are inevitable given the technical characteristics of imagery.

Thresholding the texture information into a change map and generalising the information into a gridded change map is crucial to interpret the results. The threshold, applied to the change map, simplifies the density of change information into binary change information. This simplification is justified by the inability to obtain fine density changes due to the relatively coarse spatial resolution of SPOT 1 imagery, but inevitably comes with a loss of information.

The aggregating of the change into 100 m grid cells has also some generalisation drawbacks. Each 100 m grid cell is labelled 'built-up' and/or 'built-up change' irrespective of the density of built-up within the cell. One cell with a fraction of built-up is treated similar to the cell entirely covered by built-up and this influences the change detection statistics. This generalisation also amplifies

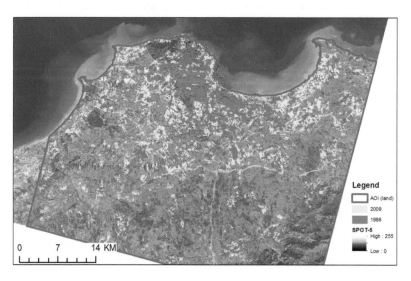

*Figure 3.1* Alger and settlements surrounding Alger over a 52 × 40 km² area. The
settlement maps are labeled red for built-up before 1986 and yellow for
built-up after 1986 (modified from Ehrlich *et al.*, 2015)

the statistics of built-up and/or its change – especially for low densities – and
should be taken into account when interpreting the results below.

The processing generates two information layers used for change analysis: a
gridded change map and a built-up map for 2009, both shown in Figure 3.1 as
yellow and red zones, respectively.

The 2009 built-up map and change maps have not been validated quanti-
tatively due to a lack of reference data both for 1986 and for 2009. However,
they have been visually inspected against the imagery from which the change
was produced (Table 3.4). The analysis shows that 339.69 km² are measured as
built-up in 1986. In only 23 years, 173.80 km² are added to the built-up land
of 1986, corresponding to an increase of 50 per cent of the built-up area. The
total built-up land over this area increases from just over 26 per cent in 1986 to
nearly 40 per cent in 2009. The statistics are based on the analysis of built-up
land computed over 100 m grid cells. In fact, different density thresholds or

*Table 3.4* Built-up and built-up change statistics over the Alger metropolitan area

| Date | 1986 | | 2009 |
|---|---|---|---|
| Built up area (km²) | 339.69 | | 513.49 |
| *Urban change (km²)* | | *173.80* | |
| Percentage of built-up area over total (%) | 26.39 | | 39.89 |
| *Urban change (%)* | | *13.50* | |

different grid cell sizes used in the analysis may provide results that differ from those provided herein. In addition, statistics may be confirmed only through a thorough validation protocol.

### Bangalore case study

A similar analysis can be performed using other imagery, for example Landsat. Here, the collections of nominal temporal signature 1975, 1990, 2000 and 2014 were used to produce a global change map. The image processing method is described in the section 'Global processing', below. In this section we describe the change analysis performed on Bangalore, the capital of the Indian state of Karnataka. Since the city population has increased from 4.3 to 8.4 million between 2001 and 2011 (ORGI, 2015), it is an interesting case study.

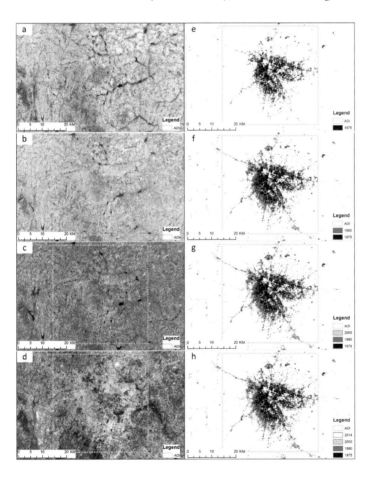

*Figure 3.2* City of Bangalore as seen from Landsat imagery: 27 Feb. 1973 (a), 14 Jan. 1992 (b), 27 Nov. 2000 (c) and 31 Mar. 2014 (d). The urban change maps encode the urban area detected in 1975 (e), 1990 (f), 2000 (g) and 2014 (h)

*Table 3.5* Built-up area per time period for the Bangalore case study, as derived from analysis of Landsat imagery collections

| Time period | 1975 | | 1990 | | 2000 | | 2014 |
|---|---|---|---|---|---|---|---|
| Built up area (km²) | 139 | | 251 | | 329 | | 520 |
| Urban change (km²) | | 112 | | 78 | | 191 | |
| Percentage of built-up area over total | 8.66 | | 15.62 | | 20.52 | | 32.42 |
| Urban change (%) | | 6.96 | | 4.9 | | 11.9 | |

All four Landsat collections were processed to generate four information layers on built-up presence. Then the layers were merged into one urban change map, which is a classification grid of 38.22 m resolution. There are four classes of urban area, which represent urban areas that appear in one of the epochs (i.e. 1975, 1990, 2000 or 2014). A class is assigned to a cell according to first occurrence of the built-up within the multi-temporal layers.

The study area is a square (40 × 40 km) that covers the present-day extent of the city of Bangalore. Figure 3.2 shows the study area on Landsat false colour images and the mapped change per epoch. Table 3.5 gathers the calculated extent of the built-up area per epoch. It can be observed that the major change occurred between 2000 and 2014 but also between 1975 and 1990. Also, here we do not have proper validation data, and we rely on visual analysis of the images. However, if assuming a positive correlation between urban extent change and population growth, the census information indicates a relevant population increase between 1970 and 1990 as well (ORGI, 2015).

## Global urban area mapping

Methods for mapping urban extension or studying its morphology and growth have radically changed with the arrival of remote sensing technology. However, urban remote sensing is very challenging due to the heterogeneity of urban areas. In some cases, the discrimination of vegetative land cover may help in the delineation of urban areas, a theory that was tested by exploiting the near-infrared band of multispectral imagery (SPOT) in Gao and Skillcorn (1998). Some researchers have used hyperspectral imagery for mapping a narrow range of urban materials (Salu, 1995; Ben-Dor *et al.*, 2001) or for analysing morphological characteristics of urban areas (Benediktsson *et al.*, 2005). Other approaches to urban area detection exploit spectral and spatial characteristics of Landsat (Guindon *et al.*, 2004; Guindon and Zhang, 2009), while Platt and Goetz (2004) found that hyperspectral AVIRIS holds advantages over Landsat ETM+ for the classification of heterogeneous and vegetated land uses (for the tested urban–rural fringe). Furthermore, SAR data (i.e. ENVISAT) have been tested as a baseline for urban mapping by means of a textural analysis (Ban *et al.*, 2015) or for urban change detection (Yousif, 2015); and night lights from DMSP-OLS data have been studied to map urban areas (Imhoff *et al.*, 1997; Small *et al.*, 2005).

Currently, the data fusion for urban area characterisation has become a common approach, and it may be done at different levels, namely multi-sensor, multiresolution (scale-space) or multi-temporal (Gamba *et al.*, 2005). There are studies that explore the integration of hyperspectral and SAR imagery for urban mapping (Hepner *et al.*, 1998; Gamba and Houshmand, 2001), the combination of DMSP-OLS data with vegetation indexes derived from MODIS (Schneider *et al.*, 2003), SPOT (Cao *et al.*, 2009) or Landsat (Zhang *et al.*, 2015). Multi-temporal Landsat data have been used to extract impervious surface time series for multi-temporal settlement mapping on Java Island (Patel *et al.*, 2015), and the analysis of annual urban dynamics in Beijing city (Li *et al.*, 2015).

Most of the methods for detecting urban areas from remote sensing imagery have been tested on some selected areas. However, it is a challenge to produce a dataset at the global scale. Currently, there are several global datasets relevant for mapping the urban extent (Elvidge *et al.*, 2009; Poterea *et al.*, 2009). There is also an ongoing project, the DLR Global Urban Footprint (GUF), which aims at mapping settlements globally at around 12 m using SAR imagery (Esch *et al.*, 2013). Additionally, there is a research team that attempts to predict the future change of urban extent for selected cities (Angel *et al.*, 2011).

In this work, we provide some details on selected datasets, namely MODIS 500m Global Urban Extent[5] (MODIS 500m), Global Land Cover 2000[6] (GLC2000), GlobCover[7] and GlobeLand30[8] (Table 3.6). Other global datasets fall in the following categories: population maps (e.g. LandScan (Bhaduri *et al.*, 2002) or WorldPop (WorldPop, 2015)), soil sealing surfaces (e.g. Global Density of Constructed Impervious Surface Areas (ISA) (Elvidge *et al.*, 2007)), nightlight-derived urban maps (Zhou *et al.*, 2015) or place-name databases (GeoNames, 2015).

MODIS 500m dataset has been created by exploiting spectral and temporal information in one year of MODIS observations (Schneider *et al.*, 2010). The global training database was created by the stratification of urban ecoregions, which have been defined via natural, physical and structural elements of urban areas. The product validation focused on 140 cities, because the method targets relatively extended settlements while neglecting sparsely urbanised areas, mainly due to the coarse resolution of the input imagery. However, MODIS 500 is an improvement over MODIS 1km, which was produced using MODIS data, DMSP-OLS dataset (1 km mosaic) and gridded population data (about 5 km). Both MODIS datasets were used to detect changes in urban areas (Mertes *et al.*, 2015).

GLC2000 is a harmonised global land cover classification database based on SPOT VGT data, created by an international partnership of 30 institutions (Bartholomé *et al.*, 2002; Bartholomé and Belward, 2005). The 'urban' class has been derived with the help of nightlight data; however, the performed validation using Landsat 7 has not targeted the 'urban' class (Bicheron *et al.*, 2008). The lessons learnt contributed to the ESA GlobCover initiative, which have delivered two global composite and land cover maps that use the same classification nomenclature. Also here, authors admit that the 'urban' class has a low accuracy, as the urban areas are underestimated and the class is not well represented in the validation dataset (i.e. points) (Bontemps *et al.*, 2011).

Table 3.6 Selected global datasets relevant for urban area mapping

| | MODIS 500m | GlobCover2005 / GlobCover2009 | GLC2000 | GlobeLand30 2000/2010 |
|---|---|---|---|---|
| Producer | University of Wisconsin–Madison | ESA | JRC (coordinator) | National Geomatics Center of China |
| Purpose | Urban area | Land cover | Land and inland water biodiversity | Land cover to support sustainable development goals |
| Input satellite data | MODIS | ENVISAT/MERIS | SPOT 4/Vegetation | LandsatTM/ETM+ / LandsatTM/ETM+ and Chinese environmental and disaster satellite (HJ-1) |
| Time consistency | 2001–2002 | 2004–2006 / 2009 | 1999–2000 | 2000 2010 |
| Type of data | Classification (1 class) | Classification (22 classes) | Classification (22 classes) | Classification (10 major classes) |
| Urban area class | Areas dominated by built environment (>50%), including non-vegetated, human-constructed elements, with minimum mapping unit >1 km$^2$ | Artificial surfaces and associated areas (urban areas >50%) | Artificial surfaces and associated areas (urban areas >50%) | Artificial cover (settlement place, industrial and mining area, traffic facilities) |
| Production method (*specific for urban class) | Supervised classification (ensemble decision-tree) | Unsupervised classification; supervised classification for urban and wetlands areas | Mainly unsupervised classification; an ad hoc classification algorithm using auxiliary data (e.g. DMSP) | POK-based approach; a supervised pixel-based classification (using spectral and texture characteristics), then segmentation, and finally visual verification and correction |
| Stratification | Urban ecoregions | Equal-reasoning areas | Continental-like | N/A |
| Urban coverage | 4.16% | 0.22% / 0.20% | 0.19% | 0.95% |
| Resolution | 500m | 300m | 1km | 30m |

Recently, a Chinese global land cover classification, GlobeLand30, has been released at 30 m as a result of a four-year effort (Chen *et al.*, 2015). The applied operational approach, called POK (the pixel-object-knowledge), mixes automatic classifiers and interactive processes (in cases of classification in complex areas and for quality control). First, each class is identified in an a priori sequence, by applying pixel- and object-based classification. Then, the results are merged through a knowledge-based interactive verification (i.e. experts using auxiliary datasets). Urban areas fall into one of the difficult cases.

When evaluating data for potential usage, we should also consider regional datasets. For example, there are multiple datasets hosted by the European Environment Agency (EEA) that can be used for urban analysis in Europe. The main datasets are HR Soil Sealing (SSL) (EEA, 2015), multi-temporal CORINE Land Cover (CLC) (EEA, 2012), and Urban Atlas (UA) (European Commission, 2011) (see Tables 3.7 and 3.8). Most of the urban area in CLC is encoded within the 'artificial surfaces' class. However, the sparse built-up structures are ignored, especially in agricultural or (semi-)natural areas (i.e. the units smaller than 25 ha are included in the dominant land cover type around or grouped in polygons labelled as 'heterogeneous'). Also, SSL underrepresents or completely omits small and dispersed rural settlements (Hurbanek *et al.*, 2010). UA offers a far more accurate picture of urban sprawl in the fringe of urban zones than CLC but it does not offer full European coverage. In practice, many research studies combine those datasets to mitigate their mutual limitations.

*Table 3.7* European datasets relevant for urban area mapping

|  | *CLC* | *SSL* | *UA* |
| --- | --- | --- | --- |
| *Input data* | Multiple imagery | SPOT 4–5 IRS P6 LISS III | Satellite imagery (SPOT 5, ALOS P/XS, RapidEye XS and QUICKBIRD), SSL, road network, topographic and cartographic maps (different scales), other ancillary data (e.g. local digital/paper maps, Bing) |
| *Method* | Photo-interpretation (MMA: 25ha) | Automatic image analysis | Photo-interpretation (MMA of 'artificial surfaces': 0.25ha) |
| *Type of data* | Land cover | Thematic gradient | Land use (extension of CLC nomenclature) |
| *Urban area definition* | 'Artificial surfaces' class | Soil sealing degree | Several classes of 'artificial surfaces' category |
| *Time consistency* | Around 1990, 2000, 2006 and 2012 | 2006 | 2005–2007 |
| *Spatial coverage* | Europe | Europe | European urban areas (>100,000 inhabitants) |
| *Spatial resolution* | Vector | 20 and 100m | Vector |

*Table 3.8* CORINE Land Cover products

|                              | CLC1990   | CLC2000    | CLC2006                  | CLC2012                  |
| ---------------------------- | --------- | ---------- | ------------------------ | ------------------------ |
| *Satellite data*             | Landsat 5 | Landsat 7  | SPOT 4–5, IRS P6 LISS III | IRS P6 LISS III RapidEye |
| *Time consistency*           | 1986–1998 | 2000 +/− 1 | 2006 +/− 1               | 2011–2012                |
| *Number of countries involved* | 26–27  | 30–35      | 38                       | 39                       |

The cost of human interaction (e.g. photo-interpretation) is usually very high when producing most global urban maps. Since there is large variety of local landscapes in different areas, there is an issue for supervised and unsupervised methods for urban areas mapping. Most classifiers are tuned into the local study area, and if the same settings are applied to other areas, the accuracy of detection may decrease significantly. Therefore, experienced researchers are needed for parsing the results, which can be very time consuming and expensive. In case of supervised methods, this cost is even higher, because training samples will be gathered per scene. Therefore, automatic methods are the best option, especially for producing time series maps at the global scale. However, existing automated classification methods have been deemed ineffective because of the low classification accuracy achievable at the global scale and at HR2 resolution, as tested at 30 m Landsat imagery in Gong *et al.* (2013).

Recently, an alternative automatic method for urban area extraction has been successfully applied at global and continental scales. The European Settlement Map[9] (ESM) was developed jointly by the Joint Research Centre (JRC) and the Directorate General for Regional Policy (DG REGIO) of the European Commission. The fundamental methodological choices followed in the processing chain are coherent with the Global Human Settlement Layer (GHSL) paradigm. The next section will outline the GHSL methodology and its applications.

## Global and regional processing

The GHSL methodology has been developed to provide an automatic image processing method for extracting built-up surfaces from remote sensing images and produce information layers of high resolution at the global scale (Pesaresi *et al.*, 2013). The methodology is able to process imagery with varying characteristics (such as diversity in spatial/spectral resolution, spatial/temporal coverage, spatial displacement errors; quality degradation that makes the calibration impossible; seasonality). As such, it qualifies as an example for the processing of remotely sensed big data (Ma *et al.*, 2014), in terms of the data volume, diversity and complexity. This is achieved through a fully automatic and computationally efficient method that is robust and general enough.

The main characteristic of the developed methodology is a scene-based processing and automatic image feature extraction by applying a multiscale learning paradigm. The multiscale learning relies on low resolution auxiliary data. The traditional approach to urban area detection for continental and global coverage (Schneider *et al.*, 2010) is based on searching for an 'urban' spectral signature (i.e. homogeneous and dominant combination of spectral characteristics) within an a priori defined local region. This approach has strong limitations when used with decametric or metric spatial resolution image data.

### Global processing

The multi-temporal medium resolution GHSL is the first geographic data-set that describes the spatial evolution of the human settlements at the global scale and along a time interval covering 40 years (from 1975 to the present). Producing spatio-temporal built-up layers is a demanding task requiring several processing steps and sophisticated modelling. Prototyping and production were fraught with several challenges such as: (1) size, diversity and quality of the input/output datasets, (2) parameterisation and fine-tuning of the information extraction and fusion techniques and (3) computational complexity.

In order to deal with this complexity in an efficient way, a new methodology has developed that is able to cope with (1) a large number of data granules (scenes), (2) imagery captured by heterogeneous sensors and (3) morphological diversity spread over different geographical areas and at different time spans. The new approach treats the image values as symbols and attempts to build associations between sequences of symbolic objects and target class values that represent the land cover semantics. The sequences can be formed by information derived either from the image bands directly or from features extracted through data-driven (statistical) or model/assumption-based (analytical) methods. Typically, this information retains a spatial consistency, yet potentially can span to time domain. The so-called Symbolic Machine Learning associative classifier, which has been defined in this context, is a supervised-learning technique that maximises the within-class similarity of the symbolic objects based on their frequent appearance in each of the classes. The classifier is controlled by very few, easily tunable parameters, and the processing chain can be modulated smoothly to any low to moderate computational infrastructure.

In the specific application of the multi-temporal medium resolution GHSL, the information was extracted from Landsat image records organised in four collections[10] corresponding to the epochs 1975, 1990, 2000 and the present time. Table 3.9 shows the type of imagery we used and quantifies the volume of data and the respective processing time. The fourth collection is composed by a set of Landsat 8 images from the years 2013 and 2014.

Both feature extraction and image classification have been implemented at the original resolution of the input images. At the final stage, the images were warped to the WGS84 Web Mercator projection at 38.22 m. The outcome of the processing is a multiclass geographic layer with the following notation: no-data (0),

*Table 3.9* Landsat imagery and GHSL processing time

|  | GLS1975 | GLS1990 | GLS2000 | Landsat-8 |
|---|---|---|---|---|
| Number of scenes | 7,588 | 7,375 | 8,756 | 4,426 (2013) 4,663 (2014) |
| Number of bands | 4 | 6 | 6 | 9 |
| Working resolution | 60m | 30m | 15m | 15m |
| Total processing time per image (sec) | 300 | 680 | 750 | 1980 |

Indicative machinery: Intel(R) Xeon(R) CPU E7420 @ 2.13GHz, 8~10GB RAM

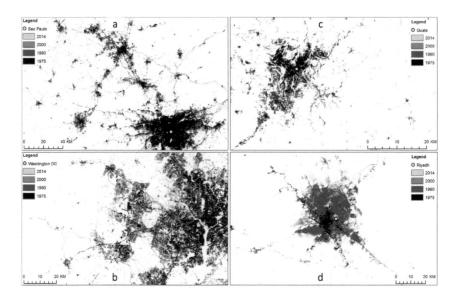

*Figure 3.3* Multi-temporal representation of four cities: Sao Paolo (a), Washington DC (b), Guate (c) and Riyadh (d)

water bodies (1), land classified as non-built-up (2), 2013/14 built-up (3), 2000 built-up (4), 1990 built-up (5) and 1975 built-up (6). Data and cloud masks are also available for each collection. Figure 3.3 shows the derived urban area change in the example of four cities.

### Continental processing for Europe

Another example of successful application of the GHSL methodology is the high-resolution regional ESM. It is produced as a built-up density map released at 100 m for the general public. It has been produced from the pan-European Copernicus (Core 003) dataset (Burger *et al.*, 2012), an image collection produced in support to the UA project, which includes multispectral SPOT 5

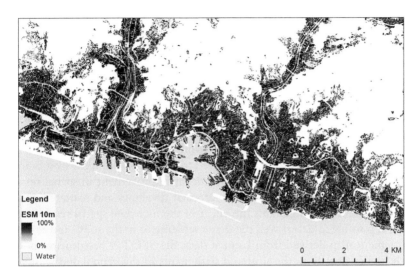

*Figure 3.4* Density of built-up depicted by the European Settlement Map for the city centre of Genoa, Italy, and its surroundings

and some SPOT 6 scenes of 2.5 m and 1.5 m spatial resolution, respectively. Although the pre-processing (i.e. pan-sharpening, histogram stretching) has caused significant spectral degradation in the data (Burger *et al.*, 2012), the method proved to be robust enough to extract meaningful information in an automatic way. In total, 2,900 SPOT images have been processed at 2.5 m resolution.

The information layer has been produced using an automatic image processing workflow (see Florczyk *et al.*, 2015). Several auxiliary datasets have been used in the production, the main being SSL, CLC and OpenStreetMap. The method combines radiometric, textural and morphological analysis in order to detect built-up structures. The produced 10 m and 100 m ESM datasets offer built-up density maps, and each pixel (i.e. 100 $m^2$ and 10,000 $m^2$ cells, respectively) represents a percentage of built-up structure within the spatial domain (i.e. cell). This approach enables a quantitative analysis of the urban area. Figure 3.4 presents an example of the city of Genoa.

### National processing for South Africa

Since 2006, the South African National Space Agency (SANSA) has been acquiring the national SPOT 5 imagery annually to support various aspects of government planning and monitoring, including mapping and monitoring of human settlements. In South Africa, the proportion of people living in urban areas increased from 52 per cent in 1990 to 62 per cent in 2011, and about 8.2 per cent of the population was living in informal settlements (Statistics South

Africa, 2011). In addition to natural population growth and the migration of people from rural areas to cities, urbanisation is also influenced by the migration of people from neighbouring and other parts of Africa (Statistics South Africa, 2011). Both cities and smaller towns are experiencing high growth rates, together with the proliferation of informal settlements around them.

To support the efforts of the government, SANSA and the JRC have developed a dedicated and fully automated workflow for the processing of SANSA's SPOT 5 imagery, based on multiscale textural and morphological image features extraction (Kemper *et al.*, 2015). In total, 485 scenes (a 2.5 m panchromatic and four 10 m multispectral bands) were processed. The SPOT 5 imagery increases the spatial detail compared to the Landsat roughly by a factor of 10 (from 30 m to 2.5 m). Such a significant improvement in spatial resolution is crucial for the monitoring of informal dwellings and scattered rural settlements. Figure 3.5 highlights the effect of the increased spatial resolution. While it is possible to detect well the dense settlements in the south and east of the settlement map derived from Landsat data, the SPOT 5-based maps show a much higher density in the scattered settlements in the central and western part of the maps.

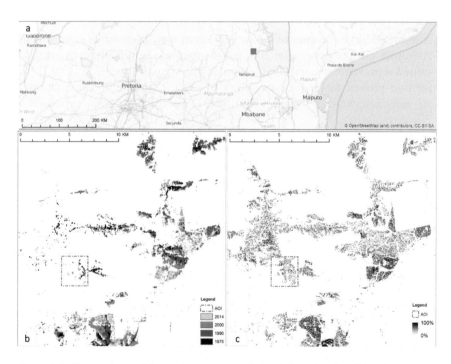

*Figure 3.5* Comparison of the settlement maps derived from Landsat (b) and SPOT 5 (c) for a selected rural area in South Africa (a). The SPOT 5 map shows building densities

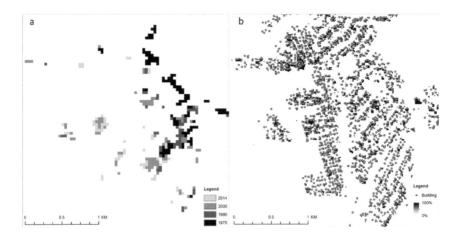

*Figure 3.6* Detail of the two settlement maps from Figure 3.5 derived from Landsat
(a) and SPOT 5 (b). The SPOT 5 settlement map (b) includes building
points derived from a visual interpretation of the imagery. Note the good
match between the SPOT 5 settlement map (black) and the building
points (red)

A closer look at these scattered settlements (Figure 3.6) confirms this
observation. The settlement map derived from SPOT 5 data is able to outline
also the built-up area of the scattered settlements, which is only partly mapped
by the Landsat data. This example illustrates clearly the need for an integrated,
multiscale concept (such as the GHSL) that is able to provide consistent information at global, regional and local scales.

In general, the GHSL datasets will provide an excellent basis for measuring
changes in built-up in the future. Also, they will be the baseline that will be
used to hindcast built-up change across large areas.

## Discussion

Mapping urban area change from optical remotely sensed data at the global
scale poses several challenges. The first challenge in processing the Landsat GLS
collections was cloud coverage. For example, 12.6 per cent and 6.6 per cent
of the land masses were not covered in GLS1975 and GLS1990 respectively
due to cloud cover. Furthermore, 5 per cent of the GLS2000 images processed
had more than 10 per cent of cloud cover (Gutman *et al.*, 2013). For this
reason, the images that made up one collection in time were actually selected
from images covering a wider temporal range than the nominal year. The
second challenge to be addressed is vegetation seasonality that prevents obtaining a stable information extraction algorithm. In fact, it may cause false land
cover 'change' that is just a change in vegetation cover. The recent Landsat

8 mission's objectives include, among others, affording seasonal coverage of the global land mass for a period of no less than five and three years for reflective and thermal multispectral image data respectively (Irons *et al.*, 2012).

Another challenge is the definition of built-up when using imagery collected with different measurement scales. For example, medium resolution imagery allows identifying constructed land often referred to as impervious surface (Elvidge *et al.*, 2007). This land includes roads, parking lots, buildings, driveways, sidewalks and other man-made surfaces. Imagery with higher spatial resolution, for example the imagery that was used to produce the ESM, allows separation between buildings and open spaces that require yet another definition of built-up and another set of information extraction algorithms. The lack of appropriate and consistent multi-temporal reference datasets makes the training of algorithms and the evaluation of the precision of the produced built-up map very difficult. In fact, no proper protocol for validating built-up maps globally is available today.

The two case studies on change detection at city level test a conceptual model for quantifying built-up areas and changes in time from multi-resolution remote sensing. It is tested on VHR2/HR and MR imagery. The processing aims to provide features that are related to the building density. The features can then be modelled to provide density of built-up and, when compared in time, changes of built-up. The selection of the input datum, the processing techniques and the modelling procedure (i.e. the area at which the density is computed and the spatial rules used to compute the density) will determine the final outcome.

The visual analysis of urban area change using multi-resolution imagery is also challenging due to the difference in resolution. The visual analysis of 2.5 m imagery (i.e. SPOT 5) confirms that built-up land can be measured, because the building structure can be enumerated and the spatial arrangement of buildings can be assessed. At 10 m resolution (i.e. SPOT 1), only large buildings may be identified and mapped. With Landsat imagery, at resolution coarser than 15 m, the majority of the built-up structures cannot be identified and it is rather the density of constructed land that is detected visually.

The automatic procedure may be better suited to detect the building structures from multi-resolution imagery. However, challenges remain mostly due to the wide variety of built-up patterns (i.e. different sizes and spatial arrangements of built-up structures). The challenges in detecting built-up areas are multiplied when changes in built-up are analysed. The change detection techniques perform relative unambiguous results when the change in built-up occurs through the encroachment of built-up land into other land cover. That is when natural land is converted in dense built-up land. However, the small density changes are difficult to assess due to the characteristics of the data and to the absence of reference data.

Further refinement of the techniques and interpretation is needed. The ultimate goal of this work is to walk through the conceptual change model, test the process rather than the technique and the result. The final map and the final statistics have to evaluate based on the input imagery and the processing procedure used. Also, the map statistics need to be checked against reference data that often is not available. Each step of the procedure will be further evaluated to better understand the information content of the imagery, the techniques used to measure changes and the eventual outcome to be used in urbanisation studies.

Finally, Mertes *et al.* (2015) indicate that an urban area (or an impervious surface) is a relatively stable land type over a long period of time, and typically only positive changes occur – from natural land into built-up land. However, due to the dynamics of urban change globally, the future methods for urban area change detection should also consider negative changes, from built-up to other land cover types.

## Conclusions

This chapter addresses some of the challenges in quantifying the growth of human settlement at local, continental and global scales. It is also about semantics and terminology that are indispensable for understanding what is being mapped and for training image processing procedures. Especially in cases of continental and global change mapping, the data availability, handling large datasets, processing huge data volume in an automatic way are the main issues.

The work shows that using different satellite sensors we will produce different results, simply due to the precision of sensors' measurements. The key is to understand the limitation and the advantages of each sensor and to define transfer functions that allow comparing one with the other. VHR satellite imagery provides enough detail to map changes in the built-up environment in a systematic and thorough way. Medium resolution satellite imagery provides the unique global coverage and, most importantly, the historical records of Earth's landscape. Although the detail might not be desirable for urban change detection analysis, it can be useful for studying global change trends. Since long-term changes can only be obtained from archived imagery collections, which are the only record of past urban extent, we need to use multi-resolution change detection techniques.

This work shows examples from a medium resolution global built-up layer and derived changes computed globally over a time span of 40 years. Comparison with finer resolution SPOT datasets measured over part of South Africa shows examples of the opportunities that the SPOT GHSL product can provide. In fact, SPOT Europe provides the detail that can be used for a systematic high-resolution built-up analysis of Europe's built-up landscapes. The new forthcoming satellite imagery, such as that provided by the Copernicus service, will empower us to measure the built-up extent globally and with high quality.

The GHSL project has put in place the infrastructure that will allow us to process future satellite images collected by the Sentinel sensors and other free and open data sources and thus provide a true opportunity to monitor changes in built-up and therefore also in the loss of soil and natural landscape. The information is particularly relevant in view of the development of the composite indicators that will be used to monitor the targets of international frameworks such as the Sendai framework for disaster risk reduction and the sustainable development goals.

## Notes

1  http://landsat.usgs.gov/.
2  http://modis.gsfc.nasa.gov/.
3  www.spot-vegetation.com/index.html.
4  https://sentinels.copernicus.eu/web/sentinel/home.
5  https://nelson.wisc.edu/sage/data-and-models/schneider.php.
6  http://forobs.jrc.ec.europa.eu/products/glc2000/glc2000.php.
7  http://due.esrin.esa.int/page_globcover.php.
8  www.globallandcover.com/GLC30Download/index.aspx.
9  http://land.copernicus.eu/pan-european.
10  To download the images (http://landsat.usgs.gov/science_GLS.php), USGS provides the tool EarthExplorer at http://earthexplorer.usgs.gov/.

## References

Angel, S., Parent J., Civco, D.L., Blei, A. and Potere, D. (2011) 'The dimensions of global urban expansion: Estimates and projections for all countries, 2000–2050', *Progress in Planning*, 75, 2, 53–107.

Ardanuy, P.E., Han, D. and Salomonson, V.V. (1991) 'The moderate resolution imaging spectrometer (MODIS) science and data system requirements', *IEEE Transactions on Geoscience and Remote Sensing*, 29, 1, 75–88.

Aschbacher, J. and Milagro-Pérez, M.P. (2012) 'The European Earth monitoring (GMES) programme: Status and perspectives', *Remote Sensing of Environment*, 120, 3–8.

Ban, Y., Jacob, A. and Gamba, P. (2015) 'Spaceborne SAR data for global urban mapping at 30 m resolution using a robust urban extractor', *ISPRS Journal of Photogrammetry and Remote Sensing*, 103, 28–37.

Bartholomé, E. and Belward, A.S. (2005) 'GLC2000: A new approach to global land cover mapping from Earth observation data', *International Journal of Remote Sensing*, 26, 9, 1959–1977.

Bartholomé, E., Belward, A.S., Achard, F., Bartalev, S., Carmona-Moreno, C., Eva, H., Fritz, S., Grégoire, J.M., Mayaux, P. and Stibig, H.J. (2002) 'Global Land Cover mapping for the year 2000: Project status November 2002', European Commission, JRC, Ispra, Italy, EUR 20524 EN.

Belward, A. and Skien, J. (2015) 'Who launched what, when and why: Trends in global land-cover observation capacity from civilian earth observation satellite', *ISPRS Journal of Photogrammetry and Remote Sensing*, 103, 115–128.

Ben-Dor, E., Levin, N. and Saaroni, H. (2001) 'A spectral based recognition of the urban environment using the visible and near-infrared spectral region (0.4–1.1 mu/m): A case study over Tel-Aviv, Israel', *International Journal of Remote Sensing*, 22, 11, 2193–2218.

Benediktsson, J.A., Palmason, J.A. and Sveinsson, J.R. (2005) 'Classification of hyperspectral data from urban areas based on extended morphological profiles', *IEEE Transactions on Geoscience and Remote Sensing*, 43, 3, 480–491.

Berger, M., Moreno, J., Johannessen, J.A., Levelt, P.F. and Hanssen, R.F. (2012) 'ESA's sentinel missions in support of Earth system science', *Remote Sensing of Environment*, 120, 84–90.

Bhaduri, B.L., Bright, E.A., Coleman, P.R. and Dobson, J.E. (2002) 'LandScan: Locating people is what matters', *Geoinformatics*, 5, 2, 34–37.

Bicheron, P., Defourny, P., Brockmann, C., Schouten, L., Vancutsem, C., Huc, M., Bontemps, S., Leroy, M., Achard, F., Herold, M., Ranera, F. and Arino, O. (2008) 'GLOBCOVER products description and validation report', MEDIAS-France, December.

Bioucas-Dias, J.M., Plaza, A., Camps-Valls, G., Scheunders, P., Nasrabadi, N.M. and Chanussot, J. (2013) 'Hyperspectral remote sensing data analysis and future challenges', *IEEE Geoscience and Remote Sensing Magazine*, 1, 2, 6–36.

Bontemps, S., Defourny, P., Van Bogaert, E., Arino, O., Kalogirou, V. and Ramos-Perez, J. (2011) 'GLOBCOVER 2009 Product description and validation report', UCLouvain and ESA, February.

Burger, A., Di Matteo, G. and Astrand, P. (2012) 'Specifications of view services for GMES Core_003 VHR2 coverage', European Commission, JRC, Luxembourg, JRC Technical Report JRC70483.

Cao, X., Chen, J., Imura, H. and Higashi, O. (2009) 'A SVM-based method to extract urban areas from DMSP-OLS and SPOT VGT data', *Remote Sensing of Environment*, 113, 2205–2209.

Chen, J., Chen, J., Liao, A., Cao, X., Chen, L., Chen, X., He, C., Han, G., Peng, S., Lu, M., Zhang, W., Tong, X. and Mills, J. (2015) 'Global land cover mapping at 30 m resolution: A POK-based operational approach', *ISPRS Journal of Photogrammetry and Remote Sensing*, 103, 7–27.

Cihlar, J. (2000) 'Land cover mapping of large areas from satellites: Status and research priorities', *International Journal of Remote Sensing*, 21, 6–7, 1093–1114.

Croft, T.A. and Colvocoresses, A.P. (1979) 'The brightness of lights on earth at night, digitally recorded by DMSP satellite', U.S. Geological Survey, Palo Alto, CA, Open-File Report, 80–167.

EEA (2012) 'Implementation and achievements of CLC2006', European Environment Agency, Technical Report.

EEA (2015) 'Soil sealing data in aggregated spatial resolution (100 × 100 m)', www.eea.europa.eu/data-and-maps/data/eea-fast-track-service-precursor-on-land/-monitoring-degree-of-soil-sealing-100m, accessed 10 September 2015.

Ehrlich, D. and Bielski, C. (2011) 'Texture based change detection of built-up on SPOT panchromatic imagery using PCA', in *Joint Urban Remote Sensing Event (JURSE), 2011*, 77–80.

Ehrlich, D., Julea, A. and Pesaresi, M. (2015) 'Global spatial and temporal analysis of human settlements from Optical Earth Observation: Concepts, procedures, and preliminary results', European Commission, Joint Research Centre, Institute for the Protection and Security of the Citizen, JRC Technical Report.

Eineder, M. and Runge, H. (2002) 'Short analysis of a long-track interferometry capabilities of TerraSAR-X', DLR Memo, May.

Elvidge, C.D., Tuttle, B.T., Sutton, P.C., Baugh, K.E., Howard, A.T., Milesi, C., Bhaduri, B.L. and Nemani, R. (2007) 'Global distribution and density of constructed impervious surfaces', *Sensors*, 7, 1962–1979.

Elvidge, C.D., Sutton, P.C., Tuttle, B.T., Ghosh, T. and Baugh, K.E. (2009) 'Global urban mapping based on nighttime lights', in *Global Mapping of Human Settlement*, P. Gamba and M. Herold (eds), Taylor & Francis, Boca Raton, FL, 129–144.

ESA (1993) 'Envisat-a new ESA satellite project', *COSPAR Information Bulletin*, 1993, 127, 68–70.

Esch, T., Marconcini, M., Felbier, A., Roth, A., Heldens, W., Huber, M., Schwinger, M., Taubenböck, H., Müller, A. and Dech, S. (2013) 'Urban footprint processor: Fully automated processing chain generating settlement masks from global data of the TanDEM-X mission', *IEEE Geoscience and Remote Sensing Letters*, 10, 6, 1617–1621.

European Commission (2011) 'Urban atlas: Delivery of land use/cover maps of major European urban agglomerations', *Official Journal of the European Union*, Final Rep. (v 2.0), Call for Tenders no 2012.CE.16.BAT.066, November.

Florczyk, A.J., Ferri, S., Syrris, V., Kemper, T., Halkia, M., Soille, P. and Pesaresi, M. (2015) 'A new European settlement map from optical remotely sensed data', *IEEE Journal of Selected Topics in Applied Earth Observations and Remote Sensing*, 9, 5, 1978–1992, 10.1109/JSTARS.2015.2485662.

Gamba, P. and Houshmand, B. (2001) 'An efficient neural classification chain of SAR and optical urban images', *International Journal of Remote Sensing*, 22, 8, 1535–1553.

Gamba, P., Dell'Acqua, F. and Dasarathy, B.V. (2005) 'Urban remote sensing using multiple data sets: Past, present, and future', *Information Fusion*, 6, 4, 319–326.

Gao, J. and Skillcorn, D. (1998) 'Capability of SPOT XS data in producing detailed land cover maps at the urban-rural periphery', *International Journal of Remote Sensing*, 19, 15, 2877–2891.

GeoNames (2015) 'The GeoNames Project', www.geonames.org/, accessed 10 September 2015.

Gong, P., Wang, J., Yu, L., Zhao, Y., Zhao, Y., Liang, L., Niu, Z., Huang, X., Fu, H., Liu, S., Li, C., Li, X., Fu, W., Liu, C., Xu, Y., Wang, X., Cheng, Q., Hu, L., Yao, W., Zhang, H., Zhu, P., Zhao, Z., Zhang, H., Zheng, Y., Ji, L., Zhang, Y., Chen, H., Yan, A., Guo, J., Yu, L., Wang, L., Liu, X., Shi, T., Zhu, M., Chen, Y., Yang, G., Tang, P., Xu, B., Giri, C., Clinton, N., Zhu, Z., Chen, J. and Chen, J. (2013) 'Finer resolution observation and monitoring of GLC: First mapping results with Landsat TM and ETM+ data', *International Journal of Remote Sensing*, 34, 7, 2607–2654.

Gueguen, L., Soille, P. and Pesaresi, M. (2011) 'Change detection based on information measure', *IEEE Transactions on Geoscience and Remote Sensing*, 49, 11/2, 4503–4515.

Gueguen, L., Pesaresi, M., Ehrlich, D., Lu, L. and Guo, H. (2013) 'Urbanization detection by a region based mixed information change analysis between built-up indicators', *IEEE Journal of Selected Topics in Applied Earth Observations and Remote Sensing*, 6, 6, 2410–2420.

Guindon, B. and Zhang, Y. (2009) 'Automated urban delineation from Landsat imagery based on spatial information processing', *Photogrammetric Engineering and Remote Sensing*, 75, 7, 845–858.

Guindon, B., Zhang, Y. and Dillabaugh, C. (2004) 'Landsat urban mapping based on a combined spectral–spatial methodology', *Remote Sensing of Environment*, 92, 2, 218–232.

Gutman, G., Huang, C., Chander, G., Noojipady, P. and Masek, J.G. (2013) 'Assessment of the NASA-USGS Global Land Survey (GLS) datasets' *Remote Sensing of Environment*, 134, 249–265.

Henderson, F.M. and Lewis, A.J. (eds) (1998) *Principles and Applications of Imaging Radar: Manual of Remote Sensing: Third Edition*, volume 2, John Wiley & Sons, New York.

Henry, P., Gentet, T., Arnaud, M. and Andersson, C. (1996) 'The VEGETATION system: A global earth monitoring from SPOT satellites', *Acta Astronautica*, 38, 4–8, 487–492.

Hepner, G.F., Houshmand, B., Kulikov, I. and Bryant, N. (1998) 'Investigation of the integration of AVIRIS and IFSAR for urban analysis', *Photogrammetric Engineering and Remote Sensing*, 64, 8, 813–820.

Hurbanek, P., Atkinson, P., Pazur, R. and Rosina, K. (2010) 'Accuracy of built-up area mapping in Europe from the perspective of population surface modeling', Presented at European Forum for Geostatistics (EFGS) Conference, 5–7 October 2010, Tallinn, Estonia.

Imhoff, M.L., Lawrence, W.T., Stutzer, D.C. and Elvidge, C.D. (1997) 'A technique for using composite DMSP/OLS City Lights satellite data to map urban area', *Remote Sensing of Environment*, 61, 3, 361–370.

Irons, J.R., Dwyer, J.L. and Barsi, J.A. (2012) 'The next Landsat satellite: The Landsat Data Continuity Mission', *Remote Sensing of Environment*, 122, 11–21.

Kemper, T., Mudau, N., Mangara, P. and Pesaresi, M. (2015) 'Towards an automated monitoring of human settlements in South Africa using high resolution SPOT satellite imagery', *International Archives of the Photogrammetry, Remote Sensing and Spatial Information Sciences*, XL-7/W3, 1389–1394.

Li, X., Gong, P. and Liang, L. (2015) 'A 30-year (1984–2013) record of annual urban dynamics of Beijing City derived from Landsat data', *Remote Sensing of Environment*, 166, 1, 78–90.

Ma, Y., Wang, L., Huang, B., Ranjan, R., Zomaya, A. and Jie, W. (2014) 'Remote sensing big data computing: Challenges and opportunities', *Future Generation Computer Systems*, 15, 47–60.

Malenovský, Z., Rott, H., Cihlar, J., Schaepman, M.E., García-Santos, G., Fernandes, R. and Berger, M. (2012) 'Sentinels for science: Potential of Sentinel-1, -2, and -3 missions for scientific observations of ocean, cryosphere, and land', *Remote Sensing of Environment*, 120, 91–101.

Markham, B.L. and Helder, D.L. (2012) 'Forty-year calibrated record of earth-reflected radiance from Landsat: A review', *Remote Sensing of Environment*, 122, 30–40.

Mertes, C.M., Schneider, A., Sulla-Menashe, D., Tatem, A.J. and Tan, B. (2015) 'Detecting change in urban areas at continental scales with MODIS data', *Remote Sensing of Environment*, 158, 331–347.

Miller, S.D., Straka, W., Mills, S.P., Elvidge, C.D., Lee, T.F., Solbrig, J., Walther, A., Heidinger, A.K. and Weiss, S.C. (2013) 'Illuminating the capabilities of the Suomi National Polar-Orbiting Partnership (NPP) Visible Infrared Imaging Radiometer Suite (VIIRS) day/night band', *Remote Sensing*, 5, 12, 6717–6766.

Mucher, C.A. and de Badts, E.P.J. (2002) *Global Land Cover 2000: Evaluation of the SPOT VEGETATION Sensor for Land Use Mapping*, Alterra, Green World Research, Wageningen.

ORGI (2015) Office of the Registrar General and Census Commissioner, India (ORGI), www.censusindia.gov.in/, accessed 21 September 2015.

Patel, N.N., Angiuli, E., Gamba, P., Gaughan, A., Lisini, G., Stevens, F.R., Tatem, A.J. and Triann, G. (2015) 'Multitemporal settlement and population mapping from Landsat using Google Earth Engine', *International Journal of Applied Earth Observation and Geoinformation*, 35, B, 199–208.

Pesaresi, M., Gerhardinger, A. and Kayitakire, F. (2008) 'A robust built-up area presence index by anisotropic rotation-invariant textural measure', *IEEE Journal of Selected Topics in Applied Earth Observations and Remote Sensing*, 1, 3, 180–192.

Pesaresi, M., Guo, H., Blaes, X., Ehrlich, D., Ferri, S., Gueguen, L., Halkia, M., Kauffmann, M., Kemper, T., Lu, L., Marin-Herrera, M.A., Ouzounis, G.K., Scavazzon, M., Soille, P., Syrris, V. and Zanchetta, L. (2013) 'A global human settlement layer from optical HR/VHR RS data: Concept and first results', *IEEE Journal of Selected Topics in Applied Earth Observations and Remote Sensing*, 6, 5, 2102–2131.

Platt, R.V. and Goetz, A.H. (2004) 'A comparison of AVIRIS and Landsat for land use classification at the urban fringe', *Photogrammetric Engineering and Remote Sensing*, 70, 7, 813–819.

Poterea, D., Schneiderb, A., Angel, S. and Civcod, D.L. (2009) 'Mapping urban areas on a global scale: Which of the eight maps now available is more accurate?', *International Journal of Remote Sensing*, 30, 24, 6531–6558.

Salu, Y. (1995) 'Sub pixel localization of highways in AVIRIS images', presented at the 5th Annual JPL Airborne Geoscience Workshop, Pasadena, CA.

Schneider, A., Friedl, M.A. and Woodcock, C.E. (2003) 'Mapping urban areas by fusing multiple sources of coarse resolution remotely sensed data', in *Proceedings of IEEE International Geoscience and Remote Sensing Symposium, IGARSS '03*, volume 4, 2623–2625.

Schneider, A., Friedl, M.A. and Potere, D. (2010) 'Mapping global urban areas using MODIS 500-m data: New methods and datasets based on "urban ecoregions"', *Remote Sensing of Environment*, 114, 8, 1733–1746.

Small, C., Pozzi, F. and Elvidge, C.D. (2005) 'Spatial analysis of global urban extent from DMSP-OLS night lights', *Remote Sensing of Environment*, 96, 3–4, 277–291.

Statistics South Africa (2011) 'Statistics South Africa. Census', http://beta2.statssa.gov. za/, accessed 10 September 2015.

Tiede, D., Wania, A. and Füreder, P. (2012) 'Object-based change detection and classification improvement of time series analysis', in *Proceedings of 4th International Conference on Geographic Object Based Image Analysis (GEOBIA), Rio de Janeiro, Brazil, May*, 223–227.

WorldPop (2015) 'The WorldPop Project', www.worldpop.org.uk/, accessed 10 September 2015.

Yousif, O. (2015) 'Urban change detection using multitemporal SAR images', Doctoral Thesis in Geoinformatics, Royal Institute of Technology (KTH), Stockholm, Sweden, June.

Zhang, Q., Li, B., Thau, D. and Moore, R. (2015) 'Building a better urban picture: Combining day and night remote sensing imagery', *Remote Sensing*, 7, 9, 11887–11913.

Zhou, Y., Smith, S.J., Zhao, K., Imhoff, M., Thomson, A., Bond-Lamberty, B., Asrar, G.R., Zhang, X., He, C. and Elvidge, C.D. (2015) 'A global map of urban extent from nightlights', *Environmental Research Letters*, 10, 5, 054011.

# 4 Modelling and projecting urban land cover

*Carlo Lavalle, Filipe Batista e Silva, Claudia Baranzelli, Chris Jacobs-Crisioni, Ana Luisa Barbosa, Jean-Philippe Aurambout, Ricardo Barranco, Mert Kompil, Ine Vandecasteele, Carolina Perpiña Castillo and Pilar Vizcaino*

## Introduction

As previous chapters in this book have shown, urban expansion is an ongoing process with considerable impacts on the environment, the economy and quality of life. Europe, with its largely urban population, is no exception. To curtail the negative impacts and foster the positive effects of ongoing urban expansion, policies will have to be adjusted and harmonised. To do so an outlook of future land use and urbanisation trends is indispensable. Such an analysis of evolutions and functional profiles of European cities requires evaluating the impacts of continent-wide drivers and, at the same time, the effect of national and local strategies with their own priorities and plans.

The Directorate General Joint Research Centre (DG JRC) of the European Commission (EC) is contributing to the analysis of European regions and cities with the LUISA Territorial Modelling Platform, the aim of which is to provide an integrated methodology based on a set of spatial tools that can be used for assessing, monitoring and forecasting the development of urban and regional environments. LUISA allows quantitative and qualitative comparisons at pan-European level, among areas subject to transformation due to policy intervention. A further characteristic is that it adopts a methodology that simultaneously addresses the EU perspective on the one hand, and the regional/local dimension on the other. These features allow investigating and understanding territorial dynamics in a wider continental dimension while considering local and regional driving forces.

This chapter illustrates how European cities are evolving in the period 2010–2050, according to the reference configuration of the LUISA platform. The second section provides a sketch of the burgeoning academic field of urban land use models, while the third summarises the main technical structural characteristics of LUISA. The fourth section presents the key trends governing land use evolution in Europe for the future decades and how these influence urban developments by looking into a few key indicators. A review of conclusions and future improvements concludes the chapter.

## The role of land use modelling for urban applications: review, opportunities and limitations

Cities are complex structures characterised by specific dynamic elements that can hardly be captured with simple linear representations. Complex modelling can often be an efficient way to understand the mechanisms of urban dynamics, to evaluate current urban systems and to provide support in urban management (Schaldach and Priess, 2008). Since urban land use dynamics are the direct consequence of the action of individuals, public and private corporations acting simultaneously in time over the urban space, advanced land use models may help to build future growth scenarios and to assess possible impacts (Lambin and Geist, 2006).

Several reviews, e.g. by Berglund[1] (2014), INSIGHT (2014), Simmonds *et al.* (2013), Silva and Wu (2012), Haase and Schwarz (2009) and Schaldach and Priess (2008), present exhaustive and critical appraisals of approaches and techniques concerning directly or indirectly land use modelling for urban applications. The variety and population of such models are continually growing, hence any compilation will be necessarily incomplete. These constant developments guarantee that almost all aspects related to modelling have been or will soon be tackled by researchers and/or practitioners, hence greatly open the perspectives and potential for urban applications.

Following the classification suggested by Silva and Wu (2012), models can be described according to the following key characteristics:

- modelling approaches: mathematical or statistical, geographical, cellular automata, agent-based, rule-based and integrated
- spatial scales: regional scale, metropolitan scale, local scale and multi-scale models
- temporal resolution/span: long-term, medium-term and short-term models
- spatial emphasis: spatially or not-spatially explicit
- thematic application: land use planning, urban growth, transportation, environmental protection, impact assessment, scenario-based modelling, etc.

While there is an overall increasing acceptance of model results for the management and planning of urban areas, the applicability and usefulness of a model depends very much on the nature of the questions to be answered (Triantakonstantis and Mountrakis, 2012). Typically, urban land use models are adopted to investigate behaviours that have strictly local characteristics, but also when related to global issues, such as climate change, since the air, soil and waste emissions that occur in cities are quantified as having a direct impact on local drivers. Furthermore, because urbanisation might go along with potential environmental consequences, urban growth modelling appears to have a key role in urban planning to assist in decisions related to sustainable urban development. Very seldom urban models are employed to assess the impact of urban growth beyond the strict delineation of the urban areas of concern, such as spill-over effects or gravitational attractiveness between urban agglomerations.

A further key issue when modelling urban systems concerns the availability (in terms of both quality and quantity) of data to be fed as input into models and also to be used for calibration and validation purposes. Although this is a common and overall concern for all modelling domains, from global change to micro-economic and behavioural applications, it is particularly pronounced when, as in the urban field, there is the need to cross-correlate data from many sectors (e.g. housing, transport, environment, etc.), in many different formats (statistics, maps, surveys, time-series, etc.) and often with a varying range of accuracy and precision.

The methodological approach at the basis of the LUISA Territorial Modelling Platform hereinafter described, aims to tackle some of the above mentioned issues, in particular for what concerns the capability to resolve local features while still providing a holistic vision of continental patterns of urban development.

## Description of the LUISA Territorial Modelling Platform

### Overview

The LUISA platform has been specifically designed to assess territorial impacts of European policies (EC, 2002, 2013) by providing a vision of possible futures and quantitative comparisons between policy options. The platform accommodates multi-policy scenarios, so that several interacting and complementary dimensions of the EU are represented. At the core of LUISA is a computationally dynamic spatial model that allocates activities and services based on biophysical and socio-economic drivers. This model receives direct input from several external models covering demography, economy, agriculture, forestry and hydrology, which define the main macro assumptions that drive the model. LUISA is also compliant with given energy and climate scenarios, which are modelled further upstream and link directly to economy, forestry or hydrology models. The model was initially based on other land-use models, namely the Land Use Scanner and CLUE models (Hilferink and Rietveld, 1999; Dekkers and Koomen, 2007; Verburg and Overmars, 2009), but in its current form LUISA is the result of a continuous development effort by the JRC (Lavalle *et al.*, 2011a). The model projects future land/use cover changes, accessibility maps and gridded population distribution at the relatively fine spatial resolution of 1 hectare (100 × 100 metres) (Batista *et al.*, 2013b; Batista *et al.*, 2013c) for the time period 2010–2050, with the most relevant groups of land use/cover types being represented. LUISA is usually run for all EU countries, but can be used for more detailed case studies or, on the contrary, be expanded to cover pan-European territory.

In contrast to many other land-use models LUISA incorporates additional information on 'land functions'. Those land functions are a new concept for cross-sector integration and for the representation of complex system dynamics. They are instrumental to better understand land use/cover change processes and to better inform on the impacts of policy options. LUISA simulates future land use changes, and land functions related to the resulting land use patterns are then inferred and described by means of spatially explicit indicators. A land function can, for example, be physical (e.g. related to hydrology or topography), ecological (e.g. related to landscape or phenology), social (e.g. related to housing or recreation), economic (e.g. related to employment or production or to an infrastructural asset) or political (e.g. consequence of policy decisions). Commonly, one portion of land is perceived to exercise many functions. Land functions are temporally dynamic, depend on the characteristics of land parcels, and are constrained and driven by natural, socio-economic and technological processes. Since it is centred on this novel concept, LUISA is far beyond a single, stand-alone model. It can be best described as a platform with a land use model at its core, linked to other upstream and downstream models. LUISA was designed to yield, ultimately, a comprehensive, consistent and harmonised analysis of the impacts of environmental, socio-economic and policy changes in Europe.

As with many modelling tools, LUISA is not a forecasting model. The most meaningful and useful way to use it is by simulating two or more comparable scenarios. Typically, a 'baseline' scenario captures the policies already in place, assuming the most likely socio-economic trends and 'business-as-usual' dynamics (i.e. as observed in the recent past). Such a baseline serves as a benchmark to compare other scenarios in which future conditions or policies are assumed to change. This approach to impact assessment provides relevant elements to structure discussion and debate in a decision-making process. Two elements are crucial when performing an assessment with the LUISA integrated modelling framework: (1) the definition of a coherent multi-sector baseline scenario to be used as a benchmark for the evaluation of alternative options, (2) a consistent and comprehensive database covering socio-economic, environmental and infrastructural themes.

The baseline scenario provides the basis for comparing policy options and should ideally include the full scope of relevant policies at the European level. A comprehensive baseline integrated in a modelling platform such as LUISA serves to capture the aggregated impact of the drivers and policies that it covers. Sensitivity analysis can be helpful to identify linkages, feedbacks, mutual benefits and trade-offs between policies. The definition of the baseline should be the result of agreements between the main stakeholders and experts involved. Ideally, the baseline's assumptions should be shared and used by different models in integrated impact assessment. Since 2013, LUISA has been configured and updated to be in line with the EC's 'Reference Scenario' (Lavalle *et al.*, 2013; Baranzelli *et al.*, 2014), which has been used as a baseline in subsequent impact assessments. Various aspects of the model, such as sector forecasts and land suitability definitions, are updated whenever pertinent.

The second element refers to the wealth of data that are needed to cope with the European-wide coverage and multi-thematic nature of a territorial impact assessment. The principal input datasets required by LUISA must comply with the following set of characteristics:

- EU-wide (ideally pan-European) coverage
- geographically referenced to bring information together and infer relationships from diverse sources
- consistency of data nomenclature, quality and resolution to allow cross-country/region comparison.

LUISA is structured into three main modules: a 'demand module', a 'land use allocation module' and an 'indicator module'. The main, final output of the allocation module is a land use map. Potential accessibility and population distribution maps are also endogenously computed by the model as a result of the simulation, and are themselves important factors for the final projected land use map. From these outputs, and in conjunction with other modelling tools that have been coupled with LUISA, a number of relevant indicators can be computed in the indicator module. The indicators capture policy-relevant information from the model's outputs for specific land use functions, such as water retention or accessibility. When computed for various scenarios, differences in the indicators can be geographically identified, sensitive regions can be pinpointed and impacts can be related to certain driving factors assumed in the definition of the scenarios. In the next sections LUISA's demand and land allocation modules are elaborated upon.

### The demand module

The demand module captures top-down or macro drivers of land use change that limit the regional quantities of the modelled land use types. The demands for different land use categories are modelled by specialised upstream models. For example, regional land demands for agricultural commodities are taken from the CAPRI model (Britz and Witzke, 2008), which simulates the consequences of the Common Agricultural Policy (Lavalle *et al.*, 2011b); demographic projections from Eurostat (Eurostat, 2010) are used to derive future demands for additional residential areas in each region; and land demands for industrial and commercial areas are driven primarily by the growth of different economic sectors (Batista e Silva *et al.*, 2014). It is clear that LUISA is linked to several thematic models, and thus it also inherits the scenario configurations and assumptions of those models. Special care is therefore taken when integrating the input data from multiple source models to ensure that inputs are mutually consistent in terms of scenario assumptions.

In the case of urban, industrial and commercial areas the link between macro driving forces and land demands are modelled within LUISA's demand module. Urban land use demands are obtained from combining demand for residences and tourist accommodations. The demand for residential urban areas is a function of the number of households and a land use intensity parameter that indicates the number of households per hectare of residential urban land. The number of households is a function of the regional population and of an average household size that is assumed to converge across European regions. The land use intensity parameter can either be extrapolated from observed past trends in a business-as-usual approach, or can be modified to depend on specific urban policies. The demand for touristic land use is a function of the number of beds in a region and another land use intensity parameter that indicates the number of beds in tourist accommodations per hectare of touristic urban land. The number of beds is a function of the projected number of tourist arrivals, which are in turn obtained from the United Nations World Tourism Organization. Finally, demand for industrial, commercial and services land use (ICS) is a function of economic growth in those three sectors of activity, and, again, a land use intensity parameter that in this case indicates gross value added per hectare of ICS land (Batista e Silva *et al.*, 2014). Here the land use intensity parameter responds to GDP per capita because it has been found that economic land use intensity depends foremost on that factor.

### The land allocation module

The land use allocation module is based on the principle that competing land use classes vie for most suitable locations, given available land and the demand for various land use classes. Given that assumption, the actual allocation of land uses to space is governed by a land use optimisation approach, in which discrete land use transitions per grid cell occur in each discrete time-step. The suitability of locations for various land use types is based on both rules and statistically inferred transition probabilities that are derived from the following factors: terrain factors such as slope, orientation and elevation; socio-economic factors such as potential accessibility, accessibility to towns and distance to roads; and neighbourhood interactions between land use. The association between these factors and each land use type is obtained from past land use observations by means of statistical regressions. In addition to exogenous suitability factors, spatial planning, regulatory constraints (e.g. protected areas) and exogenous incentives influencing specific land use conversions can also be taken into account in the model. Furthermore, two matrices govern the occurrence of land use transitions. A 'transition cost matrix' informs the model on the likelihood of pair-wise transitions. This transition cost matrix is obtained from observed land use transitions recorded in the CLC time-series (1990–2006); for example indicating that in general a land use transition from agriculture to urban is more likely than from forest to urban. An 'allow matrix' informs the model on which transitions are permitted, and can also be specified to define the number of years required for a transition to take place. Both matrices can be used either as calibration or scenario parameters, and contribute, in addition to the above mentioned factors, to the overall suitability of grid cells for each land use type.

Recent developments are shifting LUISA from traditional, land-cover based modelling approaches (LC) to activity-based modelling. The foremost developments entail the endogenous computation of accessibility levels and population distributions for each grid cell as part of the land use modelling exercise; this is explained exhaustively in Batista e Silva *et al.* (2013a). Essentially these developments add that for each year, potential accessibility levels are computed given a road network and population distribution (Jacobs-Crisioni *et al.*, 2014); while the population allocation module in the model allocates people (newcomers and internal migrants) across each region based on a range of factors. With regard to population distributions the model assumes that people are, amongst others, driven by high accessibility, vicinity to other people (a proxy for economies of scale) and preferences to build housing on certain land use. A proxy for housing supply at each location limits the amount of people that can be accommodated without further development. Finally, whether one grid cell will be urban no longer depends on the discrete land use allocation process, but instead is obtained from population distributions given straightforward threshold rules.

## Urban evolutions for a baseline scenario in Europe

### *The baseline configuration*

LUISA has been configured to project a baseline scenario of land use changes up to 2050, assuming likely socio-economic trends, business as usual urbanisation processes and the effect of established European policies with direct and/or indirect land-use impacts. This baseline configuration is defined as the 'LUISA EU Reference Scenario 2014' and is described in detail in Baranzelli *et al.* (2014). Variations to that reference scenario may be used to estimate impacts of specific policies, or of alternative macro-assumptions.

LUISA includes a set of procedures that capture top-down or macro drivers of land-use change (taken from a set of upstream models) and transform them into actual regional quantities of the modelled land-use types. Regional land demands for agricultural commodities are taken from the CAPRI, which simulates market dynamics using nonlinear regional programming techniques to forecast the consequences of the Common Agricultural Policy. Demographic projections from Eurostat and tourism projections from the United Nations World Tourism Organization (UNWTO) are used to derive future demand for urban areas in each region; land demand for industrial and commercial areas are driven primarily by the economic growth as projected by the Directorate-General for Economic and Financial Affairs of the European Commission (DG ECFIN); and the demand for forest is determined by extrapolating observed trends of afforestation and deforestation rates reported under the scheme of the United Nations Framework Convention on Climate Change (UNFCCC). The demand for the different land use types is ultimately expressed in terms of acreage and defined yearly and regionally (NUTS 2).

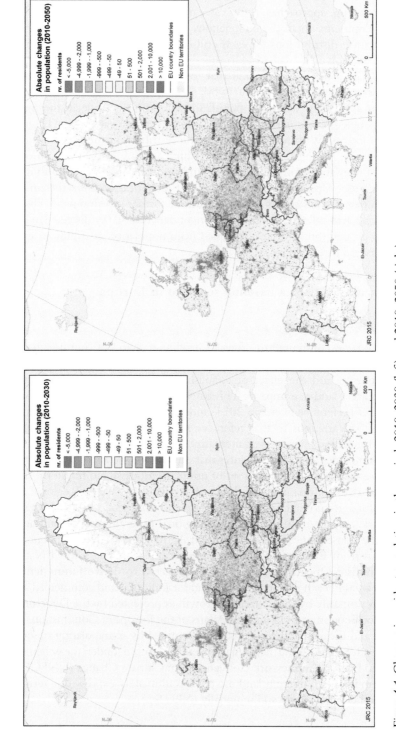

*Figure 4.1* Changes in resident population in the periods 2010–2030 (left) and 2010–2050 (right)

In the LUISA Reference Scenario 2014, the economic and demographic assumptions are consistent with the 2012 Ageing Report (EC, 2012). The demographic projections, hereinafter referred as EUROPOP2010, were produced by Eurostat, whereas the long-term economic outlook was undertaken by DG ECFIN and the Economic Policy Committee. The actual economic figures used in LUISA were taken from the GEM-E3 model, which modelled the sector composition of future economy (GVA per sector) consistently with the DG ECFIN's projections (EC, 2014). Both projections are mutually consistent in terms of scenario assumptions.

To compute the travel times that inform accessibility, a road network from the Trans-Tools transport model is used.

### Overall trends to 2030: main regional indicators

Demographic trends are amongst the main drivers of land use/cover changes, in particular for urban areas. According to the EUROPOP2010 projections, clear patterns of changes in the net population will appear in Europe in the next decades, as shown in Figure 4.1 for the periods 2010–2030 and 2010–2050. A decrease of resident population is predicted to occur in wide central and eastern areas of the European Union. Also, spots of increases are evident in some metropolitan areas, although it is worth remarking that absolute changes as those reported in Figure 4.1 are necessarily higher in densely populated areas.

The analysis of the wide picture of the evolutions of the European territory in response to such demographic patterns, and coherent economic projections, will hereinafter focus on the period 2010–2030, with a set of indicators related to urban development and accessibility at the regional level. The analysis covers variability over space and time. For this purpose, the indicators are presented at NUTS 2 level and display values for the year 2010 as well as absolute or relative changes between 2010 and 2030. A detailed list of the indicators is presented in Table 4.1.

*Table 4.1* List of indicators used to assess urban development and accessibility according to the LUISA EU Reference Scenario 2014

| THEME | INDICATOR | UNIT |
|---|---|---|
| **POPULATION** | Population density in 2010 | Person per m$^2$ |
| | Relative changes between 2010 and 2030 | % |
| **URBAN DEVELOPMENT** | Built-up area per inhabitant in 2010 | m$^2$ per person |
| | Absolute changes between 2010 and 2030 | m$^2$ per person |
| | Urban sprawl in 2010 | UPU/m$^2$ |
| | Absolute changes between 2010 and 2030 | UPU/m$^2$ |
| **ACCESSIBILITY** | Network efficiency | Dimensionless |
| | Relative changes between 2010 and 2030 | % |
| | Potential accessibility | Dimensionless |
| | Relative changes between 2010 and 2030 | % |

*Population density*

The indicator of population density is calculated as the total number of inhabitants divided by the land area in m² and is used as an ancillary indicator intended to compare the regions based on similar figures. The higher the density, the higher the concentration of population living in a specific region. The number of people is derived from EUROPOP2010 at NUTS 2 level (Eurostat, 2010). The land area corresponds to the total area of the region at NUTS 2 level (EuroBoundaryMap v81 – see Eurogeographics website, www.eurogeographics.org/).

Figure 4.2 shows the population density in 2010 and Figure 4.3 the relative changes between 2010 and 2030. According to the population projections used, Europe will diverge in terms of population density, with clear winners and losers. The change in population density also shows a high degree of autocorrelation, with large concentrations of regions with either increasing or decreasing trends.

Regions with decreasing trends in population are mostly concentrated in Central and Eastern Europe, particularly in Romania, Bulgaria, Croatia,

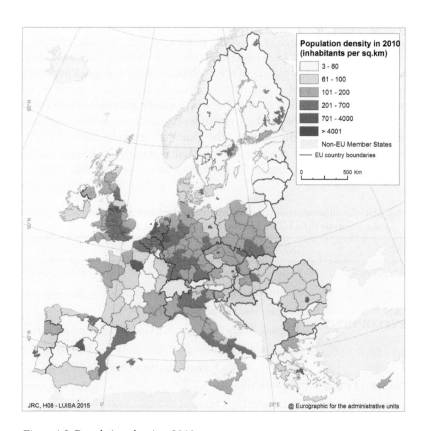

*Figure 4.2* Population density, 2010

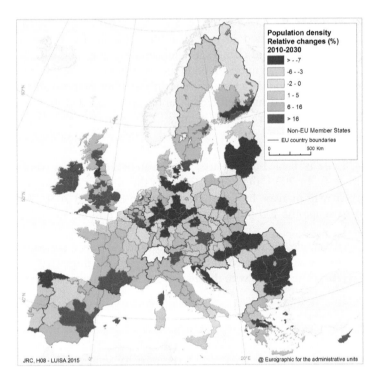

*Figure 4.3* Population density: absolute changes in percentage between 2010 and 2030

the Baltic countries and Germany. In Western Europe, only the northwest of the Iberian Peninsula is projected to show a decrease in population over the next couple of decades.

The projected population decline in most of Germany, for instance, is primarily due to negative natural growth, with immigration levels insufficient to balance population decline. In Romania and Bulgaria, on the other hand, emigration contributes to further overall population decline. However, international migration flow projections are highly uncertain due to their high volatility over time and space.

For what concerns most of the other parts of Europe, overall population growth is expected to be positive. In addition, regions with capital cities tend to stand out in terms of population growth, even in Eastern Europe. If such a scenario holds, the resulting substantial changes in regional population might generate non-negligible impacts on economy, landscape and urban dynamics.

### Built-up area per inhabitant

One indicator, built-up area per inhabitant, measures land consumption by expressing the relation between population and the size of built-up areas as the square meters of land per person. The built-up area per inhabitant is a useful tool to monitor the growth of the built-up areas and assess changes in the efficiency of land use in Europe in the period 2010–2030 according to the EU Reference Scenario 2014. The total area and changes in 'built-up areas' (i.e. land take) is a key indicator that reflects human intervention in the environment. The lower the consumption per capita of land the more efficient the use of the built-up areas

In Europe, cities use land most efficiently and population densities tend to decline the further away from city centres. This general trend can be explained by the price of land and its use, which varies with distance from the city centre (EC, 2014).

In the EU–28, the available built-up area per person in 2010 was on average 391 m² (Figure 4.4). In 2030, the model forecasts the amount of land consumed per person will increase by 6 per cent between 2010 and 2030. This implies that on average the EU population in 2030 will consume more land than in 2010.

According to the modelling results the amount of land consumed per person in 2010 is lower in the regions located in the southern part of Europe (with the exception of Cyprus). The regions with the highest land use intensity correspond

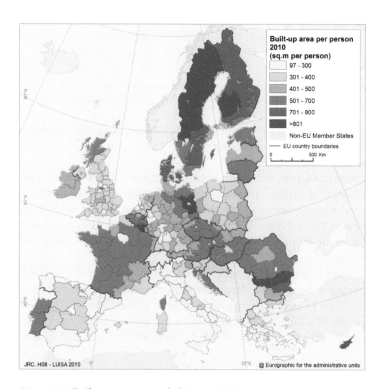

*Figure 4.4* Built-up area per inhabitant, 2010

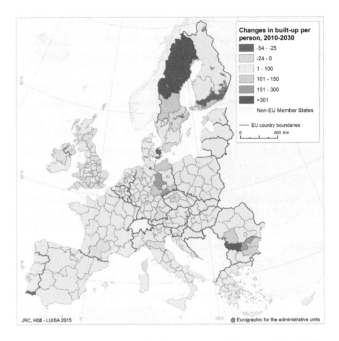

*Figure 4.5* Changes in built-up area per inhabitant, 2010–2030

to the city capitals where the land use intensity is among the highest in Europe (Figure 4.4). This pattern changes the further north one goes in Europe, with an increase in the amount of land consumed per person (Kasanko *et al.*, 2006).

Concerning the changes between 2010 and 2030, the majority of regions show an increase in the amount of land consumed per inhabitant, meaning that land use efficiency is declining over time. In this sense, the use of land will be less efficient in 2030. Countries that follow this trend are, for example, the Scandinavian countries and the eastern part of Europe (shown in red and orange hues in Figure 4.4).

There are also a few regions in the EU-28 that are expected to use land more efficiently over time. Countries that follow this trend are foreseen to decrease the land consumed per person as compared to the baseline year (red and orange hues). This is the case, for instance, in Ireland and some regions in the United Kingdom (blue hues in Figure 4.5).

*Urban sprawl*

Weighted Urban Proliferation (WUP) is an index to quantify urban sprawl, proposed by Jaeger and Schwick (2014) and implemented in LUISA (Barranco *et al.*, 2014). It is based on the following definition of urban sprawl:

the more area built over in a given landscape (amount of built-up area) and the more dispersed this built-up area in the landscape (spatial configuration), and the higher the uptake of built-up area per inhabitant or job (lower utilisation intensity in the built-up area), the higher the degree of urban sprawl.

The WUP is calculated as a combination of three different elements taking into consideration (1) the degree of urban penetration (incorporating the distance between built-up cells), (2) the building density of built-up area and (3) the population present in this built-up area. The urban sprawl is expressed in Urban Permeation Unit (UPU) per square metre (UPU/m²). The higher the UPU, the higher the urban sprawl.

In 2010, the average WUP, aggregated at NUTS 2 level for the EU-28, was 1.10 UPU/m². Much higher values were reached in capital cities such as London, Paris, Brussels and Budapest (Figure 4.6). The average WUP was projected to increase to 1.22 UPU/m² in 2020 and 1.36 UPU/m² in 2030. This increasing and accelerating trend indicates a general increase in urban sprawl across Europe but most significantly around Brussels, Prague, Vienna, London and Bucharest. This can most likely be attributed to migrations of population settling at the periphery of urban centres (Figure 4.7). In contrast, less sprawling regions can be seen all over Europe, particularly in Spain, Italy, Greece, Ireland, Scotland and in the Scandinavian countries. Some of these regions

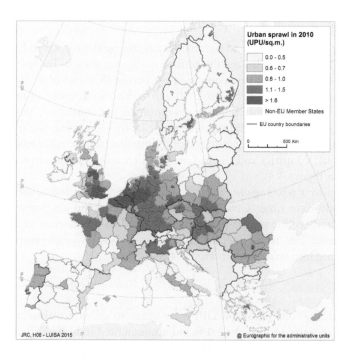

*Figure 4.6* Urban sprawl, 2010

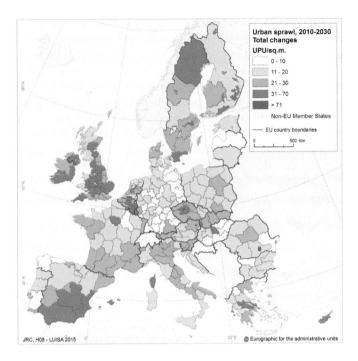

*Figure 4.7* Changes in urban sprawl, 2010–2030

also registered a significant increase in urban sprawl between 2010 and 2030, in particular in the southeast part of Spain and Ireland, most likely due to the population growth during this period (Figure 4.7).

*Accessibility*

Two indicators are shown here that measure the effects of transport network improvements on accessibility: relative network efficiency (Figures 4.8 and 4.9) and potential accessibility (Figures 4.10 and 4.11). These can be loosely linked to specific policy objectives: network efficiency measures the effectiveness of transport networks (López *et al.*, 2008); and potential accessibility measures economic opportunity (López *et al.*, 2008; Stepniak and Rosik, 2013). Both indicators are implemented in LUISA (Jacobs-Crisioni *et al.*, 2016) and are based on the shortest travel times between two municipalities and population distribution at the destination. The road network data used to obtain travel times describes the current (2006) and the expected future (2030) network; the latter takes into account the expected network improvements enabled by EU policy funding.

The analysis of accessibility maps yields common findings: for both indicators, north-western Europe has the best spatial linkages, the best network efficiency and a clearly dominant place in terms of economic opportunity. The modelled changes in accessibility levels are caused by two processes: on the one hand, changes in municipal populations modelled by LUISA; on the other hand, changes in travel times induced by transport network investments, which are taken into account in LUISA (see Batista e Silva *et al.*, 2013a; Jacobs-Crisioni *et al.*, 2016). In particular new member states are assumed to receive such network improvements. One may expect that network investments increase the accessibility provisions in currently underprovided regions. Unfortunately, as the results presented here partially show, in some cases the effects of network investments are offset by the fact that population numbers in the target regions are declining, often also with migration to more central regions that benefit from even higher accessibility levels through population growth. The results of these processes can, for example, be seen in lower network efficiency in the west of France, in the UK and in Helsinki in Finland, and poorer potential accessibility in a number of regions in the eastern part of Europe and Greece.

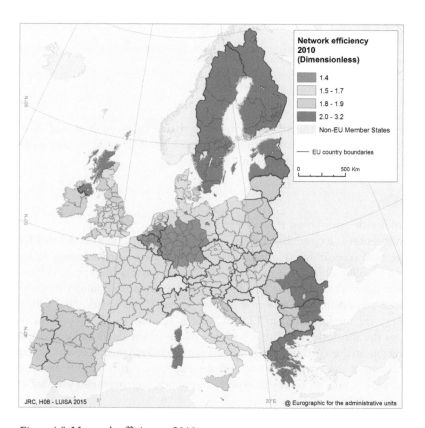

*Figure 4.8* Network efficiency, 2010

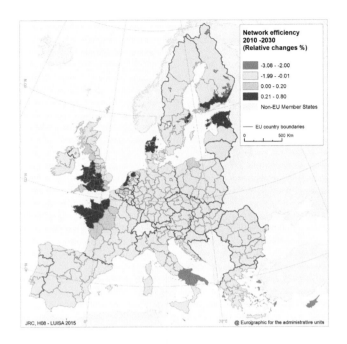

*Figure 4.9* Changes in network efficiency, 2010–2030

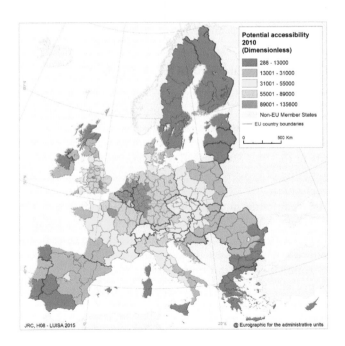

*Figure 4.10* Potential accessibility, 2010

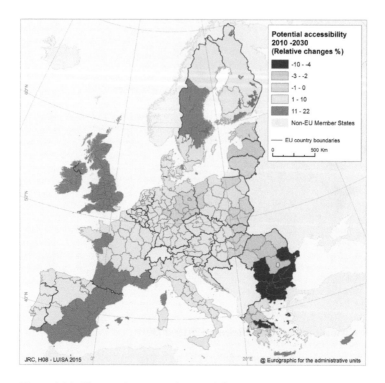

*Figure 4.11* Changes in potential accessibility, 2010–2030

## Evolutions of European Functional Urban Areas

For the purpose of analysing the evolution of individual 'urban areas', we herein adopt the definition of a city and its commuting zone as given by the OECD and the European Commission (Barranco *et al.*, 2014).

This definition identifies 828 (greater) cities with an urban centre of at least 50,000 inhabitants in the EU, Switzerland, Croatia, Iceland and Norway and allows for comparability of cross-country analysis of cities, otherwise not possible with other definitions.

Half of the European cities included in the definition are relatively small with a centre between 50,000 and 100,000 inhabitants. Only two are global cities (London and Paris). These cities host about 40 per cent of the EU population. Each city is part of its own commuting zone or a polycentric commuting zone covering multiple cities. These commuting zones are significant, especially for larger cities. The cities and commuting zones together (called 'Functional Urban Area', FUA) account for 60 per cent of the EU population.

Figure 4.12 shows the annual average population growth in the Functional Urban Areas for the period 1961–2011. With the exception of several

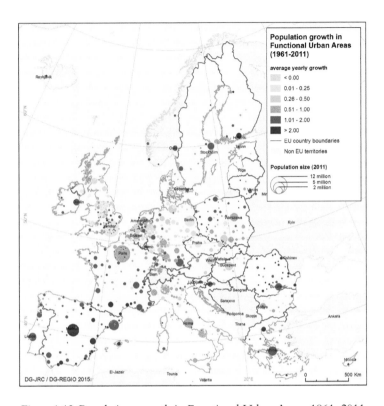

*Figure 4.12* Population growth in Functional Urban Areas, 1961–2011

cities in eastern Germany and few in northern UK, all areas present positive annual increase.

In future projections, the European Functional Urban Areas present a rather diverse picture. According to the LUISA Reference scenario, the overall increase of built-up areas in the EU for years 2030 and 2050 is 8 per cent and 13 per cent respectively when compared to the level of 2010, in spite of a population growth of respectively 4.1 per cent and 4.4 per cent. Built-up area per inhabitant sees increases of 3 per cent in 2030 and 8 per cent in 2050.

The average share (percentage vs. total surface) of built-up (i.e. artificial) surface of all FUAs per country represent a measure of the level of urbanisation around the cities, since it does not include processes of artificial development in rural areas. Figure 4.13 presents the value for the years 2010, 2030, 2050 for the 28 member states of the European Union. Bulgaria, Croatia, Germany, Hungary and Greece have an increase of less than 0.5 per cent while Italy, Lithuania, Slovakia, Cyprus, UK, Luxembourg, Malta, Romania and Belgium have increases higher than 1 per cent, with Belgium scoring for both periods more than 2.5 per cent. With few exceptions, the projections confirm that

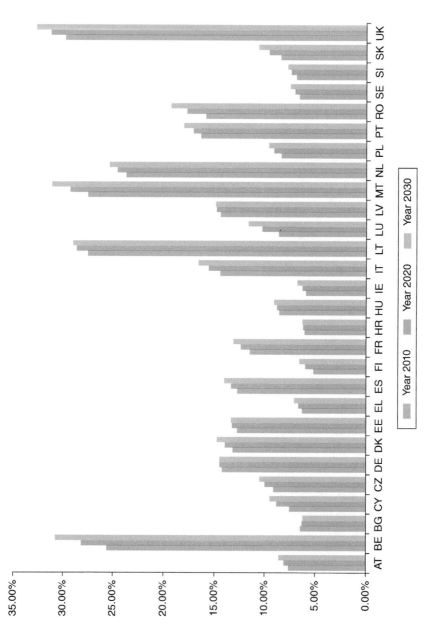

*Figure 4.13* Average share of built-up surface in FUAs per country

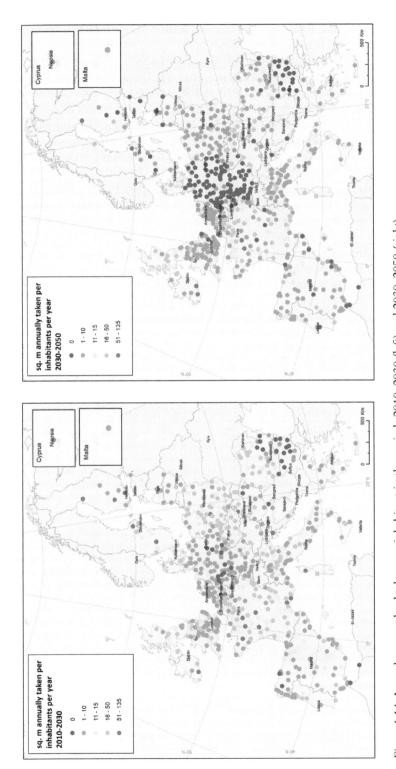

*Figure 4.14* Annual average land taken per inhabitant in the periods 2010–2030 (left) and 2030–2050 (right)

countries with a higher share of artificial areas will continue to consume more land, and to increase such behaviour.

The annual land take per inhabitant provides a measure of the rate of growth of artificial surfaces in each Functional Urban Area. Figure 4.14 gives the rate for the periods 2010–2030 and 2030–2050. The two time spans present fairly different behaviours – despite rather similar trends of demographic changes, due to the capability of LUISA to include densification phenomena (e.g. urban compactness) provoked by the various parameters used in the simulation (accessibility, suitability, attraction/repulsion rules etc.) which dynamically vary along the time.

Urban areas with highest rate of growth are spread out in Europe without a clear spatial pattern, including cities such as Le Mans, Martigues, Chartres, Nimes and Rennes in France, Namur, Leuven and Charleroi in Belgium, Tampere, Oulu and Helsinki in Finland. Large cities in Germany, namely Berlin, Stuttgart, Dresden and Frankfurt, are amongst the ones with almost null annual rate of growth.

The overall picture of urban growth in Europe can be gathered by analysing the development of built-up areas in each FUA, in direct relation to the demographic changes (Figure 4.15). The analysis reveals that 41 per cent of the FUAs are depopulating while the surface covered by built-up areas grows positively. In particular, in 23 per cent of the FUAs, artificial surfaces evidence an increase faster than population. The reverse behaviour is, however, manifested in 36 per cent of FUAs, where population grows faster than built-up areas, hinting that these cities tend to use land more efficiently, at least for what concerns the optimisation of the space employed for housing, industry, commerce and infrastructure.

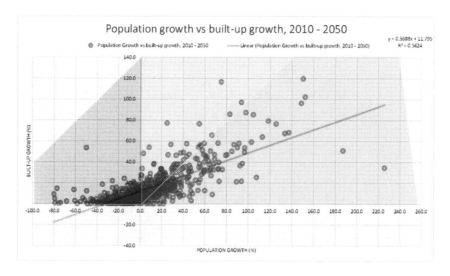

*Figure 4.15* Population growth vs. built-up growth, 2010–2050 (Lopes Barbosa, 2016)

## Conclusions

This chapter has illustrated an example of the application of advanced land use modelling for the analysis of urban development in Europe. Urban development and accessibility are important contributors to overall social and territorial cohesion. Projecting future land use according to the EU Reference Scenario 2014 gives an indication of how these two dimensions can be foreseen to evolve in the future. The modelling results show that in general, land use intensities are foreseen to decline in the EU-28 (Figure 4.16). This implies an average increase of 6 per cent of the amount of land consumed per person between 2010 and 2030. The impact on urban sprawl is much higher. On average we foresee a relative increase in urban sprawl of 23 per cent (from 1.1 UPU/m² in 2020 to 1.36 UPU/m² in 2030). This trend is particularly strong in the main capitals of the European Union.

As concluded in the study by Batista e Silva *et al.* (2013a) some of these effects can, however, be offset if adequate urbanisation policies are put in place. As such, economic growth and cohesion funds can, but do not necessarily have to be detrimental to the environment as long as appropriate spatial planning policies and recommendations are considered at different territorial scales, and more efficient land use and investment in green infrastructure is encouraged.

The coming years will see much work to improve LUISA as a comprehensive tool for evaluating the effects of various policies on land use and associated indicators. The end goal of LUISA's development should be a modelling framework that closely approximates true economic land conversions, explicitly modelling all costs and benefits that are internalised in the land use change process, while broadly taking into account both the internal and external costs and benefits of

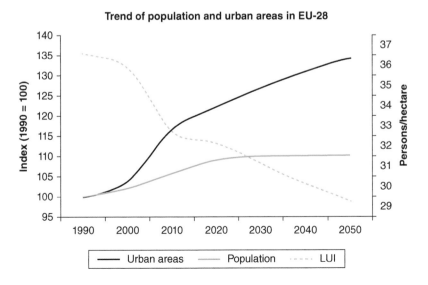

*Figure 4.16* Land use intensity in EU-28, 1990–2050

land use changes when evaluating model results. This end goal includes a better grasp of the various economic activities that drive anthropogenic land use. Lastly, a number of efforts need to be undertaken in order to better underpin the validity of the model approach, variable selection and model reliability. In the following paragraphs we discern short-term plans, for which necessary data is available, and long-term plans, which will require data sources that are currently unavailable.

One of the most important planned improvements concerning the application of LUISA in urban areas is the integration of air quality indicators. To do so, assumptions on activity levels have to be extended further from the population allocation model already in place. By integrating air quality levels in the model, the modelling platform gains a useful indicator necessary to understand the full range of external costs of land use change and also opens up possibilities to evaluate air quality improvement policies that aim at promoting behavioural changes and structural measures. Another important improvement involves redesigning the link between regional urban land use claims, the population allocation module and the discrete allocation method. Other works that will be undertaken on the short term aim to (1) underpin the conversion cost matrices currently used in the model with either empirically obtained probabilities or costs derived from an economic rationale, which serves to link more closely the model to real processes and (2) include water scarcity levels as a suitability factor for particular land uses, in order to better assess direct and indirect effects of water policies.

The frequent use of the LUISA framework in policy consultation presses the need to validate the model's output in terms of accuracy and reliability. In 2013 the JRC began a cross-validation exercise with other national and international institutes that also employs a land use model. It is expected that this validation exercise will yield useful insights into the importance of various model settings and factors that differ between the various models. Furthermore, data to do an empirical validation of the model using historical trends is finally becoming available, in the form of a historical time series of municipal population counts and historical time series land use data (EC, 2014; Barranco *et al.*, 2014). These historical data will be instrumental in empirical validation projects that are planned in the short to medium term.

Lastly, one of the most substantial improvements planned in the long term is to fully integrate an economic rationale into the land use model – based on true utilities, true costs and true willingness-to-pay data. This would better underpin the rationale of the model, and would allow inductive approaches in the model to evaluate the effect of policies on land use behaviour (i.e. not starting from an assumed overall effect, but from a clearly defined added cost or financial incentive in the utilities of particular land use conversions). In this improvement, currently unavailable data on the financial aspects of land use conversions will be critical.

## Note

1  www.nordregio.se/Global/Events/Events%202014/Attraktiva%20och%20h%C3
   %A5llbara%20stadsregioner/Report%20Review%20of%20Land-Use%20Models
   %202014-01-10.pdf, accessed 7 June 2015.

# References

Baranzelli, C., C. Jacobs, F. Batista e Silva, C. Perpiña Castillo, A. Lopes Barbosa, J. Arevalo Torres and C. Lavalle (2014) *The Reference Scenario in the LUISA Platform – Updated Configuration 2014: Towards a Common Baseline Scenario for EC Impact Assessment Procedures*, Luxembourg: Publications Office of the European Union.

Barranco, R., F. Batista e Silva, M. Marin Herrera and C. Lavalle (2014) 'Integrating the MOLAND and the urban atlas geo-databases to analyze urban growth in European cities', *Journal of Map and Geography Libraries: Advances in Geospatial Information, Collections and Archives* 10(3): 305–328.

Batista e Silva, F., C. Lavalle, C. Jacobs-Crisioni, R. Barranco, G. Zulian, J. Maes, C. Baranzelli, C. Perpiña, I. Vandecasteele, E. Ustaoglu, A. Barbosa and S. Mubareka (2013a) *Direct and Indirect Land Use Impacts of the EU Cohesion Policy: Assessment with the Land Use Modelling Platform*, Luxembourg: Publications Office of the European Union.

Batista e Silva, F., J. Gallego and C. Lavalle (2013b) 'A high-resolution population grid map for Europe', *Journal of Maps* 9(1): 16–28.

Batista e Silva, F., C. Lavalle and E. Koomen (2013c) 'A procedure to obtain a refined European land use/cover map', *Journal of Land Use Science* 8(3): 255–283.

Batista e Silva, F., E. Koomen, V. Diogo and C. Lavalle (2014) 'Estimating demand for industrial and commercial land use given economic forecasts', *PLOS ONE* 9(3): e91991.

Britz, W. and H.P. Witzke (2008) *Capri Model Documentation 2008: Version 2*, Bonn: Institute for Food and Resource Economics, University of Bonn.

Dekkers, J.E.C. and E. Koomen (2007) 'Land-use simulation for water management: application of the Land Use Scanner model in two large-scale scenario-studies', in *Modelling Land-Use Change: Progress and Applications*, edited by E. Koomen, J. Stillwell, A. Bakema and H.J. Scholten. Dordrecht: Springer, 355–374.

EC (European Commission) (2002) 'Communication from the Commission on Impact Assessment', COM(2002) 276.

EC (European Commission) (2012) 'The 2012 Ageing Report. Economic and Budgetary Projections for the 27 EU Member States (2010–2060)', European Economy, 2, 2012.

EC (European Commission) (2013) 'Assessing territorial impacts: Operational guidance on how to assess regional and local impacts within the Commission Impact Assessment System'.

EC (European Commission) (2014) 'Investment for jobs and growth – promoting development and good governance in EU regions and cities', Sixth report on economic, social and territorial cohesion, July 2014, doi 10.2776/81072.

Eurostat (2010) 'EUROPOP2010: Convergence scenario, national level 2010', http://epp.eurostat.ec.europa.eu/cache/ITY_SDDS/EN/proj_10c_esms.htm, accessed 30 July 2014.

Haase, D. and N. Schwarz (2009) 'Simulation models on human–nature interactions in urban landscapes: a review including spatial economics, system dynamics, cellular automata and agent-based approaches', *Living Reviews in Landscape Research* 3(2).

Hilferink, M. and P. Rietveld (1999) 'Land Use Scanner: an integrated GIS based model for long term projections of land use in urban and rural areas', *Journal of Geographical Systems* 1(2): 155–177.

INSIGHT (2014) 'Innovative policy modelling and governance tools for sustainable post-crisis urban development', D2.3, *Review of Urban Models: Use in Urban Policy*, www.insight-fp7.eu/, accessed 12 August 2015.

Jacobs-Crisioni, C., P. Rietveld and E. Koomen (2014) 'Evaluating the impact of land-use density and mix on spatiotemporal urban activity patterns: an exploratory study using mobile phone data', *Environment and Planning A* 46(11): 2769–2785.

Jacobs-Crisioni, C., F. Batista e Silva, C. Lavalle, C. Baranzelli, A. Barbosa and C. Perpiña Castillo (2016) 'Accessibility and territorial cohesion in a case of transport infrastructure improvements with changing population distributions', *European Transport Research Review* 8(1): 1–16.

Jaeger, J.A. and C. Schwick (2014) 'Improving the measurement of urban sprawl: Weighted Urban Proliferation (WUP) and its application to Switzerland', *Ecological Indicators* 38: 294–308.

Kasanko, M., J.I. Barredo, C. Lavalle, N. McCormick, L. Demicheli, V. Sagris and A. Brezger (2006). 'Are European cities becoming dispersed? A comparative analysis of 15 European urban areas', *Landscape and Urban Planning* 77(1): 111–130.

Lambin, E.F. and H. Geist (eds) (2006) *Land-Use and Land-Cover Change: Local Processes and Global Impacts*', Berlin: Springer.

Lavalle, C., C. Baranzelli, F. Batista e Silva, S. Mubareka, C. Rocha Gomes, E. Koomen and M. Hilferink (2011a) 'A high resolution land use/cover modelling framework for Europe: introducing the EU-ClueScanner100 model', in *Computational Science and Its Applications* – ICCSA 2011, Part I, Lecture Notes in Computer Science vol. 6782, edited by B. Murgante, O. Gervasi, A. Iglesias, D. Taniar and B.O. Apduhan. Berlin: Springer.

Lavalle, C., C. Baranzelli, S. Mubareka, C. Rocha Gomes, R. Hiederer, F. Batista e Silva and C. Estreguil (2011b) *Implementation of the CAP Policy Options with the Land Use Modelling Platform: A First Indicator-based Analysis*, Luxembourg: Publications Office of the European Union.

Lavalle, C., S. Mubareka, C. Perpiña, C. Jacobs-Crisioni, C. Baranzelli, F. Batista e Silva and I. Vandecasteele (2013) *Configuration of a Reference Scenario for the Land Use Modelling Platform*, Luxembourg: Publications office of the European Union.

Lopes Barbosa, A., Vallecillo Rodriguez, S., Baranzelli, C., Jacobs, C., Batista E. Silva, F., Perpiña Castillo, C., Lavalle, C., Maes, J. (2016) 'Modelling built-up land take in Europe to 2020: an assessment of the Resource Efficiency Roadmap measure on land', *Journal of Environmental Planning and Management*, 1–25.

López, E., J. Gutiérrez and G. Gómez (2008) 'Measuring regional cohesion effects of large-scale transport infrastructure investments: an accessibility approach', *European Planning Studies* 16(2): 277–301.

Schaldach R. and J.A. Priess (2008) 'Integrated models of the land system: a review of modelling approaches on the regional to global scale', *Living Reviews in Landscape Research* 2(1).

Silva E. and Ning Wu (2012) 'Surveying models in urban land studies', *Journal of Planning Literature* 27: 1–14.

Simmonds, D., P. Waddell and M. Wegener (2013) 'Equilibrium versus dynamics in urban modelling', *Environment and Planning B: Planning and Design* 40: 1051–1070.

Stepniak, M. and P. Rosik (2013) 'Accessibility improvement, territorial cohesion and spillovers: a multidimensional evaluation of two motorway sections in Poland', *Journal of Transport Geography* 31: 154–163.

Triantakonstantis, D. and G. Mountrakis (2012) 'Urban growth prediction: a review of computational models and human perceptions', *Journal of Geographic Information System* 4: 555–587.

Verburg, P.H. and K. Overmars (2009) 'Combining top-down and bottom-up dynamics in land use modeling: exploring the future of abandoned farmlands in Europe with the Dyna-CLUE model', *Landscape Ecology* 24(9): 1167–1181.

# 5 Drivers of urban expansion

*Stefan Fina*

## Introduction

This chapter deals with the most important drivers of urban expansion and urban sprawl according to a list of influencing factors compiled by the European Environment Agency in 2010. The structure helps to differentiate between different sectors and spatial scales that allow for a more detailed explanation of the complex aspects that drive urban expansion and urban sprawl. Examples mainly from Europe are used to give an overview over the different strands of research that are dedicated to the analysis of urbanization: economics, architecture, energy, ecology, as well as spatial planning and environmental disciplines. The emphasis is on drivers of urbanization that lead to urban expansion, not just in a sprawling sense, but in any way that impacts on the natural properties of the land. This usually leads to an increase in the two-dimensional coverage of urban and transport land, and to an increase of soil sealing and therefore loss of farmland and agricultural productive capacity. In terms of spatial coverage, the chapter will offer an overview over global and specific European drivers, and present case examples on countries, regions and local authorities mainly from Europe.

## Understanding the drivers of urban expansion

Urban expansion does not have the same immediate environmental effects that pollution has. Land consumption does neither smell nor make any noise (apart from construction activities that go along with it). It therefore does not trigger the same public levels of attention as environmental effects from emissions and other polluting activities. Public protests about urban expansion or land consumption activists are not very common in any type of environmental movement. The environmental stresses that it causes are more indirect but a time bomb nevertheless: land consumption adds to a multitude of environmental problems that are all too often irreversible. It fixes development paths to urban structures that create more land demand, for example through automobile dependencies or a segregation of land uses. And once these structures are in place, they are incredibly difficult to retrofit or change to something that could be called sustainable (Chin, 2002; Dielemann and Wegener, 2004). The worst manifestations of urban expansion in terms of these impacts are referred to as urban sprawl. Although there is no lexicographic definition for the term, most researchers agree that urban sprawl refers to a state of urban land use configurations that is very inefficient to service in terms of infrastructure, and at the same time it refers to the process leading to such configurations. The manifestations are often a combination of low–density and uniform land-expansive residential blocks detached from other land uses like industry and business, leading to long travel distances and high levels of automobile dependency (Ewing *et al.*, 2002; Wolman *et al.*, 2005). In the context of this chapter, urban expansion and urban sprawl are not being used as synonyms: urban expansion can also be a positive amalgamation of urban areas that leads to benefits to the resource efficiency of the urban compound. But in almost all instances urban expansion leads to a conversion of land that was previously used or preserved for natural or agricultural land uses.

Figure 5.1 shows a conceptual framework that illustrates how the processes of land use change are understood in the research community. The framework is closely related to the *pressure–state–response* monitoring frameworks established by the European Community in the late 1990s (European Environment Agency, 1999) and identifies the system interdependencies between the process of land use change in general with its drivers and effects and regulatory framework. Within this framework, assessment and awareness aspects are crucial, because any type of regulation needs to be based on policy strategies and possibly targets, which are ideally derived from a societal consensus and a clear political mandate. The controversies, however,

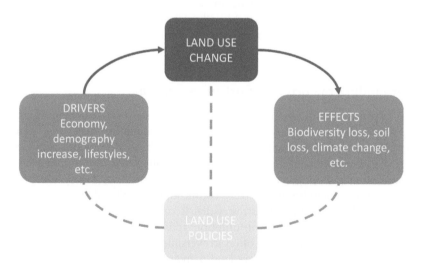

*Figure 5.1* Conceptual framework for an analysis of land use change (source: Fina *et al.*, 2014a)

start with the inherent conflicts in objectives between economically and socially desirable growth initiatives that lead to land use change, and environmental impacts and regulatory initiatives that aim to contain that land use change. Environmental impacts are often detached from the actual land use change in terms of timing and sometimes also in its spatial manifestation, for example in terms of changing floodplain patterns downstream from urbanization on a local scale, or climatic effects like heat islands when new developments block fresh air corridors.

It is therefore timely and prudent to adopt a more comprehensive monitoring and tackle the interrelationships and flow-on effects that can be attributed to land consumption. There is a large consensus amongst the research community that one needs to act on the driving forces of land consumption first, and strengthen regulatory systems in parallel. All of this in a way that caps land consumption and urban sprawl in an effective and accountable way without neglecting social and economic needs (Anthony, 2004; Bengston *et al.*, 2004; Dielemann and Wegener, 2004; Frenkel, 2004; Song and Knaap, 2004).

Christiansen and Loftsgarden (2011) have further investigated these drivers for Europe in general and Norway in particular. One of their main findings is that the multitude of drivers is very difficult to analyse if one wants to identify the most dominant one(s) in a certain geographic context. And they enhance the driving forces for urban expansion (they also speak of urban sprawl in their report) by a policy and regulatory framework group, which seems to be not a driver but a response at first glance. But looking at the actual market mechanisms in Europe, the revenue streams for territorial authorities from land sales and development are such that there is indeed an economic interest in urban expansion without other factors driving it. This is also the conclusion from a large study undertaken by Siedentop *et al.* (2009), where they looked at the land designations in urban regions exposed to demographic decline and found that one of the policy strategies against these trends was to offer more land to attract people willing to settle. The authors in this study utilize regression modelling to find the importance of driving factors on a municipality level. The results show that one needs to differentiate between demand- (population and employment, economy, transport etc.) and supply-side (land availability and pricing, infrastructure and accessibility etc.) driving factors and include development and spatial aspects in any type of comprehensive assessment. A policy climate and awareness about land consumption issues is also seen as influential: regions with a history of development pressures and associated land conflicts are likely to have developed planning processes that lead to a more efficient use of land resources.

Other strands of literature, however, describe the effect of planning and compact city policies on urban growth as minor. Angel *et al.* (2011a), for example, argue that global urban expansion will lead to massive growth in urban land cover anyway, with regional variations that are due either to population growth or to a decrease of urban density. Where in developing countries in Africa and Asia growth rates will be between 2 and 6 per cent until the year 2050, countries in Europe (and also Japan) will experience between 1 and 3 per cent growth in the scenarios supporting the analysis. One of the key assumptions for the higher scenarios is that urban densities will decrease by an average of 2 per cent, due to reasons of demographic decline and a higher demand for urban area per capita (including more living space per person and thus decreasing household sizes, and more recreational and infrastructure land per person). One of their key interpretations is that despite policy efforts to save land resources this growth will occur within the predicted ranges. The quality of the urban compound, however, is highly dependent on city planning with foresight and intelligent smart growth principles.

Other authors reflect upon the drivers of urban expansion with a view towards development that will inevitably lead to a consumption of land, but the question here is what types of land are being used (Angel *et al.*, 2011a; Hasse and Lathrop, 2003). In many ways the easiest and most frequently practised urbanization is at the cost of agricultural resources, to a very limited degree at the cost of unprotected forests, and very rarely at the cost of water dominated land uses (for example in the polder areas in the Netherlands).

In this context, case studies undertaken in German and Italian cities have shown that over 95 per cent of urbanization takes place on land previously used for agriculture (Fina *et al.*, 2014a). The conclusion here is that growth-oriented driving factors of urban expansion are not going to contain it in terms of quantity but rather manage it in terms of impact on urban structure and availability of land. According to the OECD we are living in the metropolitan century, and societies strive for urban amenities that have made cities successful – in all parts of the world (Organisation for Economic Cooperation and Development, 2015). These amenities require space and land resources in any case, but the land configuration they result in can differ significantly. Figure 5.2 shows different development paths in this respect, using CORINE land cover data for selected metropolitan regions in Europe, taking a 25 kilometre radius around the city centre for scale. They not only show that the amount of urban land differs extremely, they also show that the population base is not necessarily the reason for the difference. Some cities manage to contain urban expansion within spatial designations (Amsterdam, Munich, Manchester), designed to provide advantages in terms of service and transport structures. The opposite of that compactness we refer to as urban sprawl, where city structures are dispersed and disconnected across the region, theoretically resulting in problematic configurations to service a city efficiently with infrastructure.

It is fair to assume that these different manifestations of urban structure are the result of numerous factors. The topographic restrictions of seaside metropolitan regions like Porto or Amsterdam are an obvious one, but other effects are also known to be the result of planning strategies or compact city policies that allow for a more economic decision making with regard to land resources (Organisation for Economic Cooperation and Development, 2012). In this sense, researchers have often shown that population and economic

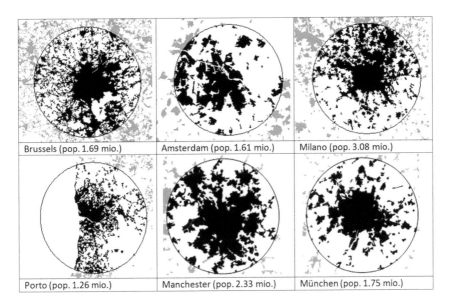

| | | |
|---|---|---|
| Brussels (pop. 1.69 mio.) | Amsterdam (pop. 1.61 mio.) | Milano (pop. 3.08 mio.) |
| Porto (pop. 1.26 mio.) | Manchester (pop. 2.33 mio.) | München (pop. 1.75 mio.) |

*Figure 5.2* Urban land use configurations in European metropolitan regions (source: Siedentop and Fina, 2012b, p. 2766)

pressures do not necessarily have to be catered for with the same amount of urban land, and that compact growth policies can be effective in terms of a more resource-efficient city development. At the same time, the debate in the scientific community and amongst decision-makers only recently started to acknowledge the value of agricultural land as public goods in this respect.

Humankind across the globe has initially settled in the most fertile regions for very obvious reasons, and if we now need to expand the urban footprints around these initial settlements for more or less good reasons, it is very likely that these fertile soils are being consumed and irreversibly being lost for cultivation. Agricultural resources are all too easily being seen as replaceable by goods from further away or from a global market, but they are actually not. It has only recently been put to the forefront of the political agenda that agricultural land actually plays a much more diverse role in the land use mix of city regions: it acts as a buffer for flood events and other forms of climate change stresses that are likely to play a role in the future (e.g. heat stress), it compensates for a multitude of infrastructure projects in environmental impact assessments, and it offers a range of other public goods that are increasingly being valued as cultural assets to a region (farmland and animal welfare, recreational and vegetation landscapes etc.; see, for example, Cooper *et al.*, 2009).

In summary it can be concluded from this brief literature analysis that there is a range of studies that allows for an identification of the main drivers for urban expansion, but their actual influence is all too often a result of their combinations. On one hand, an assessment of their importance needs to look into the geographical settings on multiple scales and systematically analyse global, regional and local trends for different sectors. On the other hand, there is an increasing need to specifically identify drivers from the impact side, i.e. which drivers are the crucial ones that lead to land consumption of agricultural resources, and which ones can be managed in a more efficient way to protect agricultural resources from urban expansion? These questions guide the structure of the subsequent sections and pick up on Figure 5.3, where the main drivers of urban expansion are illustrated in a matrix design. It was developed for the *State of the Environment* report by the European Union in 2010 and depicts the drivers from left to right on different scales of observation (global, regional, local). From top to bottom it lists the drivers that are seen as most influential for urban expansion in sectors. Drivers in bold print are seen as root causes for urban expansion, others can possibly contribute under certain circumstances. The report emphasizes that these drivers can be mutually reinforcing in some cases, in some they would level each other out in terms of the actual amount of land consumption they cause (see European Environment Agency, 2010a, pp. 22f.).

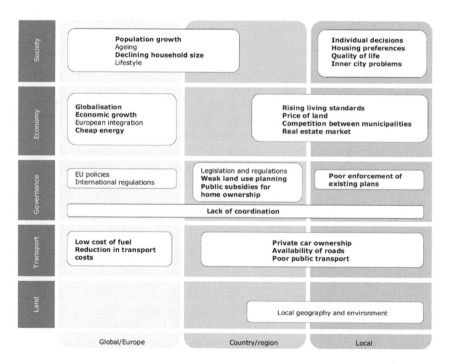

*Figure 5.3* Drivers of urban sprawl (source: European Environment Agency, 2010a, p. 23)

Note: Drivers have been organised in two dimensions: type (vertical) and spatial scale (horizontal). Demand/supply has not been differentiated. In bold: factors that drive urban sprawl; the remaining factors may become drivers under certain conditions.

## Society

For societal drivers, the European Union names *population growth* and *declining household size* as global root causes of urban expansion, and ageing and lifestyle as globally influencing variables. This assessment is very much in line with research conducted by Angel *et al.* (2011b) that sees almost no natural growth in Europe and Japan. In their models, however, urban areas will still grow between 1 and 5 per cent until 2050 (based on urban land cover in 2000), because density decline drives urban expansion in societies with a shrinking population base. This interpretation can also be substantiated by the European Union targets that aim to limit the net land take for urbanization in the European Countries to zero. The base for setting this objective was actually a net land take in the 2000s that was four times as high as population growth, leading to an unprecedented density decline in the urban compound (European Environment Agency, 2015).

The underlying dynamics of these density declines and their consequences on urban expansion have been subject to a range of research efforts in the past that explain to some degree the local root causes of urban expansion due to societal changes. A clear indication that demographic change and ageing play a role have been found in studies in Germany that looked at the housing stock build between the 1950s and 1970s. In this period the private family home became a realistic material asset to large parts of the society and led to an unprecedented suburbanization in all parts of the country. Energy prices were low and family sizes potentially high, so that family homes were built in a rather generous fashion. Today the architectural focus is on energy efficiency, healthy living and *quality of life*, aspects that the post-war building stock can only deliver with considerable modernization efforts. The remaining building stock is therefore today a burden to city planners especially in shrinking regions, with lack of investments, underutilized capacities and empty houses in private hands that are just too difficult to rejuvenate and throw whole neighbourhoods into a downward spiral of decline (Fina *et al.*, 2009; Fina *et al.*, 2012). Especially in the post-socialist countries there is massive density decline, due to population migration into the liberalized labour markets of western European countries and a new wave of suburbanization with people leaving undesirable prefabricated building blocks in the inner cities. These societal transformations and their effects on urban expansion have been documented for example by Schmidt *et al.* (2014) and Schmidt (2011).

These trends of excess capacities are often being labelled as a form of remanence, meaning that family homes only have a certain lifespan in which they cater for the intended number of people, and after a period of underutilization when the children move out ('empty-nesters') the attractiveness of these homes for the new generation depends on *individual decisions* like locational preferences, employment opportunities, but also on *housing preferences* and building characteristics. With the latter, lifestyle preferences are such that large family homes are generally not the most sought after properties anymore. Over the last few decades Western societies have seen an erosion of the traditional family model and therefore a substantial loss of potential interested parties in the existing stock of family homes. And it often proves to be more attractive and sometimes even more economic for the new generation to meet their requirements in new developments and new buildings, thus consuming new land (see for example Häußermann *et al.*, 2008).

However, these mechanisms where lifestyle and locational preferences lead to new developments are only possible if regulatory regimes and market conditions provide the settings. In that sense there are strong interdependencies between the social and economic drivers when societal demand meets economic supply. The regulatory regime has a decisive role to play in this respect, because *inner city decline* and its land use consequences can be managed on the local level with foresight and innovative ideas to some degree. There are numerous examples by now where even regions with an eroding population and economic base have successfully managed to retain inner city vibrancy, improve the environmental conditions and provide healthy environments, and attract investors to rejuvenate the building stock with successful regional development strategies. Very often these strategies rely on certain unique values, for example for recreation, the health industry or tourism, or on key businesses that have their traditional roots within the city. Despite these examples, the mainstream development strategy in shrinking regions at this point is market competition. In this respect, local communities are trying to attract families and businesses to build in new designation areas with cheap land, and neglect the long-term effects and costs for the whole community that they should be well aware of (see for example Siedentop *et al.*, 2009).

The effect that socioeconomic transformations lead to new urbanization is also visible in the many demographically stable regions in western Europe. In these cities, immigration from overseas and rural migration from the shrinking hinterlands stabilize the net loss of population that would result from natural birth rates, and offset any form of decline to the future (Siedentop and Fina, 2012a; Fina *et al.*, 2014b). The cumulative effect of demographic change is still projected to lead to a form of decline within the next decades, but for the time being there is an actual need for more housing and infrastructure. This is often true for medium-sized cities in the surrounds of the dynamic metropolitan regions, where employment opportunities are within commuting distance and a relatively stable population base has an urgent need for more housing opportunities for a more and more diverse social strata. Amongst these social classes, high-income earners continue to drive land demand for single family homes. Investors pick up on this demand for more affordable housing with block and terraced housing projects in suburban locations, with the supporting infrastructure in terms of social infrastructure (schools, kindergartens, hospitals) and shopping opportunities, leading to additional land demand.

Within the booming regions, inner cities are often exposed to gentrification processes where global players find lucrative investments in attractive markets and (re-)develop brownfield land resources or invest in the modernization and expansion of the existing building stock with a view towards profit. As a consequence certain parts of the population are driven out of the resulting overheated real estate markets, and expensive developments in the inner cities and new waves of affluent in-migrants drive the demand for affordable housing or a displacement of other inner city functions that have to relocate to suburban settings, thus driving additional land demand. Examples are university locations and research clusters, but also office parks and logistics enterprises, all of which require expansive land resources but also need to be accessible from the main city (Lüthi *et al.*, 2012). These expressions of land demand are closely linked to the next section.

## Economy

On the global and European level, the economic drivers of urban expansion need further explanation since it may not be self-explaining why globalization, cheap energy or the European integration would fuel land expansion at first glance. The most evident of these root causes seems to be *economic growth*, with all economic activities leading to a form of land demand per se. However, this relationship does not work as a very strong predictor on a national scale. Figure 5.4 shows the growth in Gross Domestic Product (GDP) for the countries of the European Union on the x-axis, and the growth of urban land cover on the y-axis. The symbols differentiate two different time periods, from 1990 to 2000 (triangles) and from 2000 to 2006 (circles). The tendency shows that economic growth and urban growth lie in the same quadrant, but in a bivariate comparison the correlation is certainly not significant. Other variables will play a role and explain the difference possibly further, but with some of the outliers visible here (Ireland, Spain, Portugal) we now know that urban expansion in the years of the observation period have created a financially troublesome real estate crisis from 2008 onwards. One conclusion here could be that economic forces do not drive urban growth on a national scale, at least not in the larger and heterogeneous countries. In this context, Vogel *et al.* (2010) describe the role of global city regions and their economic performance as the increasingly more important drivers of economic growth than the national scale.

Taking Germany again as an example, urban growth patterns and economic performance differ widely. A limited number of metropolitan regions (Munich, Stuttgart, Frankfurt, Hamburg, to some degree also Berlin) drive the economy with specific internationally visible strengths and an economically resilient backbone of industry, production and service sectors. A second tier of larger, well-connected cities is increasingly benefitting from these globally oriented economic metropolitan regions, absorbing some of the development pressures. In contrast, the remote rural areas are undergoing significant transformations in terms of their economic setup. The PLUREL project conducted by the European Union between 2007 and 2010 shows that across Europe most rural areas are struggling to compete economically on a global market. Maintaining

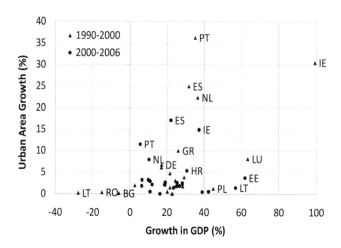

*Figure 5.4* Economic growth and urban expansion for EU countries (source: Siedentop and Fina, 2012b, p. 2780)

a form of competitiveness in times of *globalization* often means to strengthen the functional linkages between the rural hinterland and the market it services (Nilsson, 2011). And as such, a hierarchical settlement structure requires rural hubs to develop economic strategies fit to react to the requirements of a global market. This could be agricultural processing facilities in modern business parks, renewable energy power plants, or commercial centres servicing a more and more automobile-oriented client base. All of this results in a demand for land resources that becomes manifest in sprawling development patterns especially in medium-sized towns of lower central place hierarchies (= rural townships) and a form of competition for land that has worldwide effects: certain land use functions are nowadays being outsourced to land markets with cheaper agricultural resources, processing facilities or a more competitive labour market. Global trade flows allow for a timely and cost-effective delivery of even heavy products from many economic sectors, so that it is increasingly the locational advantages and global investment patterns that decide about the actual land use, rather than the local actors. In this context, Figure 5.5 shows the most striking effect of this globalization, often labelled with the negative connotation of 'land grabbing' by investors from developed countries that exploit land resources in developing countries for their production needs.

In this sense, any assessment that strives to monitor development paths towards a more sustainable land use for a certain monitoring area needs to have these downstream effects in mind and try to comprehensively include the land use that the consuming party is responsible for. This aspect has received wider attention with the concept of 'virtual water' that was established by Allan and Mallat (1995) in the mid 1990s and drew attention to the fact that imported

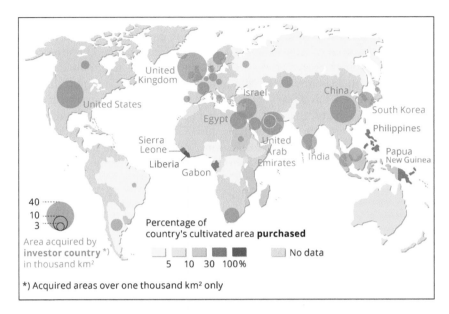

*Figure 5.5* Transnational land acquisitions, 2005–2009 (source: www.eea.europa.eu/
soer-2015/global/ecosystems, last accessed 2 June 2015, based on research
by Rulli *et al.*, 2013)

products (in his studies from the Middle East) carried an amount of water resources with them that could possibly be missing in production chains for other products in the countries of origin. The example of a cup of coffee consumed in developed countries bearing a production requirement of 11 litres of water at the production site has become a famous metaphor for this effect. Regardless of the debate about the accuracy of these numbers, the general outcome is certainly also valid for land use and land resources.

Specific structural funds dedicated to the development of rural areas in Europe may have had ambivalent effects in this respect as well: national governments and the *European integration* initiatives like the Territorial Cohesion Programme or the Common Agricultural Policy (CAP) aimed at balancing out living standards across regions, with a view towards better economic and social participation in disadvantaged areas. Evaluation of funding has shown that investments for infrastructure projects and business opportunities in some of these areas may have triggered disproportionate urban growth and sprawl without a long-term sustaining need for it (see for example Schmidt *et al.*, 2014). Another effect was that the predominant development patterns boosted automobile-oriented development and led to a form of car dependence. This was further exacerbated by the absence of any efficient type of public transport and the mutually reinforcing reliance of increasing car availability and car-oriented settlement structures. There is convincing evidence that, on average, the growth rates in the peri–urban environment and in rural townships has and will continue to have the highest growth rates when it comes to urbanization. Key explanations of this trend are the decentralization of certain types of industry, businesses and administrative functions into these areas and the increased reach of commutersheds (see also 'Transport', below). Cairncross (1997) has already pictured the 'death of distance' as a force of socioeconomic transformation, enabling businesses to operate from wherever they want without the need to be physically present in pricey locations. Now this has certainly not eventuated in general, for some industries proximity advantages within global world regions may actually have become more important in the time that has since passed. What can be taken from Cairncross' thoughts, however, is that some Silicon Valley-type new technology start-ups or business innovations could be successful from anywhere and connect via the internet to the rest of the world, with flow-on effects in terms of revenue streams and economic downstream effects that would trigger additional development, transplant new lifestyle forms into rural areas and possibly eventuate new land consumption. The predicament for such developments, next to high-speed internet connectivity, is certainly also relatively *cheap energy*, where the operation of businesses is possible without major investments into supporting infrastructure and proximity to energy facilities. Energy production and consumption patterns will also play a key role in future land use strategies and potentially transform the economic base in agricultural areas further. This is especially true in countries with high subsidies for renewable energies like Germany, where biomass production for energy has become economically so attractive that farming communities change their land use patterns in favour of such income options. This may potentially lead to a new form of land consumption that does not exhibit the same problem dimensions that we know from urbanization. There is no permanent soil sealing or loss of agricultural productivity involved as such. Nevertheless, energy production competes with food production for land resources, and the debate about food security versus sustainable forms of energy supply is only in the making.

On the regional and local levels, the European Environment Agency depicts *rising living standards* as a root driver of urban sprawl. This aspect is certainly linked with the social trends of decreasing household sizes, because this is most likely not only a question of preference and family structures, but also of affordability. In the most overheated real estate markets people would still find themselves living together in communal forms regardless of family structures due to economic reasons, sharing flats and houses. Internet platforms provide new possibilities in that respect, connecting people with no other relation than the search for affordable housing in a certain location. With decreasing economic limitations, however, large layouts of living space become a desirable lifestyle, sometimes also a second home or in some cases even multiple living locations for business or recreational reasons. At the same time, businesses will locate around the most profitable client and customer base, providing land-consuming services that are easily accessible and offer a comfortable environment to use their services. This becomes manifest in automobile oriented mall developments at the outskirts of cities or along major roads, with not only large floor-space requirements but also land needed for parking. It is also evident in shopping centres that often house very similar businesses that compete with each other, a form of redundancy that further exacerbates land needs without an actual basic reason for it. The issue of second homes, for example, has also been a particular driver of urban growth along the Mediterranean coast in Spain and in Portugal in the late 1990s and in the early 2000s, also in Denmark and other European countries where affluent city dwellers invested in houses and flats, either privately or using investors' developments that put large numbers of second homes on the market (see for example Couch *et al.*, 2007; Garcia, 2010; Christiansen and Loftsgarden, 2011).

The *price of land* is certainly a key driver in the satisfaction of demand and supply of new urban areas, especially in comparison to the price of infill development and investments into expansions (either vertically or through densification), or modernizations of the existing building stock. In this respect, some researchers observed over the last few decades that individual preferences about location and amenities favour inner city locations over new developments at the outskirts, and that many countries are entering a phase of reurbanization (see for example Siedentop, 2008). Further analysis of these initial preferences shows that the reurbanization movement can actually lead to gentrification processes that are closely linked to real estate prices in the inner cities and the options to buy land within the commutershed of the urban agglomeration. The attractiveness of such options very much depends on age and family structures and is not haphazard. It is rather a cumulative cohort effect that still drives suburbanization, possibly to a lesser extent than in the previous decades, but with a greater reach into the urban hinterland due to improved accessibility by high-speed public transport connections or other mobility options (Haag, 2002). The European Environment Agency finds in a technical report (European Environment Agency, 2010b) that tax, tax relief and urban pressure are the most influential aspects for land prices; to a lesser degree or in regional variations subsidies play a role, as well as inflation, commodity prices, land productivity and amenities. The interesting point here is that these drivers of land prices actually drive land use change. This is especially true for urban pressures where land previously used for agriculture can jump up scales of value if designated as development land in the urbanization process (e.g. in district plans or local development plans), and lends itself to all forms of speculation.

An interesting aspect in this observation is the role of the local authorities and the *competition between municipalities*. Their interest to put land on the market for residential or business developments is only partly to stabilize the demographic future and provide employment opportunities. Authors like Gutsche *et al.* (2007) have convincingly shown that in tax regimes where the local authorities benefit from property sales, business taxes lead to fiscal interests that drive urban expansions disproportionately. In some cases new developments are being marketed in a way that seems to be financially attractive in comparison to investments in the inner city, but only because maintenance costs are neglected and local infrastructure costs are financed by the commune. If development contributions and long-term maintenance would be properly included in project costs, the cost–benefit ratios would look different (see also Siedentop et al., 2009). A range of tools and initiatives have recently been disseminated by the German government to provide more accurate calculation routines to the local communities and establish stricter planning controls instruments for the designation of new urban land.

Lastly, the *real estate market* certainly has an economic interest to drive urbanization and put homes on the market. In local communities real estate agents and stakeholders in the property market can easily get involved in land use decisions, be it through lobbyism or active participation and a political mandate in the community. The same is true for the banking sector that has an economic interest in people and businesses taking up mortgages and loans to finance housing developments. Soule (2006) and his contributing authors describe these processes for the United States in their remarkable book *Urban Sprawl: A Comprehensive Reference Guide* as an actual form of subsidy, where mortgage deductions, communal development costs and low local property taxes are financed by state institutions to make the American Dream of housing property come true for large parts of the population. There are similar incentives for commercial developments, albeit with a more complex interaction of landowners, developers and businesses that rent the land or the facilities. In all cases, however, laissez-faire policies can be expected to result in land speculation under free market conditions. Self-regulation in terms of land-saving developments are not on the top of the agenda of the real estate industry, although the more serious players in that field would certainly strive for resource-efficient and environmentally sustainable neighbourhood developments, either in reaction to high levels of community awareness or as a best-practice selling point generated by market demand.

### Governance

It seems to be contradictory that the government sector is listed as a separate driver of urban expansion or urban growth when part of its duties is to care for a resource-efficient development of land use structures. The European Environment Agency starts their explanations with *EU policies* in this respect, not as a root driver, but as a factor that could be driving urban expansion under certain conditions. The key policies named here that are of influence are the transport and cohesion policies, where more than EUR 80 billion has been dedicated to structural funds, mainly for road projects. The improved accessibility in these areas made them certainly more attractive to investors for additional development, causing the following dilemma for land use planners: economic incentives in the form of infrastructure development are an effective instrument to trigger market investments, but the effects in terms of land conversion are undesirable from an environmental and sometimes also social point of view. In this respect, Figure 5.6 gives an assessment of the impact of transport policy on economic growth, researched within the ESPON framework of the European Union. It clearly shows the effects in the new member states in eastern Europe, but also in structurally weaker regions in the Mediterranean states and in remote areas of Scandinavia. It is fair to expect that these achievements have resulted in flow-on effects on land demand, be it for business or industrial enterprise along these routes or new residential developments or city expansions, although no detailed studies exist on this correlation.

*International regulations* do not seem to play that much of a role since there are no binding international laws that would influence the governance of land use-relevant policies directly. However, international agreements like the Kyoto Protocol on the reduction of $CO_2$ emissions are likely to be translated into national and regional *legislation and regulations* further downstream and can then have an effect on land use. An example is mitigation, where certain strategies aim to prevent the production of carbon emissions, for example by replacing fossil energy production chains by renewable options. This potentially transforms the

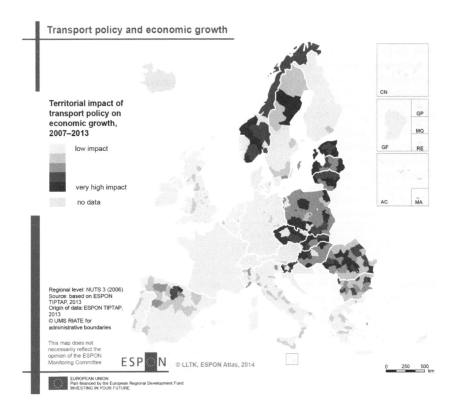

*Figure 5.6* Transport policy and economic growth (source: ESPON Atlas, http://atlas.espon.eu/, last accessed 2 June 2015)

agricultural landscape significantly and possibly leads to changing land use patterns that affect the availability of agricultural land negatively. Another example is adaptation to climate change, not necessarily triggered by international regulations, but certainly put on the forefront of the agenda by an increasing awareness about climate change risks through the work of international organizations like the Intergovernmental Panel on Climate Change (IPCC). One result of these adaptation strategies could, for example, be that floodplains are being adjusted to climate change scenarios, preventing and offsetting urbanization trends to other areas. Another effect could be that city planners place more and more value on fresh air corridors and bio-connectivity within the urban compound to preserve future options for air ventilation, water infiltration and biodiversity. Planning instruments like green belts or green wedges in regional planning serve, amongst others, such functions, but are also known to trigger increased development beyond the boundaries of the green belts. The term 'induced development' is often used in this respect, because planning instruments that contain urban development may not necessarily prevent the eventuation of new urban areas altogether. If the development pressure is very high and infill options are limited, development may be displaced to disconnected urban sprawl-like locations beyond the green belt that are highly undesirable from a compact city point of view and negatively impact upon the agricultural assets (see for example Bae and Jun, 2003; Kühn, 2003; Bengston and Youn, 2006).

In terms of *weak land use planning* as a driver of urban expansion one can only reiterate that free markets in combination with laissez-faire policies cannot be expected to deliver sustainable land use decisions. Within the PLUREL project mentioned above, Tosics *et al.* (2010) have come up with a classification of European countries in terms of their land use regulations, based on general country profiles and more in-depth case studies (see Table 5.1).

*Table 5.1* Classification of the public sector in relation to the level of control of urban development

| Control mechanism from supra-local levels of the planning system | Most important supra-local level (from land use change perspective[1] | Local level[2] | Countries |
|---|---|---|---|
| C) strongly controlled spatial policies | Large (>1) | any | |
| | Medium (0.5–1) | any | Portugal |
| | Small (<0.5) | any | Cyprus, Greece, Lithuania |
| B) medium level of control | Large (>1) | Large (>30) | Denmark, Netherlands, United Kingdom |
| | | Medium (10–30) | Belgium, France, Germany |
| | | Small (<10) | Italy, Spain |
| | Medium (0.5–1) | Large (>30) | Ireland |
| | | Medium (10–30) | |
| | | Small (<10) | Austria |

|  | Small (<0.5) |  | Large (>30) | Sweden |
|---|---|---|---|---|
|  |  |  | Medium (10–30) | Finland |
|  |  |  | Small (<10) | Estonia, Latvia, Luxembourg, Malta |
| A) Weak level of control | any |  | Large (>30) | Bulgaria |
|  |  |  | Medium (10–30) | Poland, Slovenia |
|  |  |  | Small (<10) | Czech Republic, Hungary, Romania, Slovakia |

Source: Tosics *et al.* (2010, p. 60).

Notes
1 Numbers refer to millions of inhabitants.
2 Numbers refer to thousands of inhabitants.

After grouping countries into families of spatial planning systems, the authors have looked into the different tiers of government, included hybrid approaches, and combined their results into the ranking of land use controls shown in Table 5.2, with stronger land use controls having higher values.

This classification itself does not necessarily correlate with the highest rates of urban growth since the different countries are exposed to completely different growth pressures and starting points for urbanization, especially on a regional level. Comparative studies have shown that national policies are rarely decisive in land use controls, something that Tosics *et al.* (2010) pick up on as well. They only work in combination with effective planning control systems on the regional and local levels and a planning environment with a strong mandate for compact growth strategies. In this respect, the European Environment Agency in 2006 has already presented research by authors who argue that the lack of planning coordination between the planning tiers is a key driver of urban sprawl (European Environment Agency, 2006). And even if planning controls are in place, there might still be sufficient leverage for market forces to override legislative barriers. One such example is the urban encroachment

*Table 5.2* Strength of land use controls in European countries

| Value[1] | Countries |
|---|---|
| 7 |  |
| 6 | Denmark, Netherlands, Portugal, United Kingdom |
| 5 | Belgium, Cyprus, France, Germany, Greece, Ireland, Lithuania |
| 4 | Italy, Spain, Sweeden |
| 3 | Austria, Bulgaria, Finland |
| 2 | Estonia, Latvia, Luxembourg, Malta, Poland, Slovenia |
| 1 | Czech Republic, Hungary, Romania, Slovakia |

Source: Tosics *et al.* (2010, p. 61).

Note
1 Higher values correspond to stronger control level.

of single development into green wedges in Germany, where the planning authorities had to allow some developments under special circumstances for local communities in order to be able to establish green wedges as a planning instrument in the first place. Over the years, however, the cumulative effect of all special circumstances can lead to a devaluation of the green wedge in terms of its original function. Because it is not fully natural anymore it may have lost its effectiveness as a divider between settlement structures and its ecological function as a fresh air corridor. Once this devaluation leads to lifting its status, the door is open for further development if no other restrictions are being put in place instead. In actual planning practices – these processes are quite commonplace – regulations can often not be dictated top-down to the local authorities without allowing room for development at all. Scholars in the field of urban sprawl often cite the term 'tyranny of small decisions' in this respect, which was originally coined by the economist Alfred Kahn and describes the market failures to control single decisions with a view towards their cumulative effects (Kahn, 1966). This mechanism is very true for land use decision making between different tiers of government, regardless of coordination and enforcement. The conflicts are inherent and each decision has a valid base for it to go forward. But there is a time lag difference between its initial assessment in the light of the existing land use structure, and the potential impact it may have if all other existing or yet unknown proposals to change the current land use are going to go ahead.

Another governance issue that has already been touched upon is *public subsidies for home ownership* as a driver of urban growth. It is somehow correlated to the economic and social aspects of rising living standards and housing preferences and the security that private property can provide in terms of the individual's living arrangements. Some societies have a stronger history in this respect; property levels differ widely across the world. Andrews and Sanchez (2011) have investigated home ownership rates and their driving forces for selected countries of the Organisation for Economic Cooperation and Development (OECD) in 2011.

Table 5.3 shows home ownership rates for these countries, and Figure 5.7 gives an indication of which countries offer the most public subsidies for their residents to get onto the property ladder. The latter can either be direct subsidies or tax reliefs for home ownership investments that governments use in family politics. In Germany, for example, families with children qualify for direct subsidies for residential home developments, with a view towards demographic stabilization. Other subsidies are available for investments in energy modernization in existing or new housing developments for homeowners. Overall, home ownership is changing not only due to government incentives, but also due to modified conditions for financing. Today, the loan-to-value ratios for initial down payments on housing are substantially lower than a few decades ago, albeit with some regional variations. The financial crisis of the late 2000s has not stopped this trend, quite the contrary. Because interest rates remain low and loans are easily available in the finance

*Table 5.3* Aggregate homeownership rates in selected OECD countries

| Country | 1990[1] | 2004[2] |
|---|---|---|
| Australia | 71.4 | 69.5 |
| Austria | 46.3 | 51.6 |
| Belgium | 67.7 | 71.7 |
| Canada | 61.3 | 68.9 |
| Denmark | 51.0 | 51.6 |
| Finland | 65.4 | 66.0 |
| France | 55.3 | 54.8 |
| Germany | 36.3 | 41.0 |
| Italy | 64.2 | 67.9 |
| Luxembourg | 71.6 | 69.3 |
| Netherlands | 47.5 | 55.4 |
| Spain | 77.8 | 83.2 |
| Switzerland | 33.1 | 38.4 |
| United Kingdom | 67.5 | 70.7 |
| United States | 66.2 | 68.7 |

Source: Andrews and Sanchez (2011, p. 212).

Notes
1 1987 for Austria, 1990 for Spain, 1991 for Italy, 1992 for Denmark and Switzerland, 1994 for Canada, France, Germany and Netherlands, 1995 for Australia, Belgium and Finland, 1997 for Luxembourg and United States.
2 1999 for Netherlands, 2000 for Belgium and France, 2003 for Australia, 2007 for Germany and United States.

sector, investments into real estate, including new housing developments, are more attractive than capital investments. The OECD study finds that home ownership rates are essentially driven by three groups of influencing

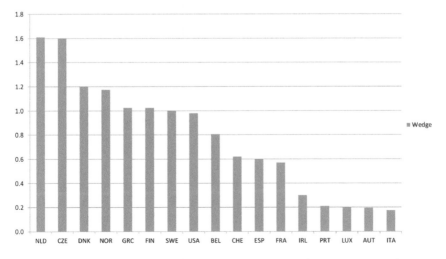

*Figure 5.7* Tax relief on debt financing cost of homeownership, 2009 (see Andrews and Sanchez, 2011, p. 216 for details on methodology and data sources)

factors: demographic factors, socioeconomic aspects and governance support in combination with mortgage markets. Their conclusion is that home ownership has effectively been boosted in some countries by public policy and financial support, which verifies the assumption that this form of governance is likely to drive urban growth as well.

The last aspect in the governance sector is the role that the *poor enforcement of existing plans* plays on the local level. This is certainly the case in many countries with difficulties in policy implementation, where building codes are either non-existent or not being enforced, or where control instruments are unfit to regulate the land market. Apart from these obvious failures, this area also covers the effective management of a cadastral system fit to support land management. It is not only the less developed countries that struggle with this task. More developed countries are also oftentimes overburdened with policy objectives that require detailed and robust information on development trends. The lack of (standardized) information in this respect can all too easily be used by interested parties to push for new development even if it goes against existing law. In cases of doubt the courts or other judicial institutions in charge of final land use decisions will often be legally bound to allow developments to go ahead. An example is the German land consumption target that was established in the early 2000s, taking cadastral information from the 1990s as the basis for the formulation of a reduction target. It states that for the whole nation, average consumption rates of 130 hectares per day in the 1990s shall be reduced to 30 hectares by the year 2020. This target was strongly supported by environmental stakeholders and interest groups as well as spatial planners, but often criticized by economic interest groups not only for its effects on the economy and possibly on the social side in terms of affordable housing, but also for its methodological problems. The base data included cadastral land use classes from inner city recreational areas and agricultural activities that were subsequently reclassified in the cadastral base. This led to a reduction of land consumption purely due to changes in the cadastre, and it was impossible to monitor the actual effect of spatial planning. Today there are many proponents of the cause that say one should have gone for a zero land consumption rate, and that due to the flawed information we now actually have a target that allows too much urbanization to still go ahead (see for example Siedentop and Fina, 2010).

### Transport

Many scholars agree that the transport sector is one of the key drivers of urban expansion and urban sprawl in modern times. Especially in the post-war period of the twentieth century, transport innovations and the advent of the private motor car have triggered suburbanization to a degree that would have been impossible to think of without modern mobility options.

Regional commutersheds are commonplace, with an expansion of metropolitan areas that allows workers to drive urbanization in locations far from their places of work and benefit from lower land prices. Some researchers have critically reflected upon transport infrastructure improvements that led to increased urban sprawl along these routes (Haag, 2002; Muniz and Galindo, 2005), others emphasize that efficient transport routes between city centres are actually quite resource efficient if serviced by different modes of transport effectively (Calthorpe and Fulton, 2001).

Apart from the accessibility gains that led to a shift in urbanization patterns, the European Environment Agency names the *low cost of fuel* as a key driver for urban expansion. This driver has to be seen in relation to household income, since fuel prices actually increase in absolute terms, but not in comparison to the growth rates in household income or gross domestic product. Tanguay and Gingras (2011) find that for Canadian metropolitan areas changes in fuel price do play a role in urban expansion and seem to be a promising policy instrument to reduce urban sprawl. However, household income in their study has a positive effect on urban sprawl, which suggests that the effects of a taxation of fuel prices could be offset by income gains. Glaeser and Kohlhase (2004) show that net transportation costs have actually decreased significantly and allowed household expenditure to be dedicated to private property and real estate investments. The same is true for freight transport and equally land-effective due to business sprawl across metropolitan regions. The authors' conclusion is that urban theories that have explained the cities of the past need to be updated. Today's cities do not need to be near natural resources anymore and efficient automobility across city environments would render city centres redundant, being remnants of a time when physical proximity was crucial. Nowadays locational aspects of quality of life are more important, not so much proximity. This argumentation has obviously some weaknesses in respect to a sustainable integration of city structures in a resource-efficient compact growth model and the authors do acknowledge that commuting times have threatened their view of cities due to increasing congestion and environmental problems that unlimited automobility causes as well. What remains uncontested, however, is that the low cost of fuel is still driving urbanization in a way that can easily become a risk to the society at large if prices increase drastically. Such scenarios of skyrocketing fuel prices have been quite widespread in the late 2000s, where the image of *peak oil* triggered substantial research efforts to test what would happen in post-fossil fuel times to urban structures and land use. A large research project in the metropolitan region of Hamburg, Germany, found that the consequences could be dramatic and advocates for spatial planning strategies that prioritize transit-oriented development patterns and compact growth initiatives along the lines of city development patterns suggested by authors like Newman (Newman and Kenworthy, 2006; Gertz *et al.*, 2015). Other fiscal instruments lead to additional *reductions in transport costs*. One example in transport policy that has been widely discussed in Germany is the regular tax breaks for commuters, where the costs of commuting can be deducted from incomes so that commuting is actually a subsidized activity. Urban growth scientists see this as a fatal misincentive and even environmental government institutions label these kinds of tax reliefs as environmentally harmful subsidies (Umweltbundesamt, 2014).

On the national/regional and local scale, high *levels of car ownership* have a reinforcing effect on urban growth. On the one hand, authors like Glaeser and Kahn (2003) argue that car ownership together with low transportation costs have triggered suburbanization and urban sprawl. On the other hand, authors like Bento *et al.* (2005) emphasize that urban structure and transport options are crucial, with suburban locations causing a form of car preference over other modes of transport and thus causing car reliance over time. Figure 5.8 shows an interesting projection of the amount of vehicle miles travelled and car ownership across world regions. One can clearly see the worrisome amount of increase in vehicle miles travelled for all regions and the correlation between the two variables. If extended to land consumption scenarios that these drivers would cause if these projections hold true, urban growth is not likely to enter any form of sustainable development path in the near future. Studies in Stuttgart, Germany, have shown that the availability of cars for the majority of the population can have flow-on effects on the urban structure in the sense that it becomes highly car-dependant, other mobility options are being phased out due to low demand, and the infrastructure to support more cars and more efficient car travel becomes a structural need (Siedentop *et al.*, 2013).

The *availability of roads* is certainly another aspect that drives the development of car-oriented urban structures. The land take for roads is a significant component of urban land use change in itself, and leads indirectly to the support of urban sprawl as described above. Figure 5.9 shows the combined effect of land take from road and rail infrastructure with rates between 1 and just over 4 per cent. On top of these direct land requirements, roads and rail lines cut through habitats and are major environmental stressors for landscape fragmentation and risks to other qualities of the natural environment.

Finally, *poor public transport* is an indirect driver of urban sprawl when it forces people to shift their mobility habits to the private motor car and therefore adds to automobile-oriented sprawling cityscapes. In this respect one needs to look at the economics of public transport in terms of their catchments and efficiencies for different urban structures. Bertraud (2004) compares the urban layout of the city of Atlanta, United States, with the city of Barcelona, Spain. Within a 800-metre radius 60 per cent of the Barcelona population live within the public transport catchments, in Atlanta it is only 4 per cent. This is certainly reflected by the number of trips made by public transport in both cities, but also in the quality of services that can be offered economically to the users.

However, public transport provision is not just a question of economics. Some countries do invest in and subsidize public transport with a view towards a more sustainable transport future, because the overall benefits are likely to outweigh profit losses when only looking at passenger ticket revenues. Road transport, after all, is also heavily subsidized with communal funds, be it for

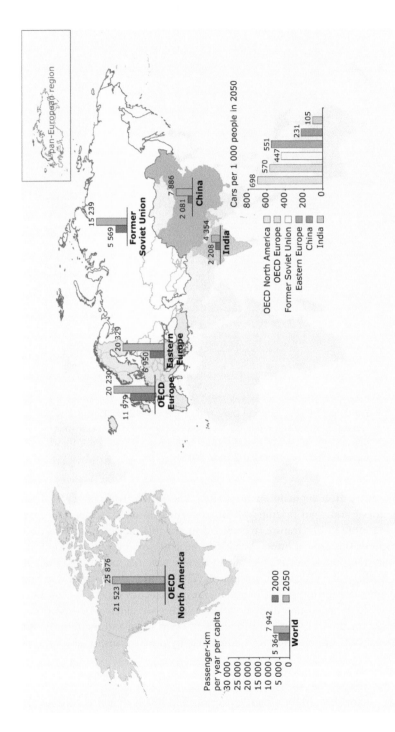

*Figure 5.8* Passenger-km per year per capita in 2000 and projected for 2050, and projected car ownership rates in 2050 (source: www. eea.europa.eu/data-and-maps/figures/passenger-km-per-year-per-capita-in-2000-and-projected-for-2050-and-projected–car-ownership-rates-in-2050/transport-outlook-map-graph.eps/image_original, last accessed 4 April 2014)

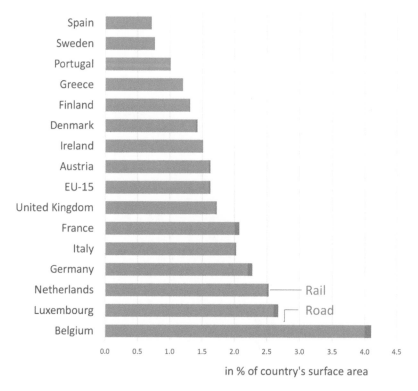

*Figure 5.9* Land take by road and rail infrastructure for the 2001 European
Union member states (source: www.eea.europa.eu/data-and-maps/
figures/eu-land-take-by-roads-and-railways-as-percentage-of-country-
surface-1998, last accessed 2 June 2015)

initial construction or maintenance costs. The whole issue of travel demand
management and road pricing systems to primarily alter transport behaviour
towards a more public transport-oriented sustainable mobility future is beyond
the scope of this contribution. What can be said, however, is that the quality
of public transport is a crucial key with a view towards future urban development. Authors like Calthorpe and Fulton (2001) and Newman and Kenworthy
(2006) advocate for integrated regional cities with centres and sub-centres connected by high-speed rail, ultimately leading to polycentric city environments
that retain quality of life options in terms of open space and medium densities,
but also capable of being serviced efficiently by public transport.

### Land

The last sector that drives urban expansion and urban sprawl according to the
European Environment Agency is the very general theme labelled 'Land'. On the

country/regional and local level, the *local geography and environment* can become a driver under certain conditions. This can be interpreted in a number of ways – the authors do not give clear directions here. Obvious drivers from the land and geography side are enclosed locations for example along the coast, where urban expansion is usually limited to the inland direction, if no land reclamation takes place. Examples are Porto and Barcelona, where the hinterland has long been a barrier to urbanization leading to particularly compact structures. The same can be said for cities surrounded by water or in mountainous regions, where building on the slopes or in floodplains would either be inconvenient and risky, or very costly. The resulting patterns are often settlement structures that are unusually compact, which can for example be seen in the valleys of the Carpathian Mountains in Romania. Apart from these topographic restrictions, there are also cultural differences that are linked to geography and history. What has been researched in a study by Siedentop and Fina (2012b) and illustrated by Hartog (2005) in his book *Europe's Ageing Cities* is that there are some countries in Europe that have a rather compact urban structure for historic reasons. The hypothesis is that, for example, climatic factors in the Mediterranean countries have favoured dense inner city building blocks that provide shade and cooling to protect from heat and allow fresh air streams to penetrate the city. In the United Kingdom and also partly in Scandinavia there is a tradition of countryside living going back to Ebenezar Howard's garden city and the health awareness at the turn of the nineteenth century, ultimately favouring a type of rural sprawl and high demands for inner city plots with gardening possibilities. In eastern Europe, the socialist regimes have long favoured dense housing units with prefabricated building blocks. And in Germany and France, traditional rural communities have developed a range of settlement structures that were adapted to the local environmental conditions and available building materials, to transport routes going back to the Roman Empire, or to fortified structures with enclosed inner cities. Very often the settlement cores were centred around particularly valuable resources, be it a river valley for water supply, agriculturally attractive areas or a combination of both. It is a historic paradox that the settlements with the most locational advantages had the best chances to prosper and expand, thus using up and destroying much of the resources that initially led to their success.

Apart from the loss of environmental resources, many of the historic and geographic predicaments mentioned above have been overrun by the massive suburbanization and modernization activities in the twentieth century, with little consideration about the underlying cultural values and environmental intelligence that are inherent to traditional settlement structures. However, the initial structures persist to some degree, and many architects and city planners rediscover and create new ways to integrate old and new settlement structures in a sustainable way. At the same time, the path that urban structures have taken differs in terms of urban sprawl, and some development paths are harder to control or to contain due to path dependencies that sprawling landscapes create for themselves. Amongst these are car dependency and a lack of public transport infrastructure. Especially the latter is incredibly difficult to retrofit if it hasn't been initially planned for and land resources haven't been set aside. The most successful public transport systems in Germany, for example, have actually benefitted from the land availability after the bombings of the Second World War, allowing planners to obtain land resources for reconstruction that would have otherwise been difficult to alter in use (see for example Diefendorf, 1993). In that sense, geographically and historically embedded development paths very often limit future options and create a form of path dependence. Any substantial changes need massive interventions and require a level of public and financial support that is very difficult to obtain a mandate for.

Figure 5.10 shows examples of cities that can be taken as representatives for such development paths in terms of urban sprawl. The categorization is based on a study conducted by Kasanko et al. (2006), where 15 European metropolitan regions have been assessed with different sprawl indicators and then aggregated to one index value.

| | Southern European cities | Eastern and central European cities | Northern and western European cities |
|---|---|---|---|
| **Sprawled** | | Udine | |
| | | Pordenone | |
| | | Dresden | Helsinki |
| | | | Copenhagen |
| | | | Dublin |
| | | | Brussels |
| | | | Grenoble |
| | Marseille | Trieste | Sunderland |
| | Porto | Vienna | Lyon |
| | | Bratislava | Tallinn |
| | | Belgrade | |
| | Iraklion | Prague | |
| | Palermo | Munich | |
| | Milan | | |
| | Bilbao | | |
| **Compact** | | | |

*Figure 5.10* Distribution of Europe's sprawling and compact cities (sources: European Environment Agency, 2006; Kasanko et al., 2006)

The result shows that the most compact cities are in southern Europe and to some degree in eastern and central Europe. The most sprawling cities are located in eastern and central Europe and also in northern and western European cities. The geographic distribution should not be overestimated in any type of interpretation, since this is only a limited number of cities. However, the geographic and topographic as well as the cultural settings have certainly a level of influence as drivers of urban sprawl that cannot be ignored.

## Conclusion

This chapter used a compilation of drivers of urban expansion and urban sprawl published by the European Environment Agency to reflect upon sector- and scale-specific influencing factors. The complexities of interactions between drivers are such that any type of overarching assessment on a national level is bound to fail – the different physical manifestations of urban expansion and urban sprawl are more characteristic on a regional observation scale.

The different driving factors have been explained based on the literature available on the respective topics, illustrated with examples mainly from across Europe. One key message is that population growth which has been driving urbanization to a large degree in the past is not the key motivation anymore. In today's stagnant or even shrinking population numbers the demand for new urban structures is nourished by social transformations and economic growth, by the preference of spacious living environments and lifestyle changes, by large-scale business developments and by the increasing importance of accessibility in metropolitan regions. Governance structures and policies have very mixed effects at best. Structural development funds and economic aid are poorly coordinated with environmental policies and contradict each other in terms of effective urban containment strategies. The planning apparatus struggles with the provision of long-term planning strategies, with the establishment of meaningful targets and appropriate monitoring systems. Information is crucial, but difficult to use in the political arena if there is disagreement on normative interpretations of development paths. Local authorities have a local mandate to do everything they can to preserve economic opportunities and stabilize the population base, often using methods that consume more and more land and will ultimately be bound to backfire economically if demographic change proceeds at the predicted rates. And economic growth on a national scale is driven by global competition, with a limited view towards sustainable land management within the own country, but also increasingly in the countries where the goods for the national market are produced. Land grabbing in other countries or remote areas occurs at scales where we only have limited insight until now, but the trend to outsource land-consumptive activities to areas that are outside the monitoring responsibilities of an institution is certainly not suitable to qualify as a move towards sustainable land management.

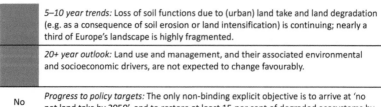

*5–10 year trends:* Loss of soil functions due to (urban) land take and land degradation (e.g. as a consequence of soil erosion or land intensification) is continuing; nearly a third of Europe's landscape is highly fragmented.

*20+ year outlook:* Land use and management, and their associated environmental and socioeconomic drivers, are not expected to change favourably.

No target   *Progress to policy targets:* The only non-binding explicit objective is to arrive at 'no net land take by 2050', and to restore at least 15 per cent of degraded ecosystems by 2020.

*Figure 5.11* Trends and outlook: land use and soil functions (source: European Environment Agency, 2015, p. 59)

Figure 5.11 and the Annex give an overview over Europe-wide (Figure 5.11) and country-specific (Annex) trends and an outlook on land use and soil functions that the European Environment Agency published in their summary *State of the Environment* report in 2015 and in 2010, respectively, with worrisome assessments. Most projections and forecasts presented in this chapter on the specific drivers are very much in line with these forecasts: urban expansion, be it in the form of urban sprawl or just as a form of land take that leads to surface sealing, reduction of farmland or natural land, is a problem area with no realistic improvement potentials under current conditions and trends.

It is the task of researchers, scholars and planners alike to disseminate this information more clearly into the decision-making arena. We can help to close the argumentation gaps where confusing information allows profiteers of urban expansion to dazzle the general public about the facts. But on top of that we also need not only to argue for new and more effective controls that influence the driving factors of urban expansion substantially, but to push them through and implement where possible, even at the cost of new economic paths to which the global community has to adapt: there is no other choice but to effectively protect land resources if sustainable land management is to be more than just talk.

## References

Allan, J.; Mallat, C. (eds) (1995) *Water in the Middle East: Legal, Political and Commercial Implications*, Tauris Academic Studies, London.

Andrews, D.; Sanchez, A. C. (2011) 'The evolution of homeownership rates in selected OECD countries: demographic and public policy influences', *OECD Journal*: Economic Studies. 2011/1. Organisation for Economic Cooperation and Development Paris.

Angel, S.; Parent, J.; Civco, D. L.; Blei, A. M. (2011a) 'Making room for a planet of cities', Policy Focus Report, Lincoln Institute of Land Policy, Cambridge, MA.

Angel, S.; Parent, J.; Civco, D. L.; Blei, A.; Potere, D. (2011b) 'The dimensions of global urban expansion: estimates and projections for all countries, 2000–2050', *Progress in Planning*, 75, 53–107.

Anthony, J. (2004) 'Do state growth management regulations reduce sprawl?', *Urban Affairs Review*, 39, 376–397.

Bae, C.-H. C.; Jun, M.-J. (2003) 'Counterfactual planning: what if there had been no greenbelt in Seoul?', *Journal of Planning Education and Research*, 23, 374–383.

Bengston, D. N.; Youn, Y.-C. (2006) 'Urban containment policies and the protection of natural areas: the case of Seoul's greenbelt', *Ecology and Society*, 11, 3.

Bengston, D. N.; Fletcher, J. O.; Nelson, K. C. (2004) 'Public policies for managing urban growth and protecting open space: policy instruments and lessons learned in the United States', *Landscape and Urban Planning*, 69, 271–286.

Bento, A. M.; Cropper, M. L.; Mobarak, A. M.; Vinha, K. (2005) 'The effects of urban spatial structure on travel demand in the United States', *The Review of Economics and Statistics*, 87, 466–478.

Bertaud, A. (2004) 'The spatial organization of cities: deliberate outcome or unforeseen consequence?', Working Paper 2004-01, Institute of Urban and Regional Development, University of California at Berkeley, Berkeley, CA.

Cairncross, F. (1997) *The Death of Distance*, Harvard Business School Press, London.

Calthorpe, P.; Fulton, W. (2001) *The Regional City: Planning for the End of Sprawl*, Island Press, Washington, DC.

Chin, N. (2002) 'Unearthing the roots of urban sprawl: a critical analysis of form, function and methodology', CASA Working Paper Series no. 47, Centre for Advanced Spatial Analysis, University College London, London.

Christiansen, P.; Loftsgarden, T. (2011) 'Drivers behind urban sprawl in Europe', TØI Report, 1136/2011, Institute of Transport Economics, Oslo.

Cooper, T.; Hart, K.; Baldock, D. (2009) *Provision of Public Goods through Agriculture in the European Union*, Report Prepared for DG Agriculture and Rural Development, Contract No. 30-CE-0233091/00-28, Institute for European Environmental Policy, London.

Couch, C.; Koeontidou, L.; Petschel-Held, G. (eds) (2007) *Urban Sprawl in Europe: Landscapes, Land-Use Change and Policy*, Blackwell Publishing, Oxford.

Diefendorf, J. M. (1993) *In the Wake of War: The Reconstruction of German Cities after World War II*, Oxford University Press, Oxford.

Dielemann, F.; Wegener, M. (2004) 'Compact city and urban sprawl', *Built Environment*, 30, 308–323.

European Environment Agency (1999) *Environmental Indicators: Typology and Overview*, Technical report no. 25, European Environment Agency, Copenhagen.

European Environment Agency (2006) *Urban Sprawl in Europe: The Ignored Challenge*, EEA Report no. 10/2006, European Environment Agency, Copenhagen.

European Environment Agency (2010a) *The European Environment: State and Outlook 2010: Land Use*, SOER 2010, European Environment Agency, Copenhagen.

European Environment Agency (2010b) *Land in Europe: Prices, Taxes and Use Patterns*, Technical report no. 4/2010, European Environment Agency, Copenhagen.

European Environment Agency (2015) *The European Environment: State and Outlook 2015: Synthesis Report*, European Environment Agency Copenhagen.

Ewing, R.; Pendall, R.; Chen, D. (2002) 'Measuring sprawl and its impact', *Smart Growth America* [Online]. Available at: www.smartgrowthamerica.org/documents/MeasuringSprawl.PDF [Last viewed on 8 February 2013].

Fina, S.; Planinsek, S.; Zakrzewski, P. (2009) 'Suburban crisis? Demand for single family homes in the face of demographic change', *Europa Regional*, 17, 2–14.

Fina, S.; Planinsek, S.; Zakrzewski, P. (2012) 'Germany's post-war suburbs: perspectives of the ageing housing stock', in Ganser, R. and Piro, R. (eds) *Parallel Patterns of Shrinking Cities and Urban Growth: Spatial Planning for Sustainable Development of City Regions and Rural Areas*. Ashgate, London, 111–124.

Fina, S.; Pileri, P.; Siedentop, S.; Maggi, M. (2014a) 'Strategies to reduce land consumption: a comparison between Italian and German city regions', *Archivo di Studi Urbani e Regionali*, 108, 37–56.

Fina, S.; Schmitz-Veltin, A.; Siedentop, S. (2014b) 'Räumliche Muster der internationalen Migration im Zeitverlauf am Beispiel Stuttgart: vom Wanderungsziel zum Migrationsknoten?', in Gans, P. (ed.) *Räumliche Auswirkungen der internationalen Migration*. Akademie für Raumforschung und Landesplanung, Hannover, 381–401.

Frenkel, A. (2004) 'The potential effect of national growth-management policy on urban sprawl and the depletion of open spaces and farmland', *Land Use Policy*, 21, 357–369.

Garcia, M. (2010) 'The breakdown of the Spanish urban growth model: social and territorial effects of the global crisis', *International Journal of Urban and Regional Research*, 34, 967–980.

Gertz, C.; Maaß, J.; Guimaraes, T. (eds) (2015), *Auswirkungen von steigenden Energiepreisen auf die Mobilität udn Landnutzung in der Metropolregion Hamburg: Ergebnisse des Projekts €LAN – Energiepreisentwicklung und Landnutzung*, Schriftenreihe des Instituts für Verkehrsplanung und Logistik, no. 13, Technische Universität Hamburg-Harburg, Institut für Verkehrsplanung und Logistik, Hamburg.

Glaeser, E. L.; Kahn, M. E. (2003) *Sprawl and Urban Growth*, Harvard Institute of Economic Research, Harvard University, Cambridge, MA.

Glaeser, E. L.; Kohlhase, J. (2004) 'Cities, regions and the decline of transport costs', *Regional Science*, 83, 197–228.

Gutsche, J.-M.; Schiller, G.; Siedentop, S. (2007) *Von der Außen- zur Innenentwicklung in Städten und Gemeinden: Das Kostenparadoxon der Baulandentwicklung*, Texte 31/2009 [Online]. Available at: www.umweltbundesamt.de/uba-infomedien/ mysql_medien.php?anfrage=Kennummer&Suchwort=3858 [Last viewed on 8 February 2013].

Haag, G. (2002) *Sprawling Cities in Germany*, Franco Angeli, Milano.

Hartog, R. (2005) *Europe's Ageing Cities*, Verlag Müller und Busmann KG, Wuppertal.

Hasse, J. E.; Lathrop, R. G. (2003) 'Land resource impact indicators of urban sprawl', *Applied Geography*, 23, 159–175.

Häußermann, H.; Läpple, D.; Siebel, W. (2008) *Stadtpolitik*, Suhrkamp, Frankfurt am Main.

Kahn, A. E. (1966) 'The tyranny of small decisions: market failures, imperfections, and the limits of economics', *Kyklos*, 19, 23–47.

Kasanko, M.; Barredo, J. I.; Lavalle, C.; Mccormick, N.; Demicheli, L.; Sagris, V.; Brezger, A. (2006) 'Are European cities becoming dispersed? A comparative analysis of 15 European urban areas', *Landscape and Urban Planning*, 77, 111–130.

Kühn, M. (2003) 'Greenbelt and green heart: separating and integrating landscapes in European city regions', *Landscape and Urban Planning*, 64, 19–27.

Lüthi, S.; Thierstein, A.; Bentlage, M. (2012) 'The relational geography of the knowledge economy in Germany: on functional urban hierarchies and localised value chain systems', *Urban Studies*, doi: 10.1177/0042098012452325.

Muniz, I.; Galindo, A. (2005) 'Urban form and the ecological footprint of commuting: the case of Barcelona', *Ecological Economics*, 55, 499–514.

Newman, P.; Kenworthy, J. R. (2006) 'Urban design to reduce automobile dependence', *Opolis*, 2, 35–52.

Nilsson, K. (2011) *Peri-Urban Land Use Relationships – PLUREL Project: Publishable Final Activity Report*, Danish Centre for Forest, Landscape and Planning, University of Copenhagen, Copenhagen.

Organisation for Economic Cooperation and Development (2012) *Compact City Policies: A Comparative Assessment*, OECD Green Growth Studies [Online]. Available at: www.oecd-ilibrary.org/urban-rural-and-regional-development/compact-city-policies_9789264167865-en [Last viewed on 8 February 2013].

Organisation for Economic Cooperation and Development (2015) *The Metropolitan Century: Understanding Urbanisation and Its Consequences*, OECD, Paris.

Rulli, M. C.; Saviori, A.; D'Odorico, P. (2013) 'Global land and water grabbing', *PNAS*, 110/3, 892–897.

Schmidt, S. (2011) 'Sprawl without growth in eastern Germany', *Urban Geography*, 32/1, 105–128.

Schmidt, S.; Fina, S.; Siedentop, S. (2014) 'Post-socialist sprawl: a cross-country comparison', *European Planning Studies*, doi: 10.1080/09654313.2014.933178.

Siedentop, S. (2008) 'Die Rückkehr der Städte? Zur Plausibilität der Reurbanisierungs hypothese', *Informationen zur Raumentwicklung*, 3, 193–210.

Siedentop, S.; Fina, S. (2010) *Datengrundlagen zur Siedlungsentwicklung. Gutachten im Auftrag des Ministeriums für Wirtschaft, Mittelstand und Energie des Landes Nordrhein-Westfalen*, Institut für Raumordnung und Entwicklungsplanung, Stuttgart.

Siedentop, S.; Fina, S. (2012a) '"Eine neue Geographie der Segregation?" Entwicklung der ethnischen und generativen Segregation in der Landeshaupstadt Stuttgart', *Statistik und Informationsmanagement*, Monatsheft 10/2012, 346–357.

Siedentop, S.; Fina, S. (2012b) 'Who sprawls most? Exploring the patterns of urban growth across 26 European countries', *Environment and Planning B*, 44, 2765–2784.

Siedentop, S.; Junesch, R.; Strasser, M.; Zakrzewski, P.; Samaniego, L.; Weinert, J. (2009) *Einflussfaktoren der Neuinanspruchnahme von Flächen*, Research Notebook no. 139, Bundesamt für Bauwesen und Raumordnung, Bonn.

Siedentop, S.; Roos, S.; Fina, S. (2013) 'Ist die "Autoabhängigkeit" städtischer Siedlungsgebiete messbar? Entwicklung und Anwendung eines Indikatorenkonzepts in der Region Stuttgart', *Raumforschung und Raumordnung*, 71, 329–341.

Song, Y.; Knaap, G.-J. (2004) 'Measuring urban form: is Portland winning the war on sprawl?', *Journal of the American Planning Association*, 70, 210–225.

Soule, D. C. (2006) *Urban Sprawl: A Comprehensive Reference Guide*, Greenwood Press, Westport, CT.

Tanguay, G.; Gingras, I. (2011) 'Gas Prices Variations and Urban Sprawl: An Empirical Analysis of the 12 Largest Canadian Metropolitan Areas', Cirona Scientific Series 2011s-37, Cirano, Montreal.

Tosics, I.; Szemző, H.; Illés, D.; Gertheis, A.; Lalenis, K.; Kalergis, D. (2010) *National Spatial Planning Policies and Governance Typology*, Peri-Urban Land Use Relationships – Strategies and Sustainability Assessment Tools for Urban-Rural Linkages, PLUREL Report no. 2.2.1, Copenhagen.

Umweltbundesamt (2014) *Umweltschädliche Subventionen in Deutschland: Aktualisierte Ausgabe 2014*, Fachbroschüre. Umweltbundesamt, Dessau-Roßlau.

Vogel, R. K.; Savitch, H. V.; Xu, J.; Yeh, A. G. O.; Wu, W.; Sancton, A.; Kantor, P.; Newman, P.; Tsukamoto, T.; Cheung, P. T. Y.; Shen, J.; Wu, F.; Zhang, F. (2010) 'Governing global city regions in China and the West', *Progress in Planning*, 73, 1–75.

Wolman, H.; Galster, G.; Hanson, R.; Ratcliffe, M.; Furdell, K.; Sarzynski, A. (2005) 'The fundamental challenge in measuring sprawl: which land should be considered?', *The Professional Geographer*, 57/1, 94–105.

# Annex

*Annex Table* Land cover change in EEA member and collaborating countries: total changes for 1990–2000 and 2000–2006, and examples of specific trends for 2000–2006

| Country | Annual land cover change, % of total area | | Characteristic land cover changes, 2000–2006 | | |
|---|---|---|---|---|---|
| | 1990–2000 | 2000–2006 | Artificial areas | Agricultural areas | Forest and nature |
| Albania | – | 0.18 | Very high rate of residential sprawl | Loss of agricultural land | Forest: gains from agriculture, losses to urbanization |
| Austria | 0.03 | 0.08 | Expansion of sport, leisure and recreation sites | Agricultural land uptake by artificial areas | Accelerated decrease of alpine glacier area |
| Belgium | 0.17 | 0.10 | Slow-down of land uptake | Slow-down of change dynamics, land uptake by artificial areas | Internal forest conversions, formation of water bodies |
| Bosnia and Herzegovina | – | 0.12 | Diffuse residential sprawl | Loss of pasture/mosaics, vineyards and orchards | Semi-natural land transitions, fires |
| Bulgaria | 0.11 | 0.09 | Urban sprawl accelerates | Overall stabilization, loss of pasture/mosaics, vineyards and orchards | Forest management has replaced forest expansion |
| Croatia | 0.19 | 0.17 | Accelerated artificial sprawl driven by highway construction | Uptake of pasture by arable and complex cultivation land | Forest management, loss of open spaces, re-growth of burnt areas |
| Cyprus | – | 0.49 | Diffuse sprawl of residential areas, sport and leisure facilities | Consumption of agricultural land | Transitional woodland formation over burnt areas |
| Czech Republic | 0.81 | 0.33 | Urban sprawl accelerates, driven by construction | Slow-down, continued conversion from arable land to pasture | Stabilization in natural landscapes, some loss of natural grasslands |

| | | | | | |
|---|---|---|---|---|---|
| Denmark | 0.13 | 0.13 | Diffuse residential sprawl accelerated | Consumption of arable land | Forest creation, changes in wetlands and water bodies |
| Estonia | 0.44 | 0.38 | Doubled sprawl of artificial areas: mines and construction | Slow-down of changes, conversion from pasture to arable land | Exchange between mineral extraction sites and forested land |
| Finland | – | 0.35 | Sprawl of housing and recreation | Conversion from forest and wetlands to arable land | Forest management, net loss of forest and wetlands |
| France | 0.20 | 0.11 | Continued urban expansion | Reduced agricultural transitions, loss of different farmland types | Slowed changes of natural areas, forest management, fires |
| Former Yoguslav Republic of Macedonia | – | 0.14 | Residential sprawl, development of mineral extraction | Transitions of different land types, loss of vineyards and orchards | Forest management, new water bodies, loss of natural grasslands |
| Germany | 0.24 | 0.10 | Land uptake slows down | Decreased change dynamics, conversion of pasture to arable land | Forest and water bodies created on open spaces and former mining areas |
| Greece | – | – | – | – | – |
| Hungary | 0.56 | 0.48 | Expansion of construction and mineral extraction | Withdrawal of farming, some conversion of pasture to arable land | Transitional woodland creation over former farmland and grasslands |
| Iceland | – | 0.10 | Land take driven by construction | Loss of pastures to artificial land uptake | Decrease of permanent snow and glaciers, new transitional woodlands |
| Ireland | 0.79 | 0.38 | Continued expansion of artificial areas on agricultural land | Rapidly reduced agriculture dynamics, withdrawal of farming | Transitional woodland over open natural and farmed areas |
| Italy | 0.13 | 0.10 | Growth of economic sites and recycling of urban land | Loss of farmland, less farming withdrawal and arable/pasture transition | Reduced expansion on to farmland, transitions of natural land cover |

*(continued)*

| Country | Annual land cover change, % of total area | | Characteristic land cover changes, 2000–2006 | | |
| --- | --- | --- | --- | --- | --- |
| | 1990–2000 | 2000–2006 | Artificial areas | Agricultural areas | Forest and nature |
| Kosovo under UNSCR 1244/99 | – | 0.16 | Dominance of residential sprawl | Loss of farmland and conversion from pasture to arable/crop land | Forest transitions, re-vegetation of burnt areas |
| Latvia | 0.78 | 0.38 | Faster artificial sprawl in surroundings of capital city | Slowed agricultural transitions, accelerated loss of farmland | Recent forest transitions, loss of pastures/mosaics to transitional woodland |
| Liechtenstein ★ | – | – | Steady increase of artificial areas | Continued decrease of agricultural land | Observed impacts of natural disturbances |
| Lithuania | 0.48 | 0.25 | Faster sprawl, driven by development of construction sites | Rapid slowdown of internal agriculture conversions | Natural land transitions, loss of pastures/mosaics to transitional woodland |
| Luxembourg | 0.15 | 0.23 | Slow-down of sprawl of housing and recreation facilities | Accelerated consumption of pasture, formation of arable land | Transitional woodland becoming forest, some loss to economic sites |
| Malta | 0.07 | 0.00 | No change in urban areas | No change in agricultural land cover | Natural areas almost without change |
| Montenegro | 0.02 | 0.04 | Extension of construction sites and residential areas | Loss of pastures and mosaics to artificial surfaces | Forest transitions, loss of natural areas to economic sites, fires |
| Netherlands | 0.30 | 0.27 | Increased construction and urban land management | Agricultural land uptake by development of artificial areas | Growth of natural areas, e.g. grasslands, withdrawal of farming |
| Norway | – | 0.10 | Extension of sport and leisure facilities, residential sprawl | Low intensity of agricultural changes | Forest transitions, some loss of natural areas, fires, decrease of glaciers |
| Poland | 0.10 | 0.10 | Increased sprawl of economic sites, highway construction | Loss of agricultural land (mostly arable) | Transitional woodland on former farmland, new water bodies |

| Portugal | 0.78 | 1.43 | Development driven by construction around key areas | Slow-down of agricultural transitions, farmland abandonment | Forest transitions, new forested land and water bodies, fires |
| Romania | 0.16 | 0.05 | Residential sprawl accelerates around main cities | Slow-down of agricultural transitions, loss of pastures | Recent felling and land transition, some loss of natural open areas. |
| Serbia | 0.11 | 0.07 | Slower residential sprawl, doubled extension of mines | New formation of arable land, loss of pasture/ mosaics, fruit and berry | Low forest formation, loss of grasslands, new water bodies |
| Slovakia | 0.51 | 0.25 | Slow-down of residential land take | Slow-down of changes, loss of agricultural land | Forest creation after withdrawal of farming |
| Slovenia | 0.02 | 0.03 | New construction sites drive future land take | Limited changes, loss of agricultural land | Limited changes, forest felling and loss of land, new water bodies |
| Spain | 0.34 | 0.29 | Urban extension, faster sprawl of construction and transport land | Loss of arable land to, olive groves, vineyards, orchards, construction | Forest transitions, afforestation of dry semi-natural land, fires |
| Sweden | – | 0.49 | Dynamic development of artificial land cover | Loss of arable land | Forest transitions, some uptake of forested areas by economic sites |
| Switzerland ** | – | – | Slower urban and infrastructure extension | Decline in arable land, increase in pasture, withdrawal of farming | Remote area reverting to wild vegetation, glacier retreat |
| Turkey | – | 0.08 | Development mostly driven by construction and mining | Increased arable land e.g. irrigated lands, loss of pasture/ mosaics | Loss of natural open land to transitional woodland/shrub |
| United Kingdom | – | – | – | – | – |

Source: European Environment Agency (2010a), based on Corine land cover data.

Notes
* Land cover changes in Liechtenstein remained below the detection level of the Corine land cover change methodology; land cover trends are assessed from the national contribution to SOER 2010.
** Land cover trends for Switzerland are assessed from the national contribution to SOER 2010.

# Part II

# Impact of land take and soil sealing on soil-related ecosystem services

# 6 The effects of urban expansion on soil health and ecosystem services

## An overview

*Mitchell Pavao-Zuckerman and Richard V. Pouyat*

## Introduction

Cities are often thought of as open systems with large, extractive footprints that are dependent on the productivity and ecosystem services of surrounding hinterlands (Rees, 1997). The ecosystem service concept was developed in part to help describe ecosystems in ways that might help conserve them in the face of land use change (Daily *et al.*, 1997, Setälä *et al.*, 2014). Despite this perspective on cities, many ecologists now recognize that cities can be thought of as ecosystems, that have internal structures and functions that generate ecosystem services (Pickett *et al.*, 2001, Adler and Tanner, 2013, Davies *et al.*, 2011, Grimm *et al.*, 2000). Moreover, the well-being of urban residents may be improved by the provision of urban ecosystem services (McPhearson *et al.*, 2014, Andersson *et al.*, 2014, Barthel *et al.*, 2010). This recognition of cities as urban ecosystems has led to a paradigm shift and focus on urban ecosystem services in research and planning perspectives, with new attention now being paid to the supply and demand for urban ecosystem services (Ernstson *et al.*, 2010, Gill *et al.*, 2008, Gomez-Baggethun and Barton, 2013, Pataki *et al.*, 2011).

As cities develop they have many potential environmental impacts that alter soils and their ecosystem services. Notably, urban development can have a significant impact on soil formation factors, altering the trajectories of soil development (Pickett and Cadenasso, 2009). Urban soil "parent material" is often partially comprised of building debris, trash, and imported fill materials, also affecting soil formation (Effland and Pouyat, 1997). These impacts on urban soil formation are critical for understanding urban ecosystem services, yet there are a host of urban impacts on soils that can be characterized as either direct or indirect impacts (Pavao-Zuckerman, 2012, Pavao-Zuckerman, 2008). Direct impacts result from the physical process of urbanization and the process of development. Relative to the process of soil development, direct impacts tend to be rapid and short duration. Indirect impacts result from the presence of built surfaces, the functioning of the city, and the actions of people in urban management (that are not directly impacting soil physical properties). Indirect impacts tend to occur over longer periods of time and their effects may be cumulative. Ecologists differentiate disturbances as "pulses" or "presses" by their temporal impacts. Pulse disturbances are short-term and temporary, where the system may recover easily from the disturbance once the pulse event ceases. On the other hand, press disturbances are longer-term disturbances with long-term consequences for an ecosystem. Press disturbances often drive the system into a new equilibrium state, resulting in changes in ecosystem structure and processes. In theory an indirect effect is "reversible," if that pressure is altered (i.e., if you mediate the urban heat island), whereas a direct effect is a structural change (a press).

Soils are a critical ecosystem component underlying or directly supporting the majority of terrestrial ecosystem services. Soil can be seen as a form of "natural capital" that supports the provision of ecosystem services (Figure 6.1; Dominati *et al.*, 2010, Robinson *et al.*, 2009). Soil health (the ability of soil to function and provide desired services and maintain environmental quality) is a critical component of a soil's role as natural capital and an important property supporting the provision of ecosystem services (Doran and Parkin, 1994). It is the knowledge of the interaction of physical, chemical, and biological properties of soils that underlies management of soil health through conservation, restoration, and design practice (Heneghan *et al.*, 2008, Pavao-Zuckerman, 2008). In this chapter, we explore the direct and indirect effects of cities on soils (Pavao-Zuckerman 2008, 2012) to explore the provision, degradation, and restoration of ecosystem services in cities. The goal of the paper is to discuss the implications for soil health and ecosystem services following the process of urbanization. Our focus is primarily on what happens to soils within cities through the process of urbanization, while other chapters in this volume cover the implications for land outside cities as the process of urbanization occurs. The direct/indirect framework facilitates a concrete understanding of the drivers and mechanisms of urban influences on soils and ultimately support the management of urban ecosystem services in a way a more generalized discussion of "urbanization" would not. In this chapter, we discuss (1) the direct and indirect effects of urban expansion on soils, (2) the effects of urban expansion on soil health and ecosystem services, and (3) approaches to enhance ecosystem services in cities through restoration and mitigation approaches for urban soils.

## Direct effects of urban expansion

The characteristics of urban soils vary widely and are dependent on both direct and indirect effects resulting from urban land use change (Figure 6.1). Examples of direct effects include soil disturbances such as grading (McGuire, 2004, Pitt and Lantrip, 2000, Trammell *et al.*, 2011), management inputs such as irrigation (Zhu *et al.*, 2006, Tenenbaum *et al.*, 2006) and compaction through trampling (Godefroid and Koedam, 2004), while indirect effects include environmental changes such as the urban heat island effect (Savva *et al.*, 2010) and atmospheric deposition (Lovett *et al.*, 2000, Rao *et al.*, 2014). Here we address the direct effects of urban land use expansion on native soils.

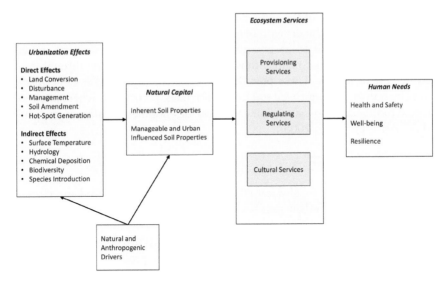

*Figure 6.1* Direct and indirect effects of urbanization influence on both the natural capital of soils and the generation of ecosystem services (source: adapted from Dominati *et al.*, 2010)

### Urban disturbance

When land is converted to urban uses, both initial and post-development factors that physically disrupt soil or result from horticultural management, e.g., fertilization and irrigation, can have profound effects on soil characteristics (Pouyat *et al.*, 2010). Nevertheless, for most urban land use conversions human-caused disturbance is more pronounced during rather than after the land-development process. The initial phase of urban development typically includes the clearing of existing vegetation, grading of soil, and the building of structures, which collectively result in a drastic alteration of the C, N, and water cycles in the resultant landscape. In turn, the extent and magnitude of these initial disturbances is dependent on infrastructure requirements (e.g., stormwater retention areas), topography, and other site limiting factors. As an example, a topographic change analysis of 30 development projects in Baltimore County, USA showed that the total volumetric change of soil per development was positively correlated with mean slope of the site (r = 0.54, p = 0.002) (McGuire, 2004).

Very little is known about C and N losses from urban soils although we do know that large-scale development projects can physically impact large volumes of surface soil. Using the topographic change analysis in McGuire (2004), Pouyat *et al.* (2007b) estimated that the potential amount of SOC that was disturbed during a 2,600 m$^2$ development project with an average depth of 3 m was roughly 2.7 × 10$^4$ kg SOC. How much SOC that actually gets lost during the development process is unknown and depends partly on the type of soil and the ultimate fate of the surface soil layers, which in the USA are typically sold as "topsoil" for the development of lawns (Pouyat *et al.*, 2007a). Soil losses of N potentially tend to be greater directly after a site is developed and are reduced as soil organic matter, thus C concentrations, increase through time (Golubiewski, 2006). These post-development effects on soil organic matter can also translate to improved soil infiltration rates, particularly where attention is paid to the use of soil amendments in post-development management (Chen *et al.*, 2014, Pitt *et al.*, 2008).

## Urban management

In addition to direct or physical effects to soil during urban land-use change, humans supplement urban soils with various amendments including fertilizer, compost, mulch, lime, and irrigated water. Ironically, these supplements are required to make up for the loss of SOC and nutrients that were lost during the initial disturbance of native soils in the development process.

Results in the literature suggest that turf-grass systems can accumulate SOC to levels that are comparable to or exceed other grassland and forested systems. In comparing results from studies of managed lawns in California, Maryland, and Wisconsin, Falk (1980) estimated that the range for net primary productivity was about 1.0 to 1.7 kg ha yr-1 in temperate climates, most of which was below-ground. Other studies have shown somewhat lower productivity rates for lawns (0.6 to 0.7 kg ha yr-1) (Blancomontero *et al.*, 1995, Jo and McPherson, 1995). In measuring C sequestration rates in turf-grass soils using C14 analysis, Qian *et al.* (2010) found rates of accumulation between 0.32 and 0.78 Mg ha-1 yr-1 during the first four years after turf establishment. These rates are similar in range to 0.9 to 1.0 Mg ha-1 yr-1 during the first 25 years (Bandaranayake *et al.*, 2003).

To manage turf grasses typically associated with lawns, homeowners and institutional land managers in the USA apply about 16 million kg of pesticides each year (Aspelin, 1997) as well as fertilizers at rates similar to or exceeding those of cropland systems (Talbot, 1990). Moreover, lawns are typically clipped on a regular basis during the growing season and depending on the practice, can result in a significant amount of N on an annual basis (Templer *et al.*, 2015). Depending on the state of recovery of the turf-grass system after development and the prevailing climate, the effect of fertilizer, pesticides, and irrigated water on lawn productivity will vary from region to region and due to the age of the development (Selhorst and Lal, 2013).

Although managed turf-grass systems have shown a high capacity to sequester C, flux rates of C from these systems appear to be higher than the native systems replaced. For instance, measurements in permanent forest and lawn plots in the Baltimore metropolitan area indicate that fluxes from turf-grass plots generally were higher than at forested sites (Groffman and Pouyat, 2009, Groffman *et al.*, 2009). Other soil-atmosphere exchanges of greenhouse gases, especially nitrous oxide and methane, also have been altered by turf management. For example, trace-gas measurements in the Baltimore metropolitan area showed that turf-grass soils have a reduced rate of methane uptake and increased nitrous oxide fluxes compared to rural forest soils (Groffman and Pouyat, 2009; Groffman *et al.*, 2009). Similarly, in the Colorado Front Range, turf-grass systems had reduced methane uptake and increased nitrous oxide fluxes relative to native short-grass steppe in that region (Kaye *et al.*, 2005). The specific mechanism for elevated $CO_2$ and nitrous oxide fluxes and a reduced methane sink in turf-grass systems has not been determined, though a possible explanation is that higher atmospheric concentrations of $CO_2$, N inputs from fertilization, and elevated atmospheric and soil temperatures play significant roles in these soil-flux responses (Yesilonis and Pouyat, 2012).

### Sealed surfaces

Sealed or impervious surfaces can partially constrain distributions of plant species, trace gas fluxes, water infiltration as well as the movement of nutrients and contaminants in urban ecosystems (Pouyat *et al.*, 2007a). The disconnection of the soil and atmosphere "short circuits" the below-ground from the above-ground ecosystem, which diminishes an ecosystem's overall ability to buffer changes in water, nutrient, and contaminant inputs. As a result, the ecosystem's capacity to retain or process these materials is altered. For totally sealed soil surfaces, soil C stocks can be half of vegetated soils; more data are reported in Chapter 10 of this book. In addition, atmospherically derived contaminant and nutrient inputs can accumulate on impervious surfaces and be washed off repeatedly by small rainfall events onto nearby exposed soil or into surface waters (Gobel *et al.*, 2007, Lee and Bang, 2000, Lee *et al.*, 2002). In addition, gaseous exchanges between the atmosphere and the soil-plant continuum should be diminished, again short circuiting the ability of the below-ground ecosystem to assimilate C or gas-phase contaminants. However, we are unaware of any such measurements of sealed soils in the literature.

The tendency of the built environment and human activity to concentrate flow paths and chemical inputs can result in the development of "hotspots" in the landscape. Hotspots are areas or patches that show disproportionately high reaction rates relative to the surrounding area or matrix (McClain *et al.*, 2003). The concept of hotspots developed from studies of N processing in soil cores (Parkin, 1987) and riparian zones that showed that anoxic microsites with high C content were zones of elevated denitrification rates. Generally, hotspots are sites where reactants for specific biogeochemical reactions coincide in an environment conducive for the reaction to take place (McClain *et al.*, 2003). Human activities and the introduction of built structures provide such conditions in urban landscapes at various scales. Examples include septic systems, horticultural beds, golf greens, stormwater retention basins, and compost piles. In all these examples, the potential for N leaching or trace gas emissions is higher than in other soil patches found in urban landscapes. Urban soil hotspots also can be sinks for contaminants, nutrients, or C.

## Indirect effects of urban expansion

### Physical effects

Urban ecosystems are characterized by an alteration of energy, water, and material fluxes that stem from disturbance, management, and other physical alterations to the environment (Kaye et al., 2006). Cities therefore can indirectly impact soils through these direct processes associated with urbanization. Here we address indirect influences of urbanization on soils through changes in surface temperature, hydrology, chemical inputs, and ecological structures.

Urban heat islands are a ubiquitous pattern of environmental impacts of cities. Ambient air temperatures differ across urban landscapes and in comparison to rural land due to the presence and percentage of built surfaces, with urban cores tending to be warmer than surrounding areas. Using analysis of satellite imagery, Buyantuyev and Wu (2010) observed a strong heat island in the Phoenix metropolitan area. In addition to the general warming pattern, they also observed strong variability in surface temperatures that were driven by intra-urban socioeconomic drivers, particularly median family income (Buyantuyev and Wu, 2010). The observed variability of surface temperatures across the urban landscape is a reflection of direct urban influences on physical space through local processes of management and development. Urban heat islands in turn can affect soil moisture status and impact rates of soil biogeochemical cycling (McDonnell et al., 1997). Elevated rates of litter decomposition and nutrient cycling have been attributed to urban heat island effects in several urban areas (McDonnell et al., 1997, Pavao-Zuckerman and Coleman, 2005, Pouyat et al., 1997).

The process of urbanization alters elements of the hydrologic cycle and resulting water balance, including evapotranspiration, infiltration, and surface runoff, at local and broader spatial scales. The ultimate urban water balance at various scales ultimately depends on the proportion of a catchment or watershed that is covered by impervious surfaces and the extent of surface sealing of soils (Wessolek, 2008), but a general pattern is that increases in imperviousness generate reduction in infiltration rates and increases in surface runoff (Paul and Meyer, 2001). Even patches of the urban landscape that are not impervious may have reduced infiltration rates due to soil compaction, exacerbating surface runoff dynamics in cities (Gregory et al., 2006). Water repellency of soils is found to increase with rates of urbanization (McDonnell, 1997), and this increase in hydrophobicity has been shown to reduce rates of water infiltration into soils (Doerr and Ritsema, 2005). Recent studies have turned to the interaction of soil condition and characteristics to hydrologic processes in cities. For example, Ossola et al. (2015) observed that different approaches to park management in Melbourne that reflected habitat complexity (Byrne, 2007) led to variation in litter and soil surface characteristics. These properties altered infiltration rates and saturated hydraulic conductivity in a way that suggests that management of parks with an eye to soil properties and habitat complexity may support better practices for stormwater management (Ossola et al., 2015).

## Chemical effects

Cities are also characterized by elevated concentrations and fluxes of chemicals than surrounding areas that derive from both point (power plants, industrial combustion, heating) and non-point sources (vehicle traffic) (Bilos *et al.*, 2001, Schauer *et al.*, 1996). Additionally, local gradients of land-use and the specific management of parcels in cities may in fact drive broader urbanization gradients of pollutant deposition patterns (Tanner and Fai, 2000). A general trend is for concentration and fluxes of anions and cations to decrease with increasing distance from an urban core (Lovett *et al.*, 2000). Indices of vehicular usage (such as $CO_2$ emissions) and the urban core are strongly correlated with inorganic nitrogen inputs to forest soils (Rao *et al.*, 2014). A recent study in Gold Coast, Australia links atmospheric deposition of Zn, Cd, Ni, and Cu with local vehicle traffic drivers, and importantly makes the connection between atmospheric deposition and stormwater runoff pollution, which may serve as another transport mechanism by which pollutants move into soils (Gunawardena *et al.*, 2013).

Elevated wet and dry deposition can result in high amounts of nitrogen, sulfur, and heavy metal deposition on urban soils (McDonnell *et al.*, 1997). Elevated N deposition has been shown to indirectly alter soil carbon dynamics by shifting extracellular enzyme activities due to alterations in litter chemistry (Waldrop *et al.*, 2004). Deposition dynamics in an urban region may be well understood by examining the source fingerprint and spatio-temporal dynamics of atmospheric pollutants (Azimi *et al.*, 2005). Rao *et al.* (2014) observed significant $NO_3$ leaching from sites that correlated with inorganic N deposition rates, and concluded that this leaching did not require saturation of aboveground and below-ground N pools.

## Ecological effects

Finally, urbanization can indirectly impact biodiversity and community structures, playing a key role in driving urban ecosystem services. The influence of urbanization on biodiversity is complex, varying greatly by taxa, climate, land use and management. Despite their perception as being biologically "inert" urban places can harbor high levels of soil biodiversity. Recent community analysis of the soil microbiota in New York City's Central Park describes a high level of microbial diversity, with an interesting degree of endemism and novelty that matches those found in non-urban ecosystems (Ramirez *et al.*, 2014). Urbanization gradient studies showed reductions in microbial populations but some settings indicated that urbanization may alter the functional composition of soil food webs, but not overall diversity levels (Pavao-Zuckerman and Coleman, 2005, Pouyat *et al.*, 1994). The role of environmental drivers and habitat conditions are important for soil microfaunal abundances, indicating that urban management can strongly affect biodiversity in soils (Byrne, 2007, Pavao-Zuckerman and Byrne, 2009). Again, urbanization can lead to unexpected results with respect to these drivers. For example, Tuhackova *et al.* (2001) demonstrated gradients of polycyclic aromatic hydrocarbons (PAHs) in soils that were driven by proximity to highways. These gradients of PAHs served as energy sources for microbes and resulted in strong increases in abundance of both bacteria and fungi close to highways. Importantly, ecosystem services are driven by the function of organisms across spaces, yet the impact of urbanization on the functional and physiological ecology of organisms is a critical issue and an emerging frontier of research (Hahs and Evans, 2015).

Invasive species that are typically introduced into urban areas can have a strong influence on soil health in urban or urban–rural interface areas (Pouyat *et al.*, 2010). For example, in the northeastern and mid-Atlantic United States where native earthworm species are rare or absent, urban areas are important foci of invasive earthworm introductions, especially Asian species from the genus Amynthas, which are expanding their range to outlying forested areas (Groffman and Bohlen, 1999, Steinberg *et al.*, 1997, Szlavecz *et al.*, 2006). Invasions by earthworms into forests have resulted in altered C and N cycling processes, sometimes this can lead to increased losses of N through trace gas and leaching fluxes (Carreiro *et al.*, 2009, Bohlen *et al.*, 2004, Hale *et al.*, 2005). Invasive earthworms may have compounding indirect effects through influences on soil properties that drive microbial function, as their casts have higher moisture contents than the soils that house them. Similarly, invasions by exotic plant species can impact C and N losses, which in some cases can facilitate the colonization of additional invasive species, further exacerbating the turnover of N in the soil (Pavao-Zuckerman, 2008). Examples of plant invasions in urban metropolitan areas that have been shown to alter C and N cycles include species of the shrub *Berberis thunbergii*, the tree *Rhamnus cathartica*, and the grass *Microstegium vimineum* (Ehrenfeld *et al.*, 2001, Heneghan *et al.*, 2002, Kourtev *et al.*, 2002).

## Urban expansion effects on soil health and ecosystem services

### Direct and indirect effects on soil health and quality

As mentioned previously, urban land use change can affect soils indirectly through changes caused in environmental factors and directly through physical or management effects with the former having the potential to influence soils beyond the boundaries of what is considered urban land use (Pouyat *et al.*, 2007b). For example, forest soils within or near urban areas have been shown to receive high amounts of heavy metals, organic compounds, and acidic compounds in atmospheric deposition. Lovett *et al.* (2000) quantified atmospheric nitrogen inputs over two growing seasons in oak forest stands along an urbanization gradient in the New York City metropolitan area. They found that the urban remnant forests received up to a two-fold greater deposition of nitrogen than in similar rural oak forests. Similar results were found in Louisville, KY, USA, the San Bernardino Mountains in the Los Angeles metropolitan area, CA, USA, the city of Oulu, Finland, and the city of Kaunas, Lithuania, where N deposition rates into urban and forest patches were higher than in rural forest patches (Bytnerowicz *et al.*, 1999, Carreiro *et al.*, 1999, Fenn and Bytnerowicz, 1993, Juknys *et al.*, 2007, Ohtonen and Markkola, 1991).

Evidence of a similar depositional pattern has been found for heavy metals along urbanization gradients in the New York City, Baltimore, and Budapest, Hungary, metropolitan areas. Pouyat *et al.* (2008) found up to a twofold to threefold increase in contents of lead, copper, and nickel in urban than in rural forest remnants. A similar pattern but with greater differences was found by Inman and Parker (1978) in the Chicago, IL, USA, metropolitan area, where levels of heavy metals, particularly lead and copper, were more than five times higher in urban than in rural forest patches. Other urbanization gradient studies have shown a similar pattern (Sawicka-Kapusta *et al.*, 2003, Watmough *et al.*, 1998), although cities having more compact development patterns exhibited less of a difference between urban and rural remnant forests (Pavao-Zuckerman, 2003, Pouyat *et al.*, 2008, Carreiro *et al.*, 2009). Besides heavy metals, Wong *et al.* (2004) found more than a twofold higher gradient of Polycyclic Aromatic Hydrocarbons (PAHs) concentrations in forest soils in the Toronto, Canada, metropolitan area, with concentrations decreasing with distance from the urban center to the surrounding rural area. Similarly, Jensen *et al.* (2007) and Zhang *et al.* (2006) found significantly higher concentrations of PAHs in surface soils of Oslo, Norway, and Hong Kong, China, respectively, than in surrounding rural areas.

How these pollutants affect the health of soil is uncertain, but results thus far suggest that the effects are variable and depend on other urban factors (Lorenz and Lal, 2009, Pouyat *et al.*, 2007a, Pouyat *et al.*, 2007b, Carreiro *et al.*, 2009). For example, Inman and Parker (1978) found a negative effect of soil contamination of Cu (76 mg kg-1) and Pb (400 mg kg-1) on leaf litter decay rates in urban stands suggesting a negative effect from pollution in the Chicago metropolitan area. Similarly, Pouyat *et al.* (1994) found an inverse relationship between litter fungal biomass and fungivorous invertebrate abundances with heavy metal concentrations along an urbanization gradient in the New York City metropolitan area. However, responses of soil invertebrates along urbanization gradients in Europe were not related to urban environmental effects but rather local factors, such as habitat connectivity or patch size (Niemela *et al.*, 2002). In fact, where heavy metal contamination of soil is moderate to low relative to other atmospherically deposited pollutants elements such as N, biological activity may actually be stimulated. Decay rates, soil respiration, and soil N-transformation increased in forest patches near or within major metropolitan areas of the USA in southern California (Fenn and Dunn, 1989, Fenn, 1991), Ohio (Kuperman, 1999), southeastern New York (McDonnell *et al.*, 1997, Carreiro *et al.*, 2009), and Maryland (Groffman *et al.*, 2006, Szlavecz *et al.*, 2006).

*Expanded spatial scales and development patterns on ecosystem services*

Urban soils or native soils that have been influenced by urban environmental conditions are generally thought of as having lower quality than native soils found in a particular region (Craul and Klein, 1980, Jim, 1993, Patterson et al., 1980, Short et al., 1986). However, more recent studies show a greater variety of soil conditions that are often more favorable for plant growth than the preexisting native soil (Davies and Hall, 2010, Edmondson et al., 2012, Hope et al., 2005, Pouyat et al., 2007b). For example, most literature assumes that the conversion of native soil types to urban uses results in losses of C (Lorenz and Lal, 2009, Pouyat et al., 2010, Scharenbroch et al., 2005). Yet, depending on the climate and native soils, C has been shown to accumulate in soils of urban landscapes to a level that is greater than that in the native soil replaced (Pouyat et al., 2015). The assumed cause for increasing C storage in what were once disturbed soils is the supplementation of water and nutrients, which in native soil and climate conditions would otherwise have limiting conditions for plant growth. Thus, an important characteristic of urban land use change with respect to C and N cycles is the replacement of native cover types with lawn cover, which often requires added nutrients and water (Kaye et al., 2005, Milesi et al., 2005, Golubiewski, 2006, Pouyat et al., 2009). In North America, the estimated amount of lawn cover for the conterminous USA is 163,800 km$^2$ $\pm$ 35,850 km$^2$, or 73 percent of all irrigated cultivated lands (excluding lawn cover) (Lubowski et al., 2006). Moreover, it is estimated that roughly half of all residences apply fertilizers (Law et al., 2004, Osmond and Hardy, 2004), which can approach or exceed rates applied in cropland systems, e.g., > 200 kg ha-1 yr-1 (e.g., Morton et al., 1988).

## Restoration and mitigation of direct and indirect effects of urban expansion

*Planning, design, restoration approaches to enhance services in cities, organized by how approaches address direct effects and indirect effects*

Urbanization represents a complicated mix of scales with respect to ecosystem service provision and sustainability. The growth of cities often comes at the expense of agricultural and natural ecosystems and the services that they provide (Setälä et al., 2014). At the same time, a greater proportion of the Earth's population resides in cities, and the provision and flow of ecosystem services to these urban dwellers is an important consideration for their well-being, health, and the overarching sustainability and resilience of urban places. From a cost–benefit perspective, restoration of ecosystem services in cities may meet environmental, social, and economic goals for sustainable development (Elmqvist et al., 2015). The recognition that cities are capable of providing urban ecosystem services may reduce the reliance of non-urban ecosystem service flows. For example, cities may produce 15–20 percent of the world's food supplies through urban agricultural practices (Smit *et al.*, 1996), with many local instances (especially in developing nations) contributing a great proportion of local food needs (Beniston and Lal, 2012). However, as discussed above, urbanization has many direct and indirect impacts on ecosystems and soils that alter their ability to provide ecosystem services (Pavao-Zuckerman, 2012). Here we discuss approaches to mitigate these effects of urbanization through soil management, restoration, planning, design, and policy in order to enhance ecosystem service provision within cities by improving soil quality.

Urban ecosystem service provision can be enhanced through soil management approaches that address direct and indirect impacts of urbanization. However, prescribing a universal solution to urban soil issues is problematic because urban soils are very heterogeneous in nature. Thus the ability to generalize an urban soil "condition" is limited. This has significant implications for urban soil management and restoration of ecological processes in cities. Assessment of local conditions that drives site-specific strategies to mitigate urban impacts on soils is critical. Therefore, a key focus for soil remediation in cities looks to soil management practices at local scales (i.e., lot, parcel). Soil carbon management is another critical approach to improving urban soil quality. Increasing the soil carbon pool has many direct and indirect ecosystem benefits for improving soil structure, enhancing infiltration rates, and increasing populations of soil biota (Lal, 2007). This can be achieved through composts and mulches and biochar, where repeated application can improve soil physical properties affected by urbanization, such as bulk density, infiltration rates, and soil water-holding capacity (Cogger, 2005). Indirect benefits of compost applications on soil properties may help to also alleviate urbanization impacts on plant productivity (Scharenbroch, 2009), which may have additional indirect benefits for urban soil quality through root and litter production. Cities often produce large quantities of organic waste materials that could be redirected for soil amendment and management (Beniston and Lal, 2012), further enhancing the localized production of ecosystem service benefits.

Urban ecosystem service provision can also be enhanced through approaches that increase the effective unsealed soil surface in a city. These approaches largely fall within the general scope of green infrastructure, and serve to increase the urban surface area of soils that interact with hydrologic flows or support enhanced biological activity in urban soils. Large-scale tree planting efforts have been viewed as a strategy to restore ecosystem services in cities due to the many benefits that trees provide (Nowak and Crane, 2002, Oldfield *et al.*, 2014). In the initial stages of an afforestation project in New York City, Oldfield *et al.* (2014) report that site preparation and soil amendment improves the health of urban soils. Specifically, they observed reductions in bulk density, increases water-holding capacities, increased microbially available carbon, and enhanced carbon storage. However, it should be noted that site preparation itself (weeding, rototilling to ~15cm, and surface mulching) dominated treatment effects (compost amendment) in the early stages of the afforestation project (Oldfield *et al.*, 2014). Low-impact development approaches (such as, biofilters, bioretention basins, bioswales, rain gardens) seek to manage and control urban stormwater by increasing retention times and the duration that water interacts with soils in the urban landscape (Askarizadeh *et al.*, 2015, Fletcher *et al.*, 2014). These systems can be designed to provide ecosystem services related to hydrologic and water quality goals. For example, soil media depth, composition, mulching, basin geometry, and vegetation composition can all be adjusted to reduce peak flows, enhance infiltration, sequester pathogens, nutrients, and metals, and support plant growth along streets, lots, and parking areas in cities (Hunt *et al.*, 2012). Some low-impact development projects that seek to restore or enhance ecosystem services in cities can be viewed as forms of novel ecosystems (Kowarik, 2011). For example, green roofs are assemblages of soil (often constructed) and plants that are uniquely designed ecological systems that take advantage of vast amounts of novel rooftop spaces in urban landscape (Oberndorfer *et al.*, 2007). The added soil surface of green roofs may enhance rainfall retention, reduce runoff rates, and reduce nutrient concentrations in runoff, although there is a great deal of variability in green roof performance that is driven by design parameters (slopes, age, soil depth, etc.) (Berndtsson, 2010). The promise of green infrastructure for affecting urban sustainability and resilience through ecosystem service provision is high (Andersson *et al.*, 2014, Tzoulas *et al.*, 2007), the specific performance of green infrastructure *in situ* and its ability to affect ecosystem services at the scale of neighborhoods, watersheds, cities, and regions remains an important research direction (Pataki *et al.*, 2011, Berndtsson, 2010)

Vacant lots hold great promise for providing many types of ecosystem services in the urban fabric. In an assessment of hydrologic properties of vacant residential lots in Cleveland, OH, Shuster *et al.* (2014, 2015b) suggest that policies and procedures for vacant lot management may positively impact soil properties such that these lots become part of a green infrastructure that addresses stormwater management issues. Moreover, similar lot-scale management and processes may allow lots to function as stormwater harvesting green infrastructure in semi-arid cities to help relieve irrigation burdens for landscaping (Shuster *et al.*, 2015a). A review by Beniston and Lal (2012) indicates that agriculture on vacant urban land has the potential to significantly address human health and economic issues centering on localized food production. Kremer *et al.* (2013) conducted a social-ecological assessment of vacant lot utilization in New York City and found a range of uses (including gardening, park space, parking, and athletic activities). Importantly, they found that whether and how residents used lots was a localized phenomenon, and was influenced by socio-economic factors (Kremer *et al.*, 2013). This suggests that planning and management of vacant lots for ecosystem services that takes into account local conditions and demand for services might better contribute to urban sustainability (McPhearson *et al.*, 2014).

Ecosystem services play an interesting role in soil policy and management in that they can serve as a communication tool to move conservation policies forward, while at the same time be a beneficiary of such policies

(Breure *et al.*, 2012). Ecosystem services can thus be a nexus in urban soil management, functioning as both a driver of and response to policy and planning initiatives (Hough *et al.*, in review, Burkhard *et al.*, 2014). A host of programs and policies directly addressing soil from ecological, societal, and economic perspectives has emerged (largely in Europe), but broader initiatives focused on ecosystem services also indirectly link to urban soil ecosystem services (i.e., TEEB—The Economy of Ecosystems and Biodiversity, the Millennium Ecosystem Assessment, IBPES—Intergovernmental Platform on Biodiversity and Ecosystem Services). At more local scales policy quickly becomes complex when local policy actors and stakeholders may value services and perceive disservices differently (Otte *et al.*, 2012). The transition away from policies that derive from soil degradation paradigms to those of sustainable use (Figure 6.2, Breure *et al.*, 2012) have greater application in urban areas because of the nature of urban soils and their potential role in ecosystem service provision (Pavao-Zuckerman, 2008). This represents a

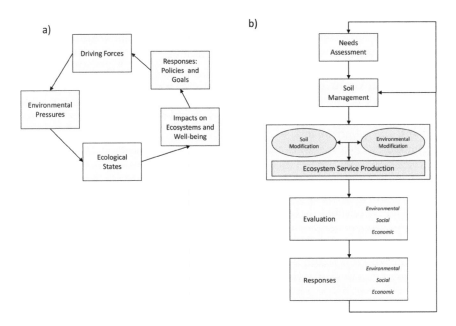

*Figure 6.2* Two paradigms for policy and management related to urbanization influences on soils and ecosystem services. (a) Through the DPSIR (Driver – Pressure – State – Impact – Response) Framework, urbanization is viewed from a degradation perspective (source: modified from Kristensen, 2004). (b) An alternative view of urban soil management that explicitly focuses on supporting the generation of ecosystem services, rather than only the mitigation of environmental degradation (source: modified from Breure *et al.*, 2012)

pragmatic approach that recognizes that in cities soil function and ecosystem service provision may be constrained by direct and indirect urbanization effects. Multiple stakeholder perspectives on ecosystem service values (both monetary and non-monetary) allow the setting of realistic goals for service provision within the constraints of urban environments that would be supported by management and policy, and ideally, coupled with evaluation and adaptive management programs (Breure *et al.*, 2012). The interface of science and policy through policy-effective and actionable research generated through a cycle of adaptive policy development through phases of assessment, implementation, and evaluation is ideally suited for urban landscapes (Otte *et al.*, 2012). The necessity of local-scale assessment and site characterization, guarantees that policies will be rooted in local settings. The experimental nature of environmental planning and design in cities helps to facilitate an iterative and adaptive policy process that uses the city as a living laboratory for the provision of ecosystem services (Felson and Pickett, 2005, Felson *et al.*, 2013).

## Conclusions

The process of urbanization is a dominant and significant transformation of ecosystem structure and function. Urbanization is in strong conflict with other land uses that provide the ecosystem services that society relies upon for resources, environmental regulation, and overall well-being. At the same time, the majority of the world's population now lives in cities, and this population also relies on urban ecosystems to provide ecosystem services in cities, towns, and suburban areas to contribute to environmental, social, health, and economic well-being. Soils are a critical form of natural capital for ecosystem service provision, yet in cities, direct and indirect environmental impacts can limit their ability to provide ecosystem services to urban residents. Management choices, development patterns, and localized land-use approaches and patterns ultimately determine these direct and indirect impacts on soils and their ability to provide ecosystem services in cities. Despite these impacts on soils and ecosystem services, mitigation approaches, restoration, ecological design and planning all show significant promise for enhancing urban soil for the purpose of ecosystem service provision.

## References

Adler, F. R. and Tanner, C. J. (2013). *Urban Ecosystems: Ecological Principles for the Built Environment*. Cambridge, Cambridge University Press.

Andersson, E., Barthel, S., Borgström, S., Colding, J., Elmqvist, T., Folke, C. and Gren, Å. (2014). Reconnecting cities to the biosphere: Stewardship of green infrastructure and urban ecosystem services. *Ambio*, 43, 445–453.

Askarizadeh, A., Rippy, M. A., Fletcher, T. D., Feldman, D. L., Peng, J., Bowler, P., Mehring, A. S., Winfrey, B. K., Vrugt, J. A., AghaKouchak, A., Jiang, S. C., Sanders, B. F., Levin, L. A., Taylor, S. and Grant, S. B. (2015). From rain tanks to catchments: use of low-impact development to address hydrologic symptoms of the urban stream syndrome. *Environmental Science & Technology*, 49, 11264–11280.

Aspelin, A. L. (1997). *Pesticide Industry Sales and Usage: 1994 and 1995 Market Estimates.* Washington, DC, US EPA, Biological and Economic Analysis Division.

Azimi, S., Rocher, V., Muller, M., Moilleron, R. and Thevenot, D. R. (2005). Sources, distribution and variability of hydrocarbons and metals in atmospheric deposition in an urban area (Paris, France). *Science of the Total Environment*, 337, 223–239.

Bandaranayake, W., Qian, Y. L., Parton, W. J., Ojima, D. S. and Follett, R. F. (2003). Estimation of soil organic carbon changes in turfgrass systems using the CENTURY model. *Agronomy Journal*, 95, 558–563.

Barthel, S., Folke, C. and Colding, J. (2010). Social–ecological memory in urban gardens: retaining the capacity for management of ecosystem services. *Global Environmental Change*, 20, 255–265.

Beniston, J. and Lal, R. (2012). Improving soil quality for urban agriculture in the north central U.S. IN Lal, R. and Augustin, B. (eds) *Carbon Sequestration in Urban Ecosystems*. Dordrecht, Springer, 279–313.

Berndtsson, J. C. (2010). Green roof performance towards management of runoff water quantity and quality: a review. *Ecological Engineering*, 36, 351–360.

Bilos, C., Colombo, J. C., Skorupka, C. N. and Presa, M. J. R. (2001). Sources, distribution and variability of airborne trace metals in La Plata City area, Argentina. *Environmental Pollution*, 111, 149–158.

Blancomontero, C. A., Bennett, T. B., Neville, P., Crawford, C. S., Milne, B. T. and Ward, C. R. (1995). Potential environmental and economic-impacts of turfgrass in Albuquerque, New Mexico (USA). *Landscape Ecology*, 10, 121–128.

Bohlen, P. J., Pelletier, D. M., Groffman, P. M., Fahey, T. J. and Fisk, M. C. (2004). Influence of earthworm invasion on redistribution and retention of soil carbon and nitrogen in northern temperate forests. *Ecosystems*, 7, 13–27.

Breure, A. M., De Deyn, G. B., Dominati, E., Eglin, T., Hedlund, K., Van Orshoven, J. and Posthuma, L. (2012). Ecosystem services: a useful concept for soil policy making! *Current Opinion in Environmental Sustainability*, 4, 578–585.

Burkhard, B., Kandziora, M., Hou, Y. and Müller, F. (2014). Ecosystem service potentials, flows and demands–concepts for spatial localisation, indication and quantification. *Landscape Online*, 34, 1–32. doi: 10.3097/LO.201434.

Buyantuyev, A. and Wu, J. (2010). Urban heat islands and landscape heterogeneity: linking spatiotemporal variations in surface temperatures to land-cover and socio-economic patterns. *Landscape Ecology*, 25, 17–33.

Byrne, L. B. (2007). Habitat structure: a fundamental concept and framework for urban soil ecology. *Urban Ecosystems*, 10, 255–274.

Bytnerowicz, A., Fenn, M. E., Miller, P. R. and Arbaugh, M. J. (1999). Wet and dry pollutant deposition to the mixed conifer forest. IN Miller, P. R. and McBride, J. R. (eds) *Oxidant Air Pollution Impacts in the Montane Forests of Southern California*. New York Springer, 235–269.

Carreiro, M. M., Howe, K., Parkhurst, D. F. and Pouyat, R. V. (1999). Variation in quality and decomposability of red oak leaf litter along an urban-rural gradient. *Biology and Fertility of Soils*, 30, 258–268.

Carreiro, M. M., Pouyat, R. V. and Tripler, C. E. (2009). Carbon and nitrogen cycling in forests along urban-rural gradients in two cities. IN McDonnell, M. J., Hahs, A. and Breuste, J. (eds) *Comparative Ecology of Cities and Towns*. New York, Cambridge University Press, 308–328.

Chen, Y. J., Day, S. D., Wick, A. F. and McGuire, K. J. (2014). Influence of urban land development and subsequent soil rehabilitation on soil aggregates, carbon, and hydraulic conductivity. *Science of the Total Environment*, 494, 329–336.

Cogger, C. G. (2005). Potential compost benefits for restoration of soils disturbed by urban development. *Compost Science & Utilization*, 13, 243–251.

Craul, P. J. and Klein, C. J. (1980). Characterization of streetside soils of Syracuse, NY. *METRIA*, 3, 88–101.

Daily, G. C. (ed.) (1997). *Nature's Services: Societal Dependence on Natural Ecosystems*. Washington, DC, Island Press.

Davies, R. and Hall, S. J. (2010). Direct and indirect effects of urbanization on soil and plant nutrients in desert ecosystems of the Phoenix metropolitan area, Arizona (USA). *Urban Ecosystems*, 13, 295–317.

Davies, Z. G., Edmondson, J. L., Heinemeyer, A., Leake, J. R. and Gaston, K. J. (2011). Mapping an urban ecosystem service: quantifying above-ground carbon storage at a city-wide scale. *Journal of Applied Ecology*. doi: 10.1111/j.1365-2664.2011.02021.x.

Doerr, S. H. and Ritsema, C. J. (2005). Water movement in hydrophobic soils. *Encyclopedia of Hydrological Sciences*. doi: 10.1002/0470848944.hsa072.

Dominati, E., Patterson, M. and Mackay, A. (2010). A framework for classifying and quantifying the natural capital and ecosystem services of soils. *Ecological Economics*, 69, 1858–1868.

Doran, J. W. and Parkin, T. B. (1994). Defining and assessing soil quality. IN Doran, J. W., Coleman, D. C., Bezdicek, D. F. and Stewart, B. A. (eds) *Defining Soil Quality for a Sustainable Environment*, SSSA Special Publication no. 35. Madison, WI, Soil Science Society of America, 3–22.

Edmondson, J. L., Davies, Z. G., McHugh, N., Gaston, K. J. and Leake, J. R. (2012). Organic carbon hidden in urban ecosystems. *Scientific Reports*, 2.

Effland, W. R. and Pouyat, R. V. (1997). The genesis, classification, and mapping of soils in urban areas. *Urban Ecosystems*, 1, 217–228.

Ehrenfeld, J. G., Kourtev, P. and Huang, W. Z. (2001). Changes in soil functions following invasions of exotic understory plants in deciduous forests. *Ecological Applications*, 11, 1287–1300.

Elmqvist, T., Setälä, H., Handel, S., van der Ploeg, S., Aronson, J., Blignaut, J., Gómez-Baggethun, E., Nowak, D., Kronenberg, J. and de Groot, R. (2015). Benefits of restoring ecosystem services in urban areas. *Current Opinion in Environmental Sustainability*, 14, 101–108.

Ernstson, H., Barthel, S., Andersson, E. and Borgström, S. T. (2010). Scale-crossing brokers and network governance of urban ecosystem services: the case of Stockholm. *Ecology and Society*, 15, 28.

Falk, J. H. (1980). The primary productivity of lawns in a temperate environment. *Journal of Applied Ecology*, 17, 689–695.

Felson, A. J. and Pickett, S. T. A. (2005). Designed experiments: new approaches to studying urban ecosystems. *Frontiers in Ecology and the Environment*, 3, 549–556.

Felson, A. J., Bradford, M. A. and Terway, T. M. (2013). Promoting Earth stewardship through urban design experiments. *Frontiers in Ecology and the Environment*, 11, 362–367.

Fenn, M. (1991). Increased site fertility and litter decomposition rate in high-pollution sites in the San Bernardino Mountains. *Forest Science*, 37(4), 1163–1181.

Fenn, M. E. and Bytnerowicz, A. (1993). Dry deposition of nitrogen and sulfur to Ponderosa and Jeffrey pine in the San-Bernardino National Forest in Southern California. *Environmental Pollution*, 81, 277–285.

Fenn, M. E. and Dunn, P. H. (1989). Litter decomposition across an air-pollution gradient in the San Bernardino Mountains. *Soil Science Society of America Journal*, 53(5), 1560–1567.

Fletcher, T. D., Vietz, G. and Walsh, C. J. (2014). Protection of stream ecosystems from urban stormwater runoff : the multiple benefits of an ecohydrological approach. *Progress in Physical Geography*, 38, 543–555.

Gill, S. E., Handley, J. F., Ennos, A. R., Pauleit, S., Theuray, N. and Lindley, S. J. (2008). Characterising the urban environment of UK cities and towns: a template for landscape planning. *Landscape and Urban Planning*, 87, 210–222.

Gobel, P., Dierkes, C. and Coldewey, W. G. (2007). Storm water runoff concentration matrix for urban areas. *Journal of Contaminant Hydrology*, 91, 26–42.

Godefroid, S. and Koedam, N. (2004). The impact of forest paths upon adjacent vegetation: effects of the path surfacing material on the species composition and soil compaction. *Biological Conservation*, 119, 405–419.

Golubiewski, N. E. (2006). Urbanization increases grassland carbon pools: effects of landscaping in Colorado's front range. *Ecological Applications*, 16, 555–571.

Gomez-Baggethun, E. and Barton, D. N. (2013). Classifying and valuing ecosystem services for urban planning. *Ecological Economics*, 86, 235–245.

Gregory, J. H., Dukes, M. D., Jones, P. H. and Miller, G. L. (2006). Effect of urban soil compaction on infiltration rate. *Journal of Soil and Water Conservation*, 61, 117–124.

Grimm, N. B., Grove, J. M., Pickett, S. T. A. and Redman, C. L. (2000). Integrated approaches to long-term studies of urban ecological systems. *BioScience*, 50, 571–584.

Groffman, P. M. and Bohlen, P. J. (1999). Soil and sediment biodiversity: cross-system comparisons and large-scale effects. *Bioscience*, 49, 139–148.

Groffman, P. M. and Pouyat, R. V. (2009). Methane uptake in urban forests and lawns. *Environmental Science & Technology*, 43, 5229–5235.

Groffman, P. M., Hardy, J. P., Driscoll, C. T. and Fahey, T. J. (2006). Snow depth, soil freezing, and fluxes of carbon dioxide, nitrous oxide and methane in a northern hardwood forest. *Global Change Biology*, 12(9), 1748–1760.

Groffman, P. M., Williams, C. O., Pouyat, R. V., Band, L. E. and Yesilonis, I. D. (2009). Nitrate leaching and nitrous oxide flux in urban forests and grasslands. *Journal of Environmental Quality*, 38, 1848–1860.

Gunawardena, J., Egodawatta, P., Ayoko, G. A. and Goonetilleke, A. (2013). Atmospheric deposition as a source of heavy metals in urban stormwater. *Atmospheric Environment*, 68, 235–242.

Hahs, A. K. and Evans, K. L. (2015). Expanding fundamental ecological knowledge by studying urban ecosystems. *Functional Ecology*, 29, 863–867.

Hale, C. M., Frelich, L. E., Reich, P. B. and Pastor, J. (2005). Effects of European earthworm invasion on soil characteristics in northern hardwood forests of Minnesota, USA. *Ecosystems*, 8, 911–927.

Heneghan, L., Clay, C. and Brundage, C. (2002). Observations on the initial decomposition rates and faunal colonization of native and exotic plant species in a urban forest fragment. *Ecological Restoration*, 20, 108–111.

Heneghan, L., Miller, S. P., Baer, S., Callaham, M. A., Montgomery, J., Pavao-Zuckerman, M., Rhoades, C. C. and Richardson, S. (2008). Integrating soil ecological knowledge into restoration management. *Restoration Ecology*, 16, 608–617.

Hope, D., Zhu, W., Gries, C., Oleson, J., Kaye, J., Grimm, N. B. and Baker, L. A. (2005). Spatial variation in soil inorganic nitrogen across an arid urban ecosystem. *Urban Ecosystems*, 8, 251–273.

Hough, M., Scott, C. A. and Pavao-Zuckerman, M. A. (in review). From plant traits to social perception: ecosystem services as indicators of thresholds in social-ecological systems. *Ecosphere*.

Hunt, W. F., Davis, A. P. and Traver, R. G. (2012). Meeting hydrologic and water quality goals through targeted bioretention design. *Journal of Environmental Engineering-ASCE*, 138, 698–707.

Inman, J. C. and Parker, G. R. (1978). Decomposition and heavy metal dynamics of forest litter in northwestern Indiana. *Environmental Pollution*, 17, 34–51.

Jensen, H., Reimann, C., Finne, T. E., Ottesen, R. T. and Arnoldussen, A. (2007). PAH-concentrations and compositions in the top 2 cm of forest soils along a 120 km long transect through agricultural areas, forests and the city of Oslo, Norway. *Environmental Pollution*, 145, 829–838.

Jim, C. Y. (1993). Soil compaction as a constraint to tree growth in tropical and subtropical urban habitats. *Environmental Conservation*, 20, 35–49.

Jo, H. K. and McPherson, E. G. (1995). Carbon storage and flux in urban residential greenspace. *Journal of Environmental Management*, 45, 109–133.

Juknys, R., Zaltauskaite, J. and Stakenas, V. (2007). Ion fluxes with bulk and through-fall deposition along an urban-suburban-rural gradient. *Water Air and Soil Pollution*, 178, 363–372.

Kaye, J. P., McCulley, R. L. and Burke, I. C. (2005). Carbon fluxes, nitrogen cycling, and soil microbial communities in adjacent urban, native and agricultural ecosystems. *Global Change Biology*, 11, 575–587.

Kaye, J. P., Groffman, P. M., Grimm, N. B., Baker, L. A. and Pouyat, R. V. (2006). A distinct urban biogeochemistry? *Trends in Ecology & Evolution*, 21, 192–199.

Kourtev, P. S., Ehrenfeld, J. G. and Haggblom, M. (2002). Exotic plant species alter the microbial community structure and function in the soil. *Ecology*, 83, 3152–3166.

Kowarik, I. (2011). Novel urban ecosystems, biodiversity, and conservation. *Environmental Pollution*, 159, 1974–1983.

Kremer, P., Hamstead, Z. A. and McPhearson, T. (2013). A social-ecological assessment of vacant lots in New York City. *Landscape and Urban Planning*, 120, 218–233.

Kristensen, P. (2004). *The DPSIR Framework*. National Environmental Research Institute, Denmark.

Kuperman, R. G. (1999). Litter decomposition and nutrient dynamics in oak-hickory forests along a historic gradient of nitrogen and sulfur deposition. *Soil Biology & Biochemistry*, 31, 237–244.

Lal, R. (2007). Soil science and the carbon civilization. *Soil Science Society of America Journal*, 71, 1425–1437.

Law, N., Band, L. and Grove, M. (2004). Nitrogen input from residential lawn care practices in suburban watersheds in Baltimore County, MD. *Journal of Environmental Planning and Management*, 47(5), 737–755.

Lee, J. H. and Bang, K. W. (2000). Characterization of urban stormwater runoff. *Water Research*, 34, 1773–1780.

Lee, J. H., Bang, K. W., Ketchum, L. H., Choe, J. S. and Yu, M. J. (2002). First flush analysis of urban storm runoff. *Science of the Total Environment*, 293, 163–175.

Lorenz, K. and Lal, R. (2009). Biogeochemical C and N cycles in urban soils. *Environment International*, 35, 1–8.

Lovett, G. M., Traynor, M. M., Pouyat, R. V., Carreiro, M. M., Zhu, W.-X. and Baxter, J. W. (2000). Atmospheric deposition to oak forests along an urban-rural gradient. *Environmental Science & Technology*, 34, 4294–4300.

Lubowski, R. N., Vesterby, M., Bucholtz, S., Baez, A. and Roberts, M. J. (2006). Major uses of land in the United States, 2002. *Economic Information Bulletin*, no. 14. Washington, DC, United States Department of Agriculture.

McClain, M. E., Boyer, E. W., Dent, C. L., Gergel, S. E., Grimm, N. B., Groffman, P. M., Hart, S. C., Judson, W. H., Johnston, C. A., Mayorga, E., McDowell, W. H. and Pinay, G. (2003). Biogeochemical hot spots and hot moments at the interface of terrestrial and aquatic ecosystems. *Ecosystems*, 6, 301–312.

McDonnell, M. J., Pickett, S. T. A., Groffman, P. M., Bohlen, P., Pouyat, R. V., Zipperer, W. C., Parmelee, R. W., Carreiro, M. M. and Medley, K. (1997). Ecosystem processes along an urban-to-rural gradient. *Urban Ecosystems*, 1, 21–36.

McGuire, M. (2004). Using DTM and LiDAR data to analyze human induced topographic change. *Proceedings of ASPRS 2004 Fall Conference*. September 12–16, 2004, Kansas City, MO. Available at: http://eserv.asprs.org/eseries/source/Orders/.

McPhearson, T., Andersson, E., Elmqvist, T. and Frantzeskaki, N. (2014). Resilience of and through urban ecosystem services. *Ecosystem Services*, 12, 152–156.

Milesi, C., Running, S. W., Elvidge, C. D., Dietz, J. B., Tuttle, B. T. and Nemani, R. R. (2005). Mapping and modeling the biogeochemical cycling of turf grasses in the United States. *Environmental Management*, 36, 426–438.

Morton, T. G., Gold, A. J. and Sullivan, W. M. (1988). Influence of overwatering and fertilization on nitrogen losses from home lawns. *Journal of Environmental Quality*, 17, 124–130.

Niemela, J., Kotze, D. J., Venn, S., Penev, L., Stoyanov, I., Spence, J., Hartley, D. and de Oca, E. M. (2002). Carabid beetle assemblages (Coleoptera, Carabidae) across urban-rural gradients: an international comparison. *Landscape Ecology*, 17, 387–401.

Nowak, D. J. and Crane, D. E. (2002). Carbon storage and sequestration by urban trees in the USA. *Environmental Pollution*, 116, 381–389.

Oberndorfer, E., Lundholm, J., Bass, B., Coffman, R. R., Doshi, H., Dunnett, N., Gaffin, S., Kohler, M., Liu, K. K. Y. and Rowe, B. (2007). Green roofs as urban ecosystems: ecological structures, functions, and services. *Bioscience*, 57, 823–833.

Ohtonen, A. and Markkola, A. M. (1991). Biological activity and amount of FDA mycelium in mor humus of Scots pine stands in relation to soil properties and degree of pollution. *Biogeochemistry*, 13, 1–26.

Oldfield, E. E., Felson, A. J., Wood, S. A., Hallett, R. A., Strickland, M. S. and Bradford, M. A. (2014). Positive effects of afforestation efforts on the health of urban soils. *Forest Ecology and Management*, 313, 266–273.

Osmond, D. L. and Hardy, D. H. (2004). Characterization of turf practices in five North Carolina communities. *Journal of Environmental Quality*, 33, 565–575.

Ossola, A., Hahs, A. K. and Livesley, S. J. (2015). Habitat complexity influences fine scale hydrological processes and the incidence of stormwater runoff in managed urban ecosystems. *Journal of Environmental Management*, 159, 1–10.

Otte, P., Maring, L., De Cleen, M. and Boekhold, S. (2012). Transition in soil policy and associated knowledge development. *Current Opinion in Environmental Sustainability*, 4, 565–572.

Parkin, T. B. (1987). Soil microsites as a source of denitrification variability. *Soil Science Society of America Journal*, 51, 1194–1199.

Pataki, D. E., Carreiro, M. M., Cherrier, J., Grulke, N. E., Jennings, V., Pincetl, S., Pouyat, R. V., Whitlow, T. H. and Zipperer, W. C. (2011). Coupling biogeochemical cycles in urban environments: ecosystem services, green solutions, and misconceptions. *Frontiers in Ecology and the Environment*, 9, 27–36.

Patterson, J. C., Murray, J. J. and Short, J. R. (1980). The impact of urban soils on vegetation. *METRIA: 3, Proceedings of the Third Conference of the Metropolitan Tree Improvement Alliance.* New Brunswick, NJ, Rutgers, the State University of New Jersey, 33–56.

Paul, M. J. and Meyer, J. L. (2001). Streams in the urban landscape. *Annual Review of Ecology and Systematics,* 32, 333–365.

Pavao-Zuckerman, M. A. (2003). Soil ecology along an urban to rural gradient in the southern Appalachians. PhD Dissertation, University of Georgia, Athens, Georgia.

Pavao-Zuckerman, M. A. (2008). The nature of urban soils and their role in ecological restoration in cities. *Restoration Ecology,* 16, 642–649.

Pavao-Zuckerman, M. (2012). Urbanization, soils, and ecosystem services. IN Wall, D. H., Bardgett, R. D., Behan-Pelletier, V., Herrick, J. E., Jones, H. P., Ritz, K., Six, J., Strong, D. R. and van der Putten, W. H. (eds) *Soil Ecology and Ecosystem Services.* Oxford, Oxford University Press, 270–281.

Pavao-Zuckerman, M. A. and Byrne, L. B. (2009). Scratching the surface and digging deeper: exploring ecological theories in urban soils. *Urban Ecosystems,* 12, 9–20.

Pavao-Zuckerman, M. A. and Coleman, D. C. (2005). Decomposition of chestnut oak (Quercus prinus) leaves and nitrogen mineralization in an urban environment. *Biology and Fertility of Soils,* 41, 343–349.

Pickett, S. T. A. and Cadenasso, M. L. (2009). Altered resources, disturbance, and heterogeneity: a framework for comparing urban and non-urban soils. *Urban Ecosystems,* 12, 23–44.

Pickett, S. T. A., Cadenasso, M. L., Grove, J. M., Nilon, C. H., Pouyat, R. V., Zipperer, W. C. and Costanza, R. (2001). Urban ecological systems: linking terrestrial ecological, physical, and socioeconomic components of metropolitan areas. *Annual Review of Ecology and Systematics,* 32, 127–157.

Pitt, R. and Lantrip, J. (2000). Infiltration through disturbed urban soils. *Building Partnerships.* doi: 10.1061/40517(2000)108.

Pitt, R., Chen, S. E., Clark, S. E., Swenson, J. and Ong, C. K. (2008). Compaction's impacts on urban storm-water infiltration. *Journal of Irrigation and Drainage Engineering-ASCE,* 134, 652–658.

Pouyat, R. V., Parmelee, R. W. and Carreiro, M. M. (1994). Environmental effects of forest soil-invertebrate and fungal densities in oak stands along and urban-rural land use gradient. *Pedobiologia,* 38, 385–399.

Pouyat, R. V., McDonnell, M. J. and Pickett, S. T. A. (1997). Litter decomposition and nitrogen mineralization in oak stands along an urban-rural land use gradient. *Urban Ecosystems,* 1, 117–131.

Pouyat, R. V., Pataki, D. E., Belt, K. T., Groffman, P. M., Hom, J. and Band, L. E. (2007a). Effects of urban land-use change on biogeochemical cycles. IN Canadell, J. G., Pataki, D. E. and Pitelka, L. F. (eds) *Terrestrial Ecosystems in a Changing World.* Berlin, Springer, 45–58.

Pouyat, R. V., Yesilonis, I. D., Russell-Anelli, J. and Neerchal, N. K. (2007b). Soil chemical and physical properties that differentiate urban land-use and cover types. *Soil Science Society of America Journal,* 71, 1010–1019.

Pouyat, R. V., Yesilonis, I. D., Szlavecz, K., Csuzdi, C., Hornung, E., Korsos, Z., Russell-Anelli, J. and Giorgio, V. (2008). Response of forest soil properties to urbanization gradients in three metropolitan areas. *Landscape Ecology,* 23, 1187–1203.

Pouyat, R. V., Yesilonis, I. and Golubiewski, N. E. (2009). A comparison of soil organic carbon stocks between residential turf grass and native soil. *Urban Ecosystems*, 12, 45–62.

Pouyat, R. V., Szlavecz, K., Yesilonis, I. D., Groffman, P. M. and Schwarz, K. (2010). Chemical, physical, and biological characteristics of urban soils. IN Aitkenhead-Peterson, J. and Volder, A. (eds) *Urban Ecosystem Ecology*. Madison, WI, American Society of Agronomy, 119–152.

Pouyat, R. V., Yesilonis, I., Dombos, M., Szlavecz, K., Setälä, H., Cilliers, S., Hornung, E., Kotze, J. and Yarwood, S. (2015). A global comparison of surface soil characteristics across five cities: a test of the Ecosystem Convergence Hypothesis. *Soil Science*, in press.

Qian, Y., Follett, R. F. and Kimble, J. M. (2010). Soil organic carbon input from urban turfgrasses. *Soil Science Society of America Journal*, 74, 366–371.

Ramirez, K. S., Leff, J. W., Barberan, A., Bates, S. T., Betley, J., Crowther, T. W., Kelly, E. F., Oldfield, E. E., Shaw, E. A., Steenbock, C., Bradford, M. A., Wall, D. H. and Fierer, N. (2014). Biogeographic patterns in below-ground diversity in New York City's Central Park are similar to those observed globally. *Proceedings of the Royal Society B-Biological Sciences*, 281. doi: 10.1098/rspb.2014.1988.

Rao, P., Hutyra, L. R., Raciti, S. M. and Templer, P. H. (2014). Atmospheric nitrogen inputs and losses along an urbanization gradient from Boston to Harvard Forest, MA. *Biogeochemistry*, 121, 229–245.

Rees, W. E. (1997). Urban ecosystems: the human dimension. *Urban Ecosystems*, 1, 63–75.

Robinson, D. A., Lebron, I. and Vereecken, H. (2009). On the definition of the natural capital of soils: a framework for description, evaluation, and monitoring. *Soil Science Society of America Journal*, 73, 1904–1911.

Savva, Y., Szlavecz, K., Pouyat, R. V., Groffman, P. M. and Heisler, G. (2010). Effects of land use and vegetation cover on soil temperature in an urban ecosystem. *Soil Science Society of America Journal*, 74, 469–480.

Sawicka-Kapusta, K., Zakrzewska, M., Bajorek, K. and Gdula-Argasinska, J. (2003). Input of heavy metals to the forest floor as a result of Cracow urban pollution. *Environment International*, 28, 691–698.

Scharenbroch, B. C. (2009). A meta-analysis of studies published in Arboriculture & Urban Forestry relating to organic materials and impacts on soil, tree, and environmental properties. *Arboriculture & Urban Forestry*, 35, 221–231.

Scharenbroch, B. C., Lloyd, J. E. and Johnson-Maynard, J. L. (2005). Distinguishing urban soils with physical, chemical, and biological properties. *Pedobiologia*, 49, 283–296.

Schauer, J. J., Rogge, W. F., Hildemann, L. M., Mazurek, M. A., Cass, G. R. and Simoneit, B. R. T. (1996). Source apportionment of airborne particulate matter using organic compounds as tracers. *Atmospheric Environment*, 30, 3837–3855.

Selhorst, A. and Lal, R. (2013). Net carbon sequestration potential and emissions in home lawn turfgrasses of the United States. *Environmental Management*, 51, 198–208.

Setälä, H, Birkhofer, K., Brady, M., Byrne, L., Holt, G., de Vries, F., Gardi, C., Hotes, S., Hedlund, K., Liiri, M., Mortimer, S., Pavao-Zuckerman, M., Pouyat, R., Tsiafouli, M. and van der Putten, W. H. (2014). Urban and agricultural soils: conflicts and trade-offs in the optimization of ecosystem services. *Urban Ecosystems*, 17, 239–253.

Short, J. R., Fanning, D. S., McIntosh, M. S., Foss, J. E. and Patterson, J. C. (1986). Soils of the Mall in Washington, DC .1. Statistical summary of properties. *Soil Science Society of America Journal*, 50, 699–705.

Shuster, W. D., Dadio, S., Drohan, P., Losco, R. and Shaffer, J. (2014). Residential demolition and its impact on vacant lot hydrology: implications for the management of stormwater and sewer system overflows. *Landscape and Urban Planning*, 125, 48–56.

Shuster, W. D., Burkman, C. E., Grosshans, J., Dadio, S. and Losco, R. (2015a). Green residential demolitions: case study of vacant land reuse in storm water management in Cleveland. *Journal of Construction Engineering and Management*, 141, 06014011.

Shuster, W. D., Dadio, S. D., Burkman, C. E., Earl, S. R. and Hall, S. J. (2015b). Hydropedological assessments of parcel-level infiltration in an arid urban ecosystem. *Soil Science Society of America Journal*, 79, 398–406.

Smit, J., Ratta, A. and Nasr, J. (1996). *Urban Agriculture: Food, Jobs, and Sustainable Cities*. New York, UN Development Program.

Steinberg, D. A., Pouyat, R. V., Parmelee, R. W. and Groffman, P. M. (1997). Earthworm abundance and nitrogen mineralization rates along an urban-rural land use gradient. *Soil Biology and Biochemistry*, 29, 427–430.

Szlavecz, K., Placella, S. A., Pouyat, R. V., Groffman, P. M., Csuzdi, C. and Yesilonis, I. (2006). Invasive earthworm species and nitrogen cycling in remnant forest patches. *Applied Soil Ecology*, 32, 54–62.

Talbot, M. (1990). Ecological lawn care. *Mother Earth News*, 123, 60–73.

Tanner, P. A. and Fai, T. W. (2000). Small-scale horizontal variations in ionic concentrations of bulk deposition from Hong Kong. *Water Air and Soil Pollution*, 122, 433–448.

Templer, P. H., Toll, J. W., Hutyra, L. R. and Raciti, S. M. (2015). Nitrogen and carbon export from urban areas through removal and export of litterfall. *Environmental Pollution*, 197, 256–261.

Tenenbaum, D. E., Band, L. E., Kenworthy, S. T. and Tague, C. L. (2006). Analysis of soil moisture patterns in forested and suburban catchments in Baltimore, Maryland, using high-resolution photogrammetric and LIDAR digital elevation datasets. *Hydrological Processes*, 20, 219–240.

Trammell, T. L. E., Schneid, B. P. and Carreiro, M. M. (2011). Forest soils adjacent to urban interstates: soil physical and chemical properties, heavy metals, disturbance legacies, and relationships with woody vegetation. *Urban Ecosystems*, 14, 525–552.

Tuhackova, J., Cajthaml, T., Novak, K., Novotny, C., Mertelik, J. and Sasek, V. (2001). Hydrocarbon deposition and soil microflora as affected by highway traffic. *Environmental Pollution*, 113, 255–262.

Tzoulas, K., Korpela, K., Venn, S., Yli-Pelkonen, V., Kaźmierczak, A., Niemela, J. and James, P. (2007). Promoting ecosystem and human health in urban areas using Green Infrastructure: a literature review. *Landscape and Urban Planning*, 81, 167–178.

Waldrop, M. P., Zak, D. R., Sinsabaugh, R. L., Gallo, M. and Lauber, C. (2004). Nitrogen deposition modifies soil carbon storage through changes in microbial enzymatic activity. *Ecological Applications*, 14(4), 1172–1177.

Watmough, S. A., Hutchinson, T. C. and Sager, E. P. S. (1998). Changes in tree ring chemistry in sugar maple (Acer saccharum) along an urban-rural gradient in southern Ontario. *Environmental Pollution*, 101, 381–390.

Wessolek, G. (2008). Sealing of soils. IN Marzluff, J. M., Shulenberger, E., Endlicher, W., Alberti, M., Bradley, G., Ryan, C., Simon, U. and Zumbrunnen, C. (eds) *Urban Ecology: An International Perspective on the Interaction between Humans and Nature*. New York, Springer, 161–179.

Wong, F., Harner, T., Liu, Q. T. and Diamond, M. L. (2004). Using experimental and forest soils to investigate the uptake of polycyclic aromatic hydrocarbons (PAHs) along an urban-rural gradient. *Environmental Pollution*, 129, 387–398.

Yesilonis, I. and Pouyat, R. V. (2012). Carbon stocks in urban forest remnants: Atlanta and Baltimore as case studies. IN Lal, R. and Augustin, B. (eds) *Carbon Sequestration in Urban Ecosystems*. New York: Springer, 103–120.

Zhang, H. B., Luo, Y. M., Wong, M. H., Zhao, Q. G. and Zhang, G. L. (2006). Distributions and concentrations of PAHs in Hong Kong soils. *Environmental Pollution*, 141, 107–114.

Zhu, W.-X., Hope, D., Gries, C. and Grimm, N. B. (2006). Soil characteristics and the accumulation of inorganic nitrogen in an arid urban ecosystem. *Ecosystems*, 9, 711–724.

# 7 Impact of land take on global food security

*Ciro Gardi*

## Introduction

Food security is defined by Maxwell (1996) as existing when 'all people, at all times, have physical and economic access to sufficient, safe and nutritious food that meets their dietary needs and food preferences for an active and healthy life'. This definition underlines that food security includes four dimensions: availability, stability, safety and accessibility. Urban expansion and urbanization affects all four dimensions of food security (Matuschke, 2009), but for the scope of this book, we will concentrate on food availability.

Urban expansion occurs in most cases at the expense of agricultural land, reducing the availability of soil resources for agriculture production, limiting the availability of food. For example, in India, Saharanpur lost in 10 years (1988–1998) more than 30 per cent of its agricultural land (Fazal, 2000), and in Turkey, Kahramanmaras registered an expansion of 1100 per cent between 1950 and 2006, mainly at the expense of high quality agricultural lands (Doygun and Gurun, 2008). In Santiago (Chile) 19,600 ha, 70 per cent of which were prime agricultural land (land capability classes I, II, III and IV), were lost in 14 years (Romero and Ordenes, 2004). In China between 1990 and 2010 the potential agricultural production decreased by approximately 34.90 million tons due to urban expansion, accounting for 6.52 per cent of China's total actual production (Liu *et al.*, 2015).

## Global food security

The food crises of 2008, 2010, and 2012, as well as the continuing food price volatility, underscore the vulnerability of the world's food system. Soil sustains (directly or indirectly) more than 95 per cent of global food production (FAO, 2008). The estimated global increase of the world population from 6.8 billion in 2009 to 9.2 billion in 2050 (Speidel *et al.*, 2009) will lead to a significant increase in both food demand and land take, as consequence of urban expansion. It has been estimated that to feed 9 billion people, global agricultural production should increase by 70 per cent. At the same time, the combination of these conflicting processes has raised international concern for global food security. Global food security is affected by several factors, such as cropland shrinkage, fisheries reduction, increased wealth in countries such as China and India with the consequent increased demand for food (Godfray *et al.*, 2010), global warming (Millennium Ecosystem Assessment, 2005) and the overall intensification of agriculture and greater pressures on soils.

A total of 1.5 billion hectares of land is currently used worldwide for crop production. This represents 11 per cent of the planet's land surface (FAO, 2003). The FAO (2003) also reports that 2.7 billion hectares of land with crop production potential remain unexploited. However, these figures might over-estimate the land available for agricultural production (Bot *et al.*, 2000). Some of the land that could potentially be used for agriculture is subject to ecological constraints and pollution, while other land is protected or occupied by other land uses (e.g. forests and woodlands, human settlements). If we consider the availability of soils suitable for agricultural production, which means soils with an intrinsic fertility, we should not be surprised that only a limited portion of the planet's lands are covered by such soils (Figure 7.1).

But what exactly happened in 2008? There had been a series of events (unfavourable climate conditions, increase investments in biofuels crops, etc.) that contributed to a drastic reduction of cereals production at the global scale. At the same time demand increased for cereals for food uses (driven especially by the fast growing Asiatic economies) and for non-food use (biofuel production). Cereal prices rose rapidly, to the satisfaction of farmers (the price of wheat doubled), but with consequences for the final consumers, who experienced concomitant increases in the prices of food such as bread, pasta, etc. Another effect, less evident but probably more serious, was the drastic reduction of global cereals stocks, which are estimated at 520 millions of tonnes of

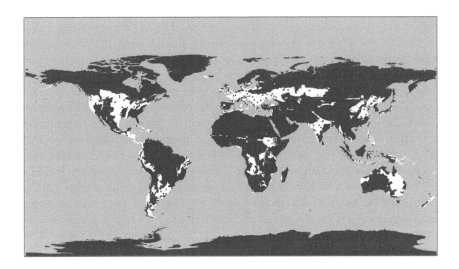

*Figure 7.1* Distribution of intrinsically fertile soils

cereals, equivalent to just months (25 per cent) of global consumption. In other words, a reduction of just 30 per cent in world cereal production will be sufficient to bring humanity to the edge of food crisis.

The magnitude and rate of the increase in Chinese demand for wheat and soybean combined with the conversion of land use in Brazil is already indicative of the pressures that are piling on cropping systems and on residual terrestrial natural and semi-natural ecosystems laying over fertile soils.

Based on this evidence and theses perspectives, it remains unclear why, at the local scale, agricultural product prices have increased so little or even decreased over the last decade, suggesting farmlands and good agricultural soils are an abundant and infinitely replaceable resources. The main reason lies precisely in the market prices of agricultural commodities that, until the recent peaks, did not guarantee profitable margins for small-scale farms and seemed to indicate the existence of agricultural systems more efficient and competitive and, above all, an unlimited availability of both food commodities and arable land. This fact, in the presence of increasing production costs for small farms and strong competitive pressure for more profitable use of the land in the short term, has encouraged the gradual abandonment of agricultural activity, for the benefit of speculative processes of urban expansion. There was an implicit (and wrong) social assumption that the loss of land and agricultural production at the local scale would be compensated in some other part the world, where agriculture is still a profitable and competitive economic activity.

The signal provided by market prices of food commodities, however, is not a reliable indicator of the absolute availability of a particular 'good', or more correctly a 'commons' such as soil, that is non-renewable, finite and degradable. Prices in reality can be an expression of the scarcity or the abundance of a resource, within a trading market, but do not account for all those people who for various reasons can not access the market and express a demand in monetary terms. In the globalization era we would think that all humanity participates in some way in the game of supply and demand that sets the prices of food commodities. In reality more than a third of the world's population is virtually cut off by this game and from the market, for the simple reason that it does not have the economic resources to express its demand or, more explicitly, because they are too poor. According to UN figures, at the beginning of the third millennium, two and a half billion people were living on less than $2 a day. Of these, 850 million were in a state of malnutrition, an increasing proportion compared to the 1990s. In addition to the people that are excluded by the market, for lack of income, we should consider those who can not participate because they do not exist – future generations. We refer to the preference and willingness to sacrifice the future of these 2.5 billion people, that by 2050, will added to the current 6.5 billion. If the legitimate food needs of those who will and those who currently are in a state of undernourishment could be expressed in terms of monetary demand discounted, the price of primary products would certainly be different, as would be the value of agricultural land and the perception of its global availability.

On the supply side there is the need to consider that the price system reflects the current productive situation. It does not provide information on the medium and long-term sustainability of current yields and, more generally, of intensive farming systems. The occurrence of a sudden fertility crisis in specific production areas of the globe, caused by soil salinization and the exhaustion of overexploited aquifers for irrigation purposes, the spread of desertification and soil erosion, such as that occurring in the Chinese loess plateau, or the consequences of climate change, could in future further increase the demand for agricultural land which may no longer be satisfied because of the irreversible changes of land use occurring in the meantime because of misleading market signals. But above all, current food commodities prices that encourage and induce those irreversible land use changes, do not take into account the expected future performance of a crucial variable for the entire system of global food supply, that is the basis of those same prices and of the market structure that makes them possible: the cost, or the availability, of fossil fuels.

The extraordinary increases in yield obtained in agriculture after the Second World War were largely due to a massive and unprecedented use of subsidiary fossil energy. We can consider that after this period modern agriculture became more oil driven than photosynthetically driven. Fossil fuel energy is used for the synthesis of chemical fertilizers and pesticides, as well as for the production and operation of agricultural machinery and irrigation pumps. It is estimated, for example, that only the production of nitrogen fertilizers covers 2 per cent of world industrial energy consumption (Hawken *et al.*, 1999). Large quantities of fossil fuels are also consumed by the transport, processing, storage, packaging and retailing of agricultural products, which is required to cover the distance from the field to the (super) market shelf. Studies conducted on the American food supply chain estimate that this consumption is four times higher than that required for the agricultural production phase. The sum of this consumption shows that, on average, every calorie contained in a food product purchased at retail costs from 1 to 10 calories as fossil fuels. Such an energy ratio suggests that with the food we are eating we are above all consuming oil.

Until a few years ago, the relatively low prices of fossil fuels meant that the cost component of food related to transport, conservation and distribution was almost irrelevant compared to other factors such as labour costs. This explains why you could find in the local market vegetables and fruits grown several thousand kilometres away with prices lower than or at least competitive with similar products from local supply chains. The possibility of importing affordable products from countries in the southern hemisphere, like New Zealand, Chile, South Africa also promotes a misleading perception of seasonality, leading to the belief that at the market it is always summer. Despite the relatively low price of oil in the present day (2014–2015) there are reasonable grounds for thinking that this sort of aberration in the food supply system, based on the low cost of energy for transport and storage, cannot last much longer, at least to this degree. Just a few years ago the steady increase in oil prices (the main energy source for transport) that exceeded the threshold of $100 per barrel (2008) revealed all the limits of a system based on cheap oil or energy prices.

## The role of bioenergy crops

The increased use of agricultural lands for bioenergy production also represents a major concern with regard to land availability for food crop production. The International Energy Agency scenarios for land use (IEA, 2004) estimated that approximately 5 per cent of the EU's cropland area would have to be converted to biofuel production in order to replace 5 per cent of its petrol supply, while 15 per cent of its cropland area would have to be converted to substitute 5 per cent of the diesel supply (Escobar *et al.*, 2009). Forecasts from the same Agency (IEA) indicate that by 2050, 27 per cent of the world transportation fuels could be provided by bioenergy crops (from the actual 2 per cent). During the last decade biofuel production has almost tripled (from 45 to 130 billions litres) (IEA, 2011). Considering an average biofuel yield of 4,500 l/ha, we can estimate that the area of agricultural land devoted to this use has risen from 10 million ha to almost 29 million ha. If the IEA's forecasts are respected, considering also the increasing demand of transport worldwide, more than 300 million hectares (20 per cent of the actual cultivated land) will have to be dedicated to biofuel production. A possible correction to this alarming situation could be provided by the third generation of biofuels, which allow the production of fuels from lingo-cellulosic materials. In other words, instead of using the main product of the crop (cereals, sugar, etc.) only the by-product of the crop would be used.

## Urban sprawl, urban expansion and land take

Sprawling cities tend to consume the best agricultural lands, forcing agriculture to move to less productive areas (Scalenghe and Marsan, 2009). The extent of agricultural land and, to a smaller extent, woodlands and semi-natural and natural areas, is decreasing due to conversion to residential, industrial or commercial areas (EEA, 2011). Urban centres often expand on the most productive land because cities are historically built mainly on fertile soils (Satterthwaite *et al.*, 2010).

Furthermore, land take causes environmental perturbations that affect agricultural ecosystems (e.g. landscape fragmentation, changes in the water cycle and reduced habitats). There is increasing evidence that European cities tend to become more dispersed, as a result of the spread of low-density settlements (urban sprawl) (Kasanko *et al.*, 2006), increasing similarities with urban areas of the US. However, differences among the urban structures and their dynamics between the US and Europe remain important, mainly because different relationships exist between central and local governments (Summers *et al.*, 1999). Land-take processes are occurring in other parts of the world at considerably higher rates than in Europe, especially in countries with rapidly growing economies. For example, 5.1 per cent of the overall territory in China was lost to manufacturing and municipal activities during the period 1996 to 2003 (Chen, 2007), and in the Beijing–Tianjin–Hebei region urban area growth expanded by 71 per cent between 1990 and 2000 (Tan *et al.*, 2005). Similar growth rates have been recorded also in India (Fazal, 2000) and in other fast growing countries. Worldwide urban areas are expanding at twice their population growth rates (Angel *et al.*, 2011).

## Land grabbing

In addition to these processes there is an increasing interest in farmland due to rising food and fuel prices, biofuel mandates, food security concerns, climate finance incentives and worries about climate change effects on scarce resources. There is concern that the remaining areas of cultivable land that are currently mostly used under customary rights are vulnerable to speculators or unscrupulous investors, who exploit smallholder farmers, herders and other local people who lack the power to stand up for their rights. This increasing interest in land has resulted in land grabbing. Land grabbing refers to large-scale land acquisitions, especially in developing countries, by domestic and transnational companies, governments and individuals. This phenomenon was also known and practised in the past, but nowadays with the term 'land grabbing' we primarily refer to large-scale land acquisitions following the 2007–2008 world food price crisis. Tens of millions of hectares have been subject to some sort of negotiation with a foreign investor. Half of the total in African countries (Ambalam, 2014; Anseeuw, 2013; Lavers, 2012; Lisk, 2013: Manji, 2012; Millar, 2015; Sulieman, 2015; Veldwisch, 2015), one-fourth in Asia (Feldman and Geisler, 2012; Kenney-Lazar, 2012; Jiao *et al.*, 2015; Semedi and Bakker, 2014; Siciliano, 2014), and not less than 10 per cent in Latin America (Borras *et al.*, 2012; Brent, 2015; Bulkan, 2014; Economist, 2011; Grajales, 2015; Grandia, 2013; Holmes, 2014; Murmis and Murmis, 2012; Oliveira, 2013; Perrone, 2013; Piñeiro, 2012; Rocheleau, 2015; Urioste, 2012). The issue suffers from a lack of transparency as contracts are often kept secret and most of these data remain undisclosed.

## Impact of land take on potential agricultural production in Europe

In 1985, the European Commission launched the CORINE (Coordination of Information on the Environment) programme. The main objectives of the CORINE Land Cover (CLC) project were to provide reliable quantitative data on land cover across Europe, and to develop one complete spatial dataset covering the EU member states (MS) plus several other European and North African countries. The datasets contain homogeneous data on land cover areas, which are represented as polygons (shapefiles), although unfortunately not all the countries involved in the programme have the same temporal coverage. Due to this limitation, CORINE Land Cover datasets from 1990, 2000 and 2006 were used to assess the extent of land take of agricultural lands in 21 EU member states.

In order to estimate the impact of land take on potential agricultural production, the regional average of winter wheat yields (NUTS2 level, 1992–2004 period – MARS, 2012), were used. These data were available for 19 of the 21 countries (Gardi et al., 2015). For each NUTS2 area, the potential agricultural production losses were calculated on the basis of the following equation:

$$PAP\_LOSSES_{NUTS2} = ALT_{NUTS2} \times AWWY_{NUTS2}$$

where:

- $PAP$ = Potential Agricultural Production
- $PAP\_LOSSES_{NUTS2}$ = Losses of PAPC at NUTS2 level (in tonnes of winter wheat)
- $ALT_{NUTS2}$ = Land take of agricultural area at NUTS2 level (ha) for the given period
- $AWWY_{NUTS2}$ = Average Winter Wheat Yields at NUTS2 level (t/ha) for the given period

Land take was calculated using CORINE Land Cover maps of 1990, 2000 and 2006. For 21 of the 27 European Union member states, agricultural land take was computed to be 752,973 ha for 1990–2000 and 436,095 ha for 2000–2006, representing 70.8 per cent and 53.5 per cent, respectively, of the total EU land take for these periods.

Table 7.1 shows the land take data, on a yearly base, expressed both in absolute and relative terms. The small countries, characterized by high population densities, present, in relative terms, the greatest loss of agricultural area due to land take. The Netherlands, for instance, experienced the highest rate of land take in relative terms, and one of the largest also when absolute values are considered (Table 7.1). This country lost almost 2.5 per cent of its agricultural land during the period 1990–2000 and 1 per cent during the period 2000–2006. The greatest land take in absolute terms, however, took place in the largest EU countries: Germany, Spain and France (1990–2000) and Spain, France and Germany (2000–2006).

The impact of this land take on the production capabilities of the agricultural sector for the period 1990–2006 for 19 of the 21 states was estimated to be equivalent to a loss of more than 6 million tonnes of wheat (0.81 per cent of the total available potential agricultural production (PAP)). From this example we can conclude that, taking a long-term perspective (e.g. 100 years), land take could be an important threat to food security in the EU. In this assessment, it was estimated that 19 EU countries lost approximately 0.81 per cent of their PAP capacity between 1990 and 2006, with large variability between regions. A more detailed analysis showed that certain regions, such as those around the largest cities, in metropolitan areas and coastal zones, experienced the greatest loss of their most fertile soils.

*Table 7.1* Absolute and relative yearly agricultural land take in 21 EU countries

| Country | Agricultural land take (ha $y^{-1}$) | | Relative land take (% $y^{-1}$) | |
|---|---|---|---|---|
| | 1990–2000 | 2000–2006 | 1990–2000 | 2000–2006 |
| Austria | 1,034 | 870 | 0.01 | 0.01 |
| Belgium | 1,579 | 426 | 0.05 | 0.01 |
| Bulgaria | 281 | 570 | 0.00 | 0.01 |
| Czech Republic | 946 | 2,011 | 0.01 | 0.03 |
| Germany | 19,097 | 9,667 | 0.05 | 0.03 |
| Denmark | 1,239 | 1,729 | 0.03 | 0.04 |
| Estonia | 148 | 366 | 0.00 | 0.01 |
| Spain | 11,872 | 17,638 | 0.02 | 0.04 |
| France | 11,570 | 12,697 | 0.02 | 0.02 |
| Hungary | 953 | 2,503 | 0.01 | 0.03 |
| Ireland | 3,120 | 3,275 | 0.04 | 0.05 |
| Italy | 7,931 | 7,735 | 0.03 | 0.03 |
| Lithuania | 52 | 551 | 0.00 | 0.01 |
| Luxembourg | 171 | 62 | 0.07 | 0.02 |
| Malta | 1 | 1 | 0.00 | 0.00 |
| Netherlands | 8,130 | 5,879 | 0.23 | 0.17 |
| Poland | 1,709 | 2,883 | 0.01 | 0.01 |
| Portugal | 4,244 | 1,838 | 0.05 | 0.02 |
| Romania | 743 | 1,396 | 0.00 | 0.01 |
| Slovenia | 12 | 70 | 0.00 | 0.00 |
| Slovakia | 1,034 | 870 | 0.01 | 0.01 |

The importance of land take as a threat to soil varies among EU countries. In countries with high land-take rates and high PAP, such as the Netherlands, land take is a particularly important issue. The same applies for most of the new member states where the agricultural land-take trend has doubled in the past few years, and for the countries affected by 'real estate bubbles', such as Spain and Ireland.

## References

Ambalam, K. (2014) 'Food sovereignty in the era of land grabbing: an African perspective', *Journal of Sustainable Development*, 7: 121–132.

Angel, S., J. Parent, D.L. Civco, A. Blei and D. Potere (2011) 'The dimensions of global urban expansion: estimates and projections for all countries, 2000–2050', *Progress in Planning*, 75(2): 53–107.

Anseeuw, W. (2013) 'The rush for land in Africa: resource grabbing or green revolution?', *South African Journal of International Affairs*, 20: 159–177.

Borras Jr, S.M., J.C. Franco, S. Gómez, C. Kay and M. Spoor (2012) 'Land grabbing in Latin America and the Caribbean', *Journal of Peasant Studies*, 39: 845–872.

Bot, A.J., F.O. Nachtergaele and A. Young (2000) *Land resource potential and constraints at regional and country levels*, World Soil Resources Reports, 90, Rome, FAO.

Brent, Z.W. (2015) 'Territorial restructuring and resistance in Argentina', *Journal of Peasant Studies*, 42: 671–694.

Bulkan, J. (2014) 'Forest grabbing through forest concession practices: the case of Guyana', *Journal of Sustainable Forestry*, 33(4): 407–434.

Chen, J. (2007) 'Rapid urbanization in China: a real challenge to soil protection and food security', *Catena*, 69: 1–15.

Doygun, H. and D.K. Gurun (2008) 'Analysing and mapping spatial and temporal dynamics of urban traffic noise pollution: a case study in Kahramanmaraş, Turkey', *Environmental Monitoring and Assessment*, 142(1–3): 65–72.

Economist (2011) 'The surge in land deals: when others are grabbing their land', *The Economist*, 399(8732). www.economist.com/node/18648855, accessed 23 June 20016.

EEA (2011) 'Land Take (CSI 014)', Assessment published February 2011. www.eea.europa.eu/data-and-maps/indicators/land-take-2/assessment, accessed 15 July 2011.

Escobar, J.C., E.S. Lora, O.J. Venturini, E.E. Yanez, E.F. Castillo and O. Almazan (2009) 'Biofuels: environment, technology and food security', *Renewable and Sustainable Energy Reviews*, 13: 1275–1287.

FAO (2003) 'World agriculture: towards 2015/2030. An FAO perspective'. www.fao.org/docrep/005/y4252e/y4252e00.htm, accessed 16 July 2013.

FAO (2008) *FAO Statistical Yearbook 2007–2008*. www.fao.org/, accessed 16 July 2013.

Fazal, S. (2000) 'Urban expansion and loss of agricultural land: a GIS based study of Saharanpur City, India', *Environment and Urbanization*, 12(2): 133–149.

Feldman S. and C. Geisler (2012) 'Land expropriation and displacement in Bangladesh', *Journal of Peasant Studies*, 39: 971–993.

Gardi, C., P. Panagos, M. Van Liedekerke, C. Bosco and D. De Brogniez (2015) 'Land take and food security: assessment of land take on the agricultural production in Europe', *Journal of Environmental Planning and Management*, 58(5): 898–912.

Godfray, H.C.J., J.R. Beddington, I.R. Crute, L. Haddad, D. Lawrence, J.F. Muir, J. Pretty and C. Toulmin (2010) 'Food security: the challenge of feeding 9 billion people', *Science*, 327: 812–818.

Grajales, J. (2015) 'Land grabbing, legal contention and institutional change in Colombia', *Journal of Peasant Studies*, 42: 541–560.

Grandia, L. (2013) 'Road mapping: megaprojects and land grabs in the Northern Guatemalan Lowlands', *Development and Change*, 44: 233–259.

Hawken, P., A. Lovins and L.H. Lovins (1999) *Natural Capitalism: Creating the Next Industrial Revolution*, Boston, MA, Little, Brown.

Holmes, G. (2014) 'What is a land grab? Exploring green grabs, conservation, and private protected areas in southern Chile', *The Journal of Peasant Studies*, 41: 547–567.

International Energy Agency (IEA) (2004) 'Biofuels for transport: an international perspective', www.iea.org, accessed 7 July 2015.

International Energy Agency (IEA) (2011) 'Technology roadmap: biofuels for transport'. www.iea.org, accessed 7 July 2015.

Jiao X., C. Smith-Hall and I. Theilade (2015) 'Rural household incomes and land grabbing in Cambodia', *Land Use Policy*, 48: 317–328.

Kasanko, M., J.I. Barredo, C. Lavalle, N. McCormick, L. Demicheli, V. Sagris and A. Brezger (2006) 'Are European cities becoming dispersed? A comparative analysis of 15 European urban areas', *Landscape Urban Plan*, 77: 111–130.

Kenney-Lazar, M. (2012) 'Plantation rubber, land grabbing and social-property transformation in southern Laos', *Journal of Peasant Studies*, 39: 1017–1037.

Lavers, T. (2012) '"Land grab" as development strategy? The political economy of agricultural investment in Ethiopia', *Journal of Peasant Studies*, 39: 105–132.

Lisk, F. (2013) '"Land grabbing" or harnessing of development potential in agriculture? East Asia's land-based investments in Africa', *Pacific Review*, 26: 563–587.

Liu, L., X. Xu and X. Chen (2015) 'Assessing the impact of urban expansion on potential crop yield in China during 1990–2010', *Food Security*, 7(1): 33–43.

Manji, A. (2012) 'The grabbed state: lawyers, politics and public land in Kenya', *Journal of Modern African Studies*, 50: 467–492.

MARS (2012) 'Monitoring of agriculture with remote sensing'. http://mars.jrc. ec.europa.eu/, accessed 8 July 2012.

Matuschke, I. (2009) 'Rapid urbanization and food security: using food density maps to identify future food security hotspots', in paper prepared for presentation at the International Association of Agricultural Economists Conference, Beijing, China, 16–22 August 2009.

Maxwell, S. (1996) 'Food security: a post-modern perspective', *Food policy*, 21(2): 155–170.

Millar, G. (2015) 'Knowledge and control in the contemporary land rush: making local land legible and corporate power applicable in rural Sierra Leone', *Journal of Agrarian Change*. doi: 10.1111/joac.12102.

Millennium Ecosystem Assessment (2005) *Ecosystems and Human Well-being: Synthesis*, Washington, DC: Island Press.

Murmis, M. and M.R. Murmis (2012) 'Land concentration and foreign land ownership in Argentina in the context of global land grabbing', *Canadian Journal of Development Studies*, 33: 490–508.

Oliveira, G. de L.T. (2013) 'Land regularization in Brazil and the global land grab', *Development and Change*, 44(2): 261–283.

Perrone, N.M. (2013) 'Restrictions to foreign acquisitions of agricultural land in Argentina and Brazil', *Globalizations*, 10: 205–209.

Piñeiro, D.E. (2012) 'Land grabbing: concentration and "foreignization" of land in Uruguay', *Canadian Journal of Development Studies*, 33: 471–489.

Rocheleau, D.E. (2015) 'Networked, rooted and territorial: green grabbing and resistance in Chiapas', *Journal of Peasant Studies*, 42: 695–723.

Romero, H. and F. Ordenes (2004) 'Emerging urbanization in the Southern Andes: environmental impacts of urban sprawl in Santiago de Chile on the Andean piedmont', *Mountain Research and Development*, 24(3): 197–201.

Satterthwaite, D., G. McGranahan and C. Tacoli (2010) 'Urbanization and its implications for food and farming', *Philosophical Transactions of the Royal Society B*, 365: 2809–2820.

Scalenghe, R. and Marsan, F.A. (2009) 'The anthropogenic sealing of soils in urban areas', *Landscape and Urban Planning*, 90(1): 1–10.

Semedi, P. and L. Bakker (2014) 'Between land grabbing and farmers' benefits: land transfers in West Kalimantan, Indonesia', *Asia Pacific Journal of Anthropology*, 15: 376–390.

Siciliano, G. (2014) 'Rural-urban migration and domestic land grabbing in China population', *Space and Place*, 20: 333–351.

Speidel, J., D. Weiss, S. Ethelston and S. Gilbert (2009) 'Population policies, programmes and the environment', *Philosophical Transactions of the Royal Society B*, 364: 3049–3065.

Sulieman, H.M. (2015) 'Grabbing of communal rangelands in Sudan: the case of large-scale mechanized rain-fed agriculture', *Land Use Policy*, 47: 439–447.

Summers, A.A., P.C. Cheshire and L. Senn (eds) (1999) *Urban Change in the United States and Western Europe: Comparative Analysis and Policy*, Washington, DC: The Urban Institute Press.

Tan, M., X. Li, H. Xie and C. Lu (2005) 'Urban land expansion and arable land loss in China: a case study of Beijing–Tianjin–Hebei region', *Land Use Policy*, 22(3): 187–196.

Urioste, M. (2012) 'Concentration and "foreignization" of land in Bolivia', *Canadian Journal of Development Studies*, 33: 439–457.

Veldwisch, G.J. (2015) 'Contract farming and the reorganisation of agricultural production within the Chókwè irrigation system, Mozambique', *Journal of Peasant Studies*. doi:10.1080/03066150.2014.991722.

# 8  Hydrological impact of soil sealing and urban land take

*Alberto Pistocchi*

## Introduction

Land take by urban development is a major hydrologic threat: it entails soil sealing and compaction, impervious surfaces may collect pollutants that are periodically washed off, and artificial drainage generally transfers runoff away much more quickly than in natural watersheds. Actually, the contemporary city is mainly built on the concept that rainwater should be evacuated rather than retained. A quick drainage of water is not an obvious and necessary need: the birth of cities in the Neolithic has been put in relation with the feminine wisdom and capacity to collect and retain (Mumford, 1961):

> Under woman's dominance, the Neolithic period is pre-eminently one of containers: it is an age of stone and pottery utensils, of vases, jars, vats, cisterns, bins, barns, granaries, houses, not least great collective containers, like irrigation ditches and villages. The uniqueness and significance of this contribution has too often been overlooked by modern scholars who gauge all technical advances in terms of the machine . . . . Wherever a surplus must be preserved and stored, containers are important. . . . But as soon as agriculture brought a surplus of food and permanent settlement, storage utensils of all kinds were essential.

Retaining and recycling water is indeed an essential urban function in arid or semi-arid environments, and the capacity to adapt landscapes to collect water and make it available for human use has been key to the development of civilizations across the world (Laureano, 2001).

Contrary to early agglomerations, however, modern cities have grown largely inconsiderate of the need to manage landscapes to retain water, and now urban drainage systems, made of underground pipes serving impervious surfaces such as roads and roofs, are regarded as 'traditional'. The mechanism of hydrological alteration operated by soil sealing and 'traditional' urban drainage is essentially a reduction of the permanence of water in the landscape, by avoiding infiltration in soils, hence shortening the hydrological pathways of runoff, and ultimately accelerating its delivery to the receiving water bodies.

The effect is exacerbated by the connection of impervious areas to receiving streams through artificial drainage networks. Sometimes, the 'connected' impervious fraction of a catchment is considered as 'effective' and looked at as a better predictor of the impacts of soil sealing (Walsh *et al.*, 2005a, as summarized by Hamel *et al.*, 2013). This causes, on the one side, the drying-up of the landscape, and on the other the reduction in duration, and increase in intensity of stream water discharges. A dryer landscape, at the same time, evaporates less water. It is customary to denote soil moisture and evapotranspiration flows as 'green water' in contrast to 'blue water' in liquid form in streams, aquifers and other water bodies, following a concept initially proposed by Falkenmark (1995). We may think of soil sealing as a process shifting water from 'green' to 'blue' (moreover available for shorter time). The consequences of this shift may be significant on the water cycle not just within the affected watershed, but also at regional scale, where reduced evapotranspiration may alter precipitation feedbacks (e.g. Rockström *et al.*, 2014).

Soil sealing should be avoided, limited, mitigated and compensated as much as possible (e.g. European Commission, 2012a). The only effective way to compensate the hydrological effects of soil sealing is through water storage, with characteristics that may vary considerably depending on the compensation target. In this contribution we discuss the impacts of soil sealing and land take from urban development in terms of water quality, water availability and floods, and we summarize options for the mitigation of impacts on floods and water availability.

## The ecological consequences of soil sealing

In USEPA's Causal Analysis/Diagnosis Decision Information System (CADDIS), urbanization is considered a specific source of stress for the aquatic environment; its impacts, due to morphological alteration of the water bodies, chemical emissions and hydrological alterations, affect water quality (with increased nutrients and toxic substances), flow regime (with flashier flows), physical habitat (with simplification of channel morphology), water temperature, and energy flow with less organic matter retention (USEPA, 2010). As a complex mixture of stressors that can be hardly disentangled, urbanization induces systematically a set of symptoms in water bodies, which have been called the 'urban stream syndrome' (Walsh *et al.*, 2005b). The CADDIS urbanization stressor module outlines the main ecological consequences on water bodies of urbanization (USEPA, 2010) as follows:

1   Urbanization can significantly modify the structure of aquatic biotic communities. The impervious cover fraction of the catchment has been put in inverse relationship with the probability of occurrence of certain fish species, and generally with ecological stream conditions. The effect becomes very strong beyond a threshold of share of the watershed or riparian area affected by urbanization (somewhere between 20 and 30 per cent).

2   The hydrological alteration induced by urbanization, with increased runoff volumes and reduced flow duration, may create conditions for incision of the stream channels, in turn reducing the water storage capacity of the alluvium. As a result, riparian vegetation may significantly change and so may the capacity of riparian areas to retain nutrients.

3   Pollution associated to urbanization is not just due to wastewater treatment effluents, but also to combined sewer overflows and wash-off of impervious surfaces where contaminants build up during inter-storm periods. The wastewater-related enrichment of streams contributes to altering macroinvertebrate diversity, while chemicals bypassing wastewater treatment or discharged with combined sewerage overflow may include endocrine disruptors or be toxic to aquatic organisms.

4   Wash-off of impervious surfaces typically carries sediments, nutrients, household pesticides, metals, polycyclic aromatic hydrocarbons (PAHs), oil and grease. The transport of contaminants is mainly associated with impervious surfaces only when they are connected with an urban drainage system. Anyway, chemical concentrations in the water bodies have been correlated to the impervious cover fraction of the watershed. Pollutants associated to urban runoff can be predicted on the basis of empirical models, among which the classic one by Heaney *et al.* (1976), although the predictability of pollutant loads with statistical methods may be rather low (e.g. Brezonik and Stadelman, 2002).

5   The alteration of riparian vegetation cover and the heat transferred from warm sealed surfaces to rainwater during wash-off tend to increase water temperature, which in turn affects the activity of microorganisms and the chemical reactivity of substances. A net effect of warmer streams is the decrease of fish and macro-invertebrate abundance and variety.

6   Sediments supply tends to decrease in the long term; this may trigger channel bank erosion and the enlargement of channels, in turn altering physical habitat availability. Road crossings also affect physical habitat through hydrodynamic alteration and scour around piles and piers. Substrates in urbanized streams may tend to increase the fine fraction of sediments and become less stable, or on the contrary be eroded and armoured.

7   Organic carbon input and metabolism are changed in urban streams: less natural carbon, and more anthropogenic carbon, is conveyed to the stream. The storage of carbon is generally reduced.

Walsh *et al.* (2012) present urban storm water runoff as a new class of environmental flow problem: contrary to 'traditional' environmental flow problems where a minimum water volume flowing in streams should be left unaffected by abstractions in order to protect ecosystems, in this case avoiding excess volumes to reach the streams generally improves ecological conditions and the services provided by water ecosystems (Figure 8.1). Therefore, urban storm water harvesting can be regarded as a win–win solution for water supply to human activities.

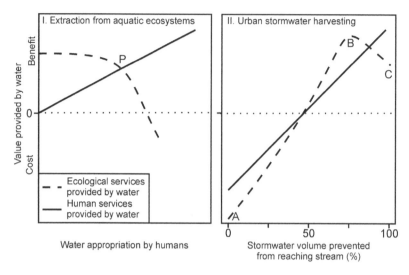

*Figure 8.1* Conceptual graphs of ecological and human value of water (source: Walsh *et al.*, 2012, under the Creative Commons Attribution licence)

## Impacts on water availability and their mitigation

Although the increase in runoff volumes due to soil sealing may facilitate storm water harvesting, flows from urbanized catchments tend to be flashier, with relatively high and short-lasting discharges. In other terms, the flow duration curve of an urbanized catchment tends to become steeper compared to undeveloped conditions, with higher high-flows of shorter duration and lower low-flows of longer duration. As a consequence, overall water availability may be reduced. This response of flow duration curves to soil sealing has exceptions (e.g. Hawley and Bledsoe, 2011) that may be due to, for example, leaking water supply infrastructure or reduced evapotranspiration. Burns *et al.* (2005) find flow recession speed to increase with the degree of development in the catchment in the Croton river basin, New York, USA, but they also find an increment of base flow. Simultaneous increase of low-flows and high-flows can be generally expected where rainfall is relatively uniform in time, and subsurface flows contribute relatively little to total discharges. This makes a generalization about the impacts on water availability difficult (Hamel *et al.*, 2013). Evidence of change of flow duration curves is reported in several cases (e.g. Schoonover *et al.*, 2006; Guo and Quader, 2009; Mejía *et al.*, 2014), and is predicted by hydrological modelling studies (e.g. Yang *et al.*, 2014).

Sustainable urban drainage systems (SUDs) and low impact development (LID) have long been advocated in order to mitigate the hydrological impacts of soil sealing. According to the SUSDRAIN platform of the construction industry research and information association (CIRIA).

Sustainable drainage is a departure from the traditional approach to draining sites. There are some key principles that influence the planning and design process enabling SuDS to mimic natural drainage by: storing runoff and releasing it slowly (attenuation); allowing water to soak into the ground (infiltration); slowly transporting (conveying) water on the surface; filtering out pollutants; allowing sediments to settle out by controlling the flow of the water.

> (www.susdrain.org/delivering-suds/using-suds/suds-principles/
> suds-principals.html, accessed 17 November 2016)

Similarly, USEPA characterizes LID as follows:

LID is an approach to land development (or re-development) that works with nature to manage stormwater as close to its source as possible. LID employs principles such as preserving and recreating natural landscape features, minimizing effective imperviousness to create functional and appealing site drainage that treat stormwater as a resource rather than a waste product. There are many practices that have been used to adhere to these principles such as bioretention facilities, rain gardens, vegetated rooftops, rain barrels, and permeable pavements. By implementing LID principles and practices, water can be managed in a way that reduces the impact of built areas and promotes the natural movement of water within an ecosystem or watershed. Applied on a broad scale, LID can maintain or restore a watershed's hydrologic and ecological functions.

> (http://water.epa.gov/polwaste/green/,
> accessed 17 November 2016)

LID may be an effective solution for both water quality and quantity alterations associated to soil sealing. For instance, Dietz and Clausen (2008) find that runoff volume and loads of contaminants from LID areas do not depend on the percentage of impervious area as strongly as those from traditional developments.

More recently, the European Commission's Blueprint to safeguard Europe's waters (European Commission, 2012b) has endorsed the use of natural water retention measures (NWRMs) as a kind of approach to retain water in the landscape, by this mitigating floods, ad improving water quality and availability. NWRMs broaden the scope of LID and SUDs:

Natural Water Retention Measures (NWRM) are multi-functional measures that aim to protect and manage water resources and address water-related challenges by restoring or maintaining ecosystems as well as natural features and characteristics of water bodies using natural means and processes. Their main focus is to enhance, as well as preserve, the water retention capacity of aquifers, soil, and ecosystems with a view to improving their status. NWRM have the potential to provide multiple benefits, including the reduction of risk of floods and droughts, water quality improvement, groundwater recharge and habitat improvement. The application of NWRM supports green infrastructure, improves or preserves the quantitative status of surface water and groundwater bodies and can positively affect the chemical and ecological status of water bodies by restoring or enhancing natural functioning of ecosystems and the services they provide. The preserved or restored ecosystems can contribute both to climate change adaptation and mitigation.

(European Commission, 2014)

SUDs, LID and NWRMs include a variety of solutions (green roofs, grassed swales, constructed wetlands, detention and retention ponds, etc.) that need to be appraised and evaluated on a case-by-case basis depending on the conditions where they have to be applied, both for new developments and for the retrofitting of existing ones. Both the SUSDRAIN platform and the European Commission's NWRM platform (nwrm.eu) aim at providing a clearinghouse of examples by collecting case studies providing evidence for the cost-effectiveness of these solutions on a continuous basis.

In order to mitigate the impact of urban development, these solutions should in principle aim at restoring the 'natural flow regime' of streams (Poff *et al.*, 1997), which requires an in-depth understanding of the hydrological behaviour of the catchments before land development, as well as hydrologically considerate design.

## Impacts on floods and their mitigation

There is evidence that soil sealing and urban development not only increase annual runoff volumes, but also flood hazards (e.g. Pitt, 2008), especially when acting together with other mechanisms. Du *et al.* (2015), for instance, show evidence of increased hazards from combined urban soil sealing and accelerated erosion produced by the displacement of agriculture. Flood hazards increase particularly in smaller catchments, where urban development tends to represent larger shares of the contributing area. Flood peaks due to extreme rainfall tend to increase because of two synergistic mechanisms: on the one side, the capacity of natural soils to infiltrate and detain rainfall is reduced by soil sealing; on the other, urban areas equipped with traditional drainage systems deliver runoff more quickly and more efficiently, with reduced detention of rainfall on urban surfaces compared to undeveloped lots. The relative importance of the two mechanisms depends on local factors. For very high return period floods, often the initial soil moisture conditions at the start of the flood event are such that soil water storage and infiltration are already quite limited. However, an extreme rainfall event may also occur after soils have had sufficient time to drain antecedent precipitation, and their infiltration capacity is near optimal conditions. In engineering practice, extreme rainfall infiltration capacity of unsealed soils is often evaluated in the range 20–70 per cent of the event rainfall volume, depending on soil characteristics and initial conditions (e.g. ASCE, 1960). On the contrary, sealed soil is often assumed to deliver 80–100 per cent of the event volume.

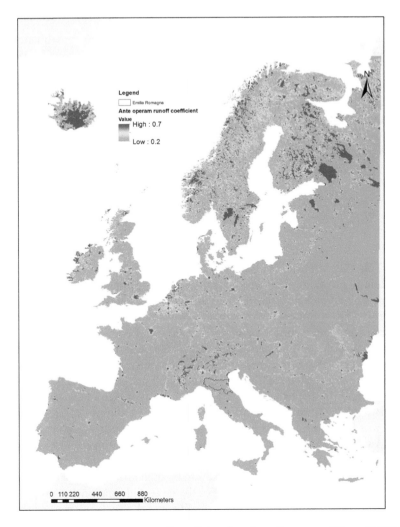

*Figure 8.2* Ante-operam runoff coefficient based on direct runoff from LISFLOOD model simulations (see text for details). The red polygon is the perimeter of Emilia Romagna

For what concerns the detention of runoff on the soil surface, typically non-sealed agricultural or natural land provides at least 50 m³/ha of volume given primarily by surface roughness and depressions. This amount is greatly reduced in urban land, e.g. by a factor of three to four (Pistocchi, 2001; Sofia *et al.*, 2014). Smaller volumes mean shorter residence time, i.e. faster concentration of runoff and consequently higher discharge peaks. Pistocchi *et al.* (2015) present an example of an area significantly affected by urban expansion in northern Italy, where soil sealing has particularly impacted the watersheds of the secondary and artificial drainage network of the plains in the region of Emilia Romagna (shown in Figure 8.2). The channels, suitable for the drainage of agricultural land as in pre-development conditions, require retrofitting in order to keep flood risks under control.

In order to avoid the need to retrofit the drainage networks in consequence of new urban developments, the local flood management plans have introduced provisions for the mitigation of flood impacts, known as the principle of hydraulic invariance of land use change (Pistocchi, 2001; Pistocchi and Zani, 2004): the hydrological impact of soil sealing should be offset by increasing the detention volume at the soil surface. The detention volume required to offset the alteration of flood peak discharges is not linearly related to the extent of urban development. For the urbanization of a share X of the catchment, the required volume is computed as (see Pistocchi *et al.*, 2015):

$$
W = W_N \left( \frac{\alpha(1-X) + \beta X}{\alpha} \right)^{1.92}
$$

where $W_N$ is the detention volume capacity of undeveloped land, respectively, while $\alpha$ and $\beta$ are the fraction of the rainfall volume during the flood event that contributes to runoff from undeveloped and developed land respectively. This volume should be made available in the catchments after urban transformation in order to offset the effects of soil sealing. We may have a first estimation of W by assuming that 90 per cent of rainfall on impervious surfaces contributes to runoff during a flood event in a small catchment ($\beta = 0.9$), and on unsealed soil this is reduced to 20 per cent ($\alpha = 0.2$), and by setting $W_N = 50$ m³/ha, as assumed by the Emilia Romagna regional flood management plans (Pistocchi, 2001). The detention volume W allows in principle to offset the increase in peak flood discharge, not the increase in total flood volume. Moreover, this volume is not effective to harvest water, but just to slow down the flow of water to the catchment outlet.

The detention volume computed above can be used as an indicator of urban development impact on peak discharges. For the sake of illustration, we propose hereafter a demonstrational calculation over Europe. A European map of the fraction of direct runoff (Burek, 2014, personal communication) prepared at 5 km resolution as part of the input parameters of the European scale hydrological model LISFLOOD (Burek *et al.*, 2013) was transformed into a map of runoff coefficients by assigning a value of 0.2 to 5 × 5 km² cells having no direct runoff contribution, a value of 0.7 to cells with 100 per cent direct runoff, and linearly distributed values in between. This map, $\Phi$, is shown in Figure 8.2 and may be considered a first approximation of present European conditions, 'ante operam'. As such, it may be used for the evaluation of impacts due to recent land take by urban expansion. For this purpose, we use the map of the changes in Urban Morphological Zones (UMZs) provided by the European Environment Agency (EEA: www.eea.europa.eu/data-and-maps/data/urban-morphological-zones-changes-2000, last accessed January 2014), a binary map of 100 m resolution indicating with 1 those grid cells turned from non-urban to urban between

2000 and 2006, and with 0 those unchanged. This map was aggregated to a 1 km resolution by assigning 1 km × 1 km cells the sum of the values of the 100 m resolution map. This sum corresponds to the percentage of the 1 km$^2$ cell turned to urban land use. The resulting map was used as a weight to a flow accumulation operation (e.g. Pistocchi, 2014, ch. 7) that yielded the contributing area upstream of each cell turned to urban land use between 2000 and 2006. The ratio of this weighted flow accumulation to the standard flow accumulation

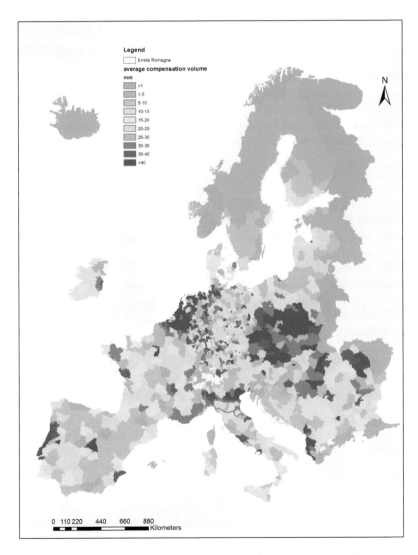

*Figure 8.3* Compensation volume (1 mm = 10 m$^3$/ha) estimated for the compensation of new urban areas built between 2000 and 2006 (see text for details). The red polygon is the perimeter of Emilia Romagna

gives the fraction of the catchment turned urban in the period, X. A map of the required offset volume (m³/ha) to offset soil sealing through 'hydraulic invariance' was estimated as:

$$W = 50\left(\frac{\left(\Phi 1 - X\right) + 0.9X}{\Phi}\right)^{1.92}$$

as we assume $W_N = 50$ m3/ha. If we average the volume values within Europe. If we average the computed values of W at the regional level within Europe, the resulting map (Figure 8.3) highlights where recent urban expansion may cause higher impacts on the flood peaks of the secondary and artificial drainage networks. The map reflects the distribution of new UMZs and does not take into account local conditions of precipitation and soil/land management other than reflected by the abovementioned direct runoff map. As such, it must be regarded as a purely illustrative example of the assessment of land take impacts on the local drainage network.

## Conclusions

We have provided an overview of the hydrological impacts of soil sealing and urban expansion. These affect the water cycle, hence ecology, as well as water availability and flood hazards in catchments. The effects of soil sealing have been identified as a significant threat to water bodies and should be appropriately addressed first of all by limiting impervious surfaces in a catchment and by avoiding their direct drainage to streams. Impervious surfaces that cannot be avoided should be accompanied by appropriate mitigation measures based on the paradigms of SUDs, LID and NWRMs, and (limited to the issue of flood peak discharges) the principle of hydraulic invariance.

## References

ASCE (American Society of Civil Engineers) (1960) *Design Manual for Storm Drainage.* New York.

Brezonik, P.L., Stadelmann, T.H. (2002) 'Analysis and predictive models of stormwater runoff volumes, loads, and pollutant concentrations from watersheds in the Twin Cities metropolitan area, Minnesota, USA', *Water Research* 36, 1743–1757.

Burek, P., van der Knijff, J., de Roo, A. (2013) '*LISFLOOD. Distributed Water Balance and Flood Simulation Model. Revised User Manual*', JRC Technical Reports – EUR 26162 EN.

Burns, D., Vitvar, T., McDonnell, J., Hassett, J., Duncan, J., Kendall, C. (2005) 'Effects of suburban development on runoff generation in the Croton River Basin, New York, USA', *Journal of Hydrology* 311, 266–281.

Dietz, M.E., Clausen, J.C. (2008) 'Stormwater runoff and export changes with development in a traditional and low impact subdivision', *Journal of Environmental Management* 87, 560–566.

Du, S., Shi, P., Van Rompaey, A., Wen, J. (2015) 'Quantifying the impact of impervious surface location on flood peak discharge in urban areas', *Natural Hazards* 76(3), 1457–1471. doi: 10.1007/s11069-014-1463-2.

European Commission (2012a) *Commission Staff Working Document. Guidelines on Best Practice to Limit, Mitigate or Compensate Soil Sealing*, SWD (2012) 101 final. European Commission, Brussels.

European Commission (2012b) *A Blueprint to Safeguard Europe's Water Resources.* Communication from the Commission COM(2012)673. http://eur-lex.europa.eu/legal-content/EN/TXT/?uri=CELEX:52012DC0673, accessed on 10 July 2015.

European Commission (2014) *EU Policy Document on Natural Water Retention Measures by the Drafting Team of the WFD CIS Working Group Programme of Measures (WG PoM)*, DG ENV, Technical Report 2014 – 082. http://ec.europa.eu/environment/water/adaptation/ecosystemstorage.htm, accessed on 10 July 2015.

Falkenmark, M. (1995) *Coping with Water Scarcity under Rapid Population Growth*, Conference of SADC Ministers, Pretoria, 23–24 November 1995.

Guo, Y., Quader, A. (2009) 'Derived flow–duration relationships for surface runoff dominated small urban streams', *Journal of Hydrological Engineering* 14(1), 42–52.

Hamel, P., Daly, E., Fletcher, T.D. (2013) 'Source-control stormwater management for mitigating the impacts of urbanisation on baseflow: a review', *Journal of Hydrology* 485, 201–211.

Hawley, R.J., Bledsoe, B.P. (2011) 'How do flow peaks and durations change in suburbanizing semi-arid watersheds? A southern California case study', *Journal of Hydrology* 405(1–2), 69–82. http://dx.doi.org/10.1016/j.jhydrol.2011.05.011, accessed on 14 June 2015.

Heaney, J., Huber, W., Nix, S.J. (1976) 'Storm water management model – Level I', Preliminary screening procedures, EPA-600/2-76-275, Cincinnati, October 1976.

Laureano, P. (2001) *The Water Atlas: Traditional Knowledge to Combat Desertification.* Bollati Boringhieri, Torino.

Mejía, A., Daly, E., Rossel, F., Jovanovic, T., Gironás, J. (2014) 'A stochastic model of streamflow for urbanized basins', *Water Resources Research* 50(3), 1984–2001.

Mumford, L. (1961) *The City in History: Its Origins, Its Transformations, and Its Prospects.* Harcourt, Brace & World, New York.

Pistocchi, A. (2001) 'La valutazione idrologica dei piani urbanistici: Un metodo semplificato per l'invarianza idraulica dei piani regolatori generali', *Ingegneria Ambientale* 30(7/8), 407–413 (in Italian).

Pistocchi, A. (2014) *GIS Based Chemical Fate Modeling: Principles and Applications.* Wiley, Hoboken, NJ.

Pistocchi, A., Zani, O. (2004) 'L'invarianza idraulica delle trasformazione urbanistiche: il metodo dell'Autorità dei bacini regionali romagnoli', *Atti XXIX Convegno di Idraulica e Costruzioni Idrauliche*, Trento, vol. 3, 107–114 (in Italian).

Pistocchi, A., Calzolari, C., Malucelli, F., Ungaro, F. (2015) 'Soil sealing and flood risks in the plains of Emilia-Romagna, Italy', *Journal of Hydrology: Regional Studies* 4, 398–409. doi: 10.1016/j.ejrh.2015.06.021.

Pitt, M. (2008) *The Pitt Review: Learning Lessons from the 2007 Floods.* Cabinet Office, London. webarchive.nationalarchives.gov.uk/20080906001345/cabinetoffice.gov.uk/thepittreview.aspx, accessed on 17 November 2016.

Poff, N.L., Allan, J.D., Bain, M.B., Karr, J.R., Prestegaard, K.L., Richter, B.D., Sparks, R.E., Stromberg, J.C. (1997) 'The natural flow regime: a paradigm for river conservation and restoration', *BioScience* 47, 769–784.

Rockström, J., Falkenmark, M., Folke, C., Lannerstad, M., Barron, J., Enfors, E., Gordon, L., Heinke, J., Hoff, H., Pahl-Wostl, C. (2014) *Water Resilience for Human Prosperity*. Cambridge University Press, Cambridge.

Schoonover, J.E., Lockaby, B.G., Helms, B.S. (2006) 'Impacts of land cover on stream hydrology in the West Georgia Piedmont, USA', *Journal of Environmental Quality* 35(6), 2123–2131.

Sofia, G., Prosdocimi, M., Dalla Fontana, G., Tarolli, P. (2014) 'Modification of artificial drainage networks during the past half-century: evidence and effects in a reclamation area in the Veneto floodplain (Italy)', *Anthropocene* 6, 48–62. http:// dx.doi.org/10.1016/j.ancene.2014.06.005, accessed on 17 November 2016.

USEPA (Environmental Protection Agency) (2010) *Causal Analysis/Diagnosis Decision Information System (CADDIS)*, Office of Research and Development, Washington, DC. www.epa.gov/caddis, last updated 23 September 2010.

Walsh, C.J., Fletcher, T.D., Ladson, A.R. (2005a) 'Stream restoration in urban catchments through redesigning stormwater systems: looking to the catchment to save the stream', *Journal of the North American Benthological Society* 24, 690–705.

Walsh, C.J., Roy, A.H., Feminella, J.W., Cottingham, P.D., Groffman, P.M., Morgan, R.P. II (2005b) 'The urban stream syndrome: current knowledge and the search for a cure', *Journal of the North American Benthological Society* 24(3), 706–723.

Walsh, C.J., Fletcher, T.D., Burns, M.J. (2012) 'Urban stormwater runoff: a new class of environmental flow problem', *PLoS ONE* 7(9): e45814. doi:10.1371/journal. pone.0045814.

Yang, J., Entekhabi, D., Castelli, F., Chua, L. (2014) 'Hydrologic response of a tropical watershed to urbanization', *Journal of Hydrology* 517, 538–546.

# 9 Impact of land take and soil sealing on biodiversity

*Geertrui Louwagie, Mirko Gregor, Manuel Löhnertz, Ece Aksoy, Christoph Schröder and Erika Orlitova*

## Introduction

Biodiversity can cover a lot of different things and refers in its simplest form to the diversity of habitats, species and genes. Thus, 'the concept covers not only overall richness of species present in a particular area but also the diversity of genotypes, functional groups, communities, habitats and ecosystems there' (Haines-Young, 2009).

Soils host many soil-dwelling species – from the very small (like fungi and bacteria) to the very large (like earthworms and moles) – and provide valued habitats for them. Soil organisms break down organic matter and transform nutrients, making them available to other plants and organisms. Soil biodiversity also controls the degradation and release of many pollutants. Soil is thus at the heart of many environmental processes and the benefits humans derive from ecosystems in general. Many of our valued habitats and rare plant species are dependent on very specific soil conditions. Soil thus has an important role in sustaining biodiversity.

Land take by the expansion of artificial areas for urban settlements and related infrastructure has been the main cause of net land cover change in Europe since at least the 1990s (when land cover monitoring in Europe started).[1] Land take differs from increase in soil sealing or imperviousness, as not all of the area mapped as artificial may actually be covered with impervious material. Nevertheless, land take is often used as a proxy for soil sealing.

In this chapter we focus on the effects of land take and soil sealing on the different aspects of biodiversity. In describing the effects, we distinguish between two types: micro- to meso-scale effects happen at or close to the concentration of soil sealing, whereas macro-scale effects occur at a distance from the soil sealing concentration. Assuming that the biggest concentration of sealing occurs in core urban areas, micro- to meso-scale effects occur in the core urban space, whereas macro-scale effects extend into the peri-urban and rural space.

## Impact

### Micro- to meso-scale effects

Sealing interrupts the contact between pedosphere and atmosphere and thus changes the gas, water and material (including nutrients) fluxes (Burghardt *et al.*, 2004), and thus directly influences biogeochemical cycling in soils.

Experiments from Beijing (China) have shown that an impervious cover (in comparison to forest and bare land) significantly affects microbial biomass, enzyme activity and nitrogen transformation processes (mineralisation and nitrification) (Zhao *et al.*, 2012). Sealing particularly affects soil characteristics in the upper 10 cm of the soil, with decreasing effects, including on soil microbial activity, with depth.

Ultimately, owing to the hampered exchanges between soil fauna and external inputs, sealing can lead to severely depleted soil biodiversity, and a slow death of most soil organisms (European Commission, 2010). Soil biota can initially survive on the moisture and organic matter that was present in the soil before sealing, until these resources are exhausted. Then, the bacteria-dominated microorganisms may enter an inactive state (dormancy) or simply die off, while small (dominantly micro-arthropods) and larger (e.g. earth-worms) invertebrates may either move away from the sealed area or, when sealing covers vast areas, die off.

Soil sealing is also a key contributor to the urban heat island effect, which is generated by the differences in heat storage of construction materials along urban–rural gradients. Soil sealing in particular influences the albedo (or reflection coefficient) of surfaces, which may lead to increased temperatures above sealed areas (micro-scale) and within urbanised areas at large (meso-scale) (Burghardt *et al.*, 2004). The urban heat island effect and the related increase in air and soil temperatures have exerted an evolutionary pressure on soil organisms (Scalenghe and Ajmone Marsan, 2009). The effect can be observed in soil fungi, organisms that cannot regulate their own temperature.

Soil sealing can eliminate a natural habitat for plants. Temperature increases owing to the heat island effect have also caused changes in plant phenology, with spring-blooming plants blooming earlier in the city than in the surrounding habitats in a variety of ecosystems in North America, Europe and China (Neil and Wu, 2006). Over time, such change (along with other factors, such as climate change or introduced species) can lead to differences in species composition.

### Macro-scale effects

#### Freshwater and related terrestrial habitats

The soil sealing pattern heavily influences water infiltration and preferential flows: rainwater cannot directly infiltrate an impervious area; thus, it either drains beside the sealed area (e.g. along a small road) or is discharged and enters the sewer system (e.g. around big buildings or parking lots). Sediment and dust particles follow the same pathway. Soil sealing also decreases plant and soil evapotranspiration. If water infiltration is not facilitated at the edges of completely sealed, larger areas, groundwater recharge may be decreased; whereas when facilitated, surface water may break through to the groundwater, with an increased risk of groundwater pollution as a consequence (Burghardt *et al.*, 2004). This so-called barrier effect at the edges of impervious areas may equally lead to erosion in adjacent areas.

In connection to these hydrologic changes and interruptions, Arnold and Gibbons (1996) reviewed literature on the effects of impervious surface coverage on stream health and connected terrestrial habitats. Research consistently shows a strong correlation between the imperviousness of a river basin and the health of the receiving stream. Stream health is among others defined by pollutant loads, habitat quality and aquatic species diversity and abundance. Stream degradation reportedly already occurs at relatively low levels of imperviousness (from 10 per cent onwards) in the watershed. The same threshold is deemed crucial for maintaining wetlands in good ecological condition. Referring to the earlier-mentioned increased erosion risk, Arnold and Gibbons (1996) also point at substantial losses of both streamside or riparian (where erosion occurs) and in-stream (where sedimentation happens) habitats.

*Ecosystem loss and fragmentation*

Development of built-up areas and transport infrastructure also leads to landscape and ecosystem loss and fragmentation, worldwide considered the main threat to biodiversity conservation.

When ecosystems are completely lost, the effects on biodiversity can be assessed by mapping the different ecosystem types affected, including their rarity (Geneletti, 2003). Rarity is a measure of how frequently an ecosystem type is found within a given area; protection of rare ecosystems is often considered the single most important function of biodiversity conservation.

Fragmentation has a number of ecological effects, such as the decline and loss of wildlife populations, an increasing endangerment of species, changed water regimes and a change in recreational quality of landscapes (EEA-FOEN, 2011) (Table 9.1). Habitat fragmentation can clearly cause remaining habitats to become too small for some organisms to persist, or too fragmented so that remaining patches may be too far apart for organisms to move between.

The fragmentation effects of land take and soil sealing have particular bearing on the liveability of cities. When cities are dense (i.e. have an increasing amount of built-up area), they are to a large extent dependent on ecosystem services from outside; the size and variety of urban green areas as well as their connection with the ecosystems surrounding the city will largely determine the biodiversity potential (Bolund and Hunhammar, 1999). However, the transport network and patches of commercial and industrial service areas around cities often lower the stabilising effect of outer core areas due to their barrier effect.

*Table 9.1* Effects of landscape fragmentation on flora and fauna

Death of animals caused by road mortality (partially due to attraction of animals by roads or railways: 'trap effect'

Higher levels of disturbance and stress, loss of refuges

Reduction or loss of habitat; sometimes creation of new habitat

Modifications of food availability and diet composition (e.g. reduced food availability for bats due to cold air build-ups along road embankments at night)

Barrier effect, filter effect to animal movement (reduced connectivity)

Disruption of seasonal migration pathways, impediment of dispersal, restriction of recolonization

Subdivision and isolation of habitats and resources, breaking up of populations

Disruption of metapopulation dynamics, genetic isolation, inbreeding effects and increased genetic drift, interruption of the processes of evolutionary development

Reduction of habitat below required minimal areas, loss of species, reduction of biodiversity

Increased intrusion and distribution of invasive species, pathways facilitating infection with diseases

Reduced effectiveness of natural predators of pests in agriculture and forestry (i.e. biological control of pest more difficult)

Source: EEA-FOEN (2011).

---

**Box 9.1   Mapping the impact of land take on soil biodiversity in Europe**

Soil sealing has already been identified as one of the major threats to soils in the 2002 European Commission's Communication 'Towards a Thematic Strategy on Soil Protection' (COM(2002) 179 final[2]). Through the disruption of cycles, soil sealing contributes to the loss of valuable soil functions, such as hosting the biodiversity pool, and thus to the loss of soil-based ecosystem services.

Pan-European maps of a number of soil functions[3] have recently become available and were used to analyse the impact of land cover changes on the potential of soils to provide those soil functions. In this specific case the impact of land take, i.e. the expansion of areas with artificial cover over areas that previously had (semi-)natural cover, on the capacity of soils to act as biodiversity pool was assessed.

Soil biodiversity throughout Europe was estimated by using critical thresholds for specific indicators that may regulate and affect the conditions of soils for biodiversity: physical and chemical soil parameters, climate, soil biomass production potential and land use/land cover. In general, the map and the underpinning model rather express the quantity (abundance) than the diversity (species richness) of soil organisms. However, as those two aspects (abundance and diversity) are often positively related, locations with a high abundance of soil organisms are expected to possess a higher species richness as well.

Figure 9.1 shows that there exist areas in the United Kingdom, Ireland and parts of central Europe (e.g. France or the Netherlands) where soils have a high potential to host the biodiversity pool. Those high levels are mainly caused by the relatively high weight that is given to the land use/ land cover parameter. In addition, grassland soils with a high biomass

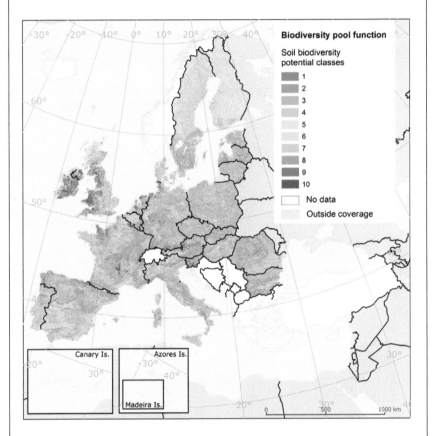

*Figure 9.1* Capacity of soils to serve as a biodiversity pool (source: ETC ULS based on texture (European Soil Database version 2.0); topsoil pH water, topsoil organic matter, total mean annual precipitation, mean annual temperature (EFSA Spatial Data version 1.1); evapotranspiration (JRC MARS); soil biomass productivity potential (JRC SoilProd model) and Corine Land Cover (version 16) data sets)

Note: the classes represent the biodiversity potential of soils, from 1 (low potential) to 10 (high potential).

*(continued)*

*(continued)*

production potential generally appear with higher soil biodiversity values than soils with high production potential under other land uses.

The impact of land take (between 2000 and 2006) is expressed per NUTS 3 area,[4] as the percentage of lost area with good soil biodiversity potential in relation to the area with good soil biodiversity potential in that NUTS 3 area (Figure 9.2). Several clusters of NUTS 3 regions with high impacts (in relative terms) can be detected: the Netherlands (along with border regions in Germany), eastern Ireland, central UK, the coastal Pays de la Loire (France), northern Portugal and the metropolitan area of Lisbon, northern Spanish coastal regions and central Spain, and the Budapest region.

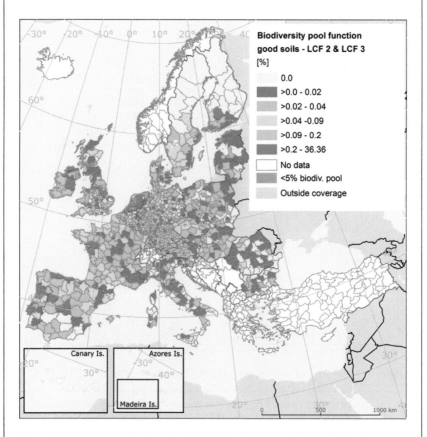

*Figure 9.2* Percentage decline (per NUTS 3 area) of land with good soil biodiversity potential due to urban residential sprawl (LCF2), and sprawl of economic sites (commercial, industrial) and infrastructure (LCF3) between 2000 and 2006 (source: ETC ULS based on soil biodiversity potential (ETC ULS, Figure 9.1) and Corine Land Cover (version 17) data sets)

In some regions these high percentage values correspond to big areas with good soil biodiversity potential: in eastern Ireland, three regions account for nearly 6,000 ha. The large expansion of commercial sites and infrastructure in the metropolitan area of Dublin (EEA, 2006) and surrounding regions particularly affected grassland areas with good soil biodiversity potential. Land take between 2000 and 2006 had a major impact in nearly all Dutch NUTS 3 regions, affecting the whole territory. The region of Utrecht (NL310) stands out with more than 1,200 ha affected by land take. Also here, large shares of grasslands can be found in the regions with the highest soil biodiversity potential. As a final example, the construction of EU-funded infrastructure (mainly motorways) in the northern Spanish regions of Asturias (ES130) and Cantabria (ES120) affected a total of 1,974 ha of soils with good potential for soil biodiversity.

There are also regions with high relative impact where the total affected area is relatively small (< 200 ha, e.g. Portuguese and central Spanish regions). These regions are also known as hotspots of land take for the period 2000–2006 (EEA Land take indicator[5]), regardless of their high potential to host soil biodiversity.

## Responses to soil sealing

### A role for planning

Landscape and habitat fragmentation can be seen as 'the negative' or opposite of connectivity. 'Green infrastructure' (GI) is a concept that is closely related to connectivity in an urban context, as it 'can be considered to comprise of all natural, semi-natural and artificial networks of multifunctional ecological systems within, around and between urban areas, at all spatial scales', emphasising the importance of quality, quantity, multi-functionality and interconnectedness of urban green spaces (Tzoulas *et al.*, 2007). The GI concept originates from spatial planning practice and highlights the role of green space in urban systems (Sandström, 2002).

Land use planning is indeed recognised as an essential instrument to curtail the negative effects of land take and soil sealing on the capacity of soil to sustain biodiversity. Local planning authorities may take account of the biodiversity effects that development projects may engender under the legislation for environmental impact assessment (as for example the Environmental Impact Assessment Directive in the European Union[6]).

In peri-urban areas (i.e. the transition zone from the core urban to the surrounding rural area), the location of new developments – implying land take and possibly soil sealing – in connection to protected nature areas is particularly interesting. In some countries or regions the planning system and designation of sites of high conservation interest may be sufficiently strong to constrain development on land with such value. In Scotland conservation status is adequate to protect land with valuable and/or rare habitats and sites of high biodiversity against development, and most of the extensive areas of valued and/or rare habitats in Scotland are not adjacent to potential development sites either (Dobbie *et al.*, 2011). However, conservation status is not always enough to prevent land take (Box 9.2). Also in Scotland's 2011 state-of-the-soil report it is recognised that conflicts between development and conservation may arise in specific cases, such as golf course developments (often including considerable built elements) or land-based renewables (e.g. hydro schemes and windfarms, often requiring significant road infrastructure).

---

**Box 9.2   Spatial patterns of land take in relation to protected areas in the European Union (EU)**

Natura 2000 is an EU-wide network of protected areas designated under the Habitats Directive[7] ('Special Areas of Conservation') and the Birds Directive[8] ('Special Protection Areas'), created with the aim of ensuring the conservation of Europe's most valuable and threatened habitats and species. The Birds and Habitats Directives restrict land use changes in Natura 2000 areas and limit the range of activities that can take place in

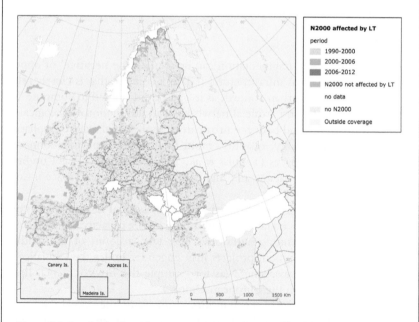

*Figure 9.3* Land take in and near nature areas protected by Natura 2000 status (source: ETC ULS based on Natura 2000 (version 2014, May 2015) and Corine Land Cover (version 18.3) data sets)

Legend: N2000 = Natura 2000; LT = land take.

Note: while the analysis only focuses on terrestrial sites, the map presents both terrestrial and marine Natura 2000 sites.

these areas. The Habitats Directive also foresees the implementation of compensating measures in case potential implications of a development are assessed as negative; nevertheless, even if evaluated negatively, projects or plans can still be carried out for 'imperative reasons of overriding public interest'.

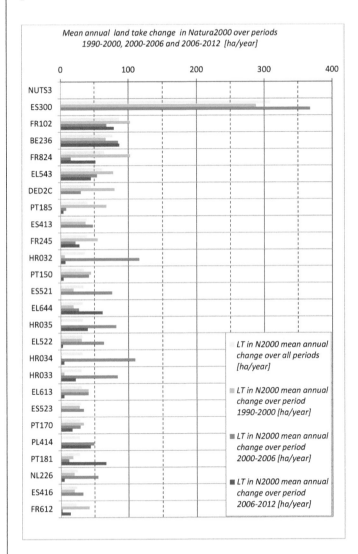

*Figure 9.4* Land take in and near Natura 2000 areas for selected NUTS 3 areas (mean annual change in hectares per year) (source: ETC ULS based on Natura 2000 (version 2014, May 2015), Corine Land Cover (version 18.3) and NUTS (2013 scale 1:1 million) data sets)

Legend: N2000 = Natura 2000; LT = land take.

*(continued)*

*(continued)*

Consisting of over 26,000 sites, the Natura 2000 network covers approximately 18.4 per cent of the EU territory,[9] and constitutes the largest protected area system worldwide (Figure 9.3). With the aim of protecting wild fauna, flora and habitats, and maintaining ecosystem services, these directives highly restrict land use changes and place certain limits on the range of activities that can take place in these areas.[10] Natura 2000 sites overlap with many nationally protected areas. However, close to half of them do not have a national designation, and thus the network provides an important expansion of protected areas. In total, about 25 per cent of land in the EU-27 is protected either by Natura 2000 sites or nationally designated areas.[11] Species listed in Annex I of the Birds Directive have been evaluated as having benefited from the nature legislation (Sanderson *et al.*, 2015).

Based on their protection status, it is expected that land take is greatly reduced or halted, potentially even reversed, in and close to Natura 2000 areas. However, this is not always the case (Figure 9.3), as also illustrated for selected NUTS 3 areas (Figures 9.4 and 9.5).

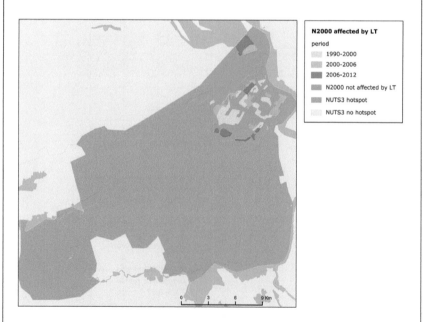

*Figure 9.5* Spatial pattern of land take in and near a Natura 2000 area in a NUTS 3 region defined as a hotspot (BE236 – Arrondissement of Sint-Niklaas) (source: ETC ULS based on Natura 2000 (version 2014, May 2015), Corine Land Cover (version 18.3) and NUTS (2013 scale 1:1 million) data sets)

Legend: N2000 = Natura 2000; LT = land take.

Note: hotspots are based on the criterion of highest mean annual increase of land take (per NUTS 3 area).

## Notes

1 EEA land take indicator: www.eea.europa.eu/data-and-maps/indicators/land-take-2/assessment-2, accessed 11 November 2016.
2 http://eur-lex.europa.eu/legal-content/EN/TXT/PDF/?uri=CELEX:52002DC0179&rid=1, accessed 11 November 2016.
3 As described in the European Commission's 'Thematic Strategy for Soil Protection' (COM(2006)231 final): http://eur-lex.europa.eu/legal-content/EN/TXT/PDF/?uri=CELEX:52006DC0231&from=EN, accessed 11 November 2016.
4 NUTS: nomenclature for territorial units for statistics in the EU – for an overview see: http://ec.europa.eu/eurostat/web/nuts/overview, accessed 11 November 2016.
5 www.eea.europa.eu/data-and-maps/indicators/land-take-2/assessment-2, accessed 11 November 2016.
6 http://ec.europa.eu/environment/eia/review.htm, accessed 11 November 2016.
7 Council Directive 92/43/EEC of 21 May 1992 on the conservation of natural habitats and of wild fauna and flora.
8 Directive 2009/147/EC of the European Parliament and of the Council of 30 November 2009 on the conservation of wild birds.
9 Natura 2000 Barometer December 2013.
10 Collectively, ecosystem services are estimated to be worth EUR 200 to EUR 300 billion a year, significantly more than the annual cost of some EUR 6 billion to manage the network (http://europa.eu/pol/env/flipbook/en/files/environment.pdf, accessed 11 November 2016).
11 www.oee.hu/upload/html/2014-02/YPEF_Educational_material_2014.pdf, accessed 11 November 2016.

## References

Arnold, Chester L. and Gibbons, C. James, 1996. Impervious surface coverage: the emergence of a key environmental indicator. *Journal of the American Planning Association*, 62 (2), pp. 243–258.
Bolund, P. and Hunhammar, S., 1999. Ecosystem services in urban areas. *Ecological Economics*, 29, pp. 293–301.
Burghardt, W., Banko, G., Hoeke, S., Hursthouse, A., de L'Escaille, T., Ledin, S., Ajmone Marsan, F., Sauer, D., Stahr, K., Amann, E., Quast, J., Nerger, M., Schneider, J. and Kuehn, K., 2004. Taskgroup 5: Sealing soils, soils in urban areas, land use and land use planning. In: Van-Camp, L., Bujarrabal, B., Gentile, A-R., Jones, R.J.A., Montanarella, L., Olazabal, C. and Selvaradjou, S-K. eds. *Reports of the Technical Working Groups Established under the Thematic Strategy for Soil Protection*, Volume VI: *Research, Sealing and Cross-cutting Issues*. EUR 21319 EN/6. Office for Official Publications of the European Communities, Luxembourg.
Dobbie, K.E., Bruneau, P.M.C and Towers, W. eds, 2011. *The State of Scotland's Soil*. Natural Scotland. [ONLINE] Available at: www.sepa.org.uk/media/138741/state-of-soil-report-final.pdf. [Accessed 16 October 2015].
EEA, 2006. *Urban Sprawl in Europe: The Ignored Challenge*. EEA Report No. 10/2006, European Environment Agency. [ONLINE] Available at: www.eea.europa.eu/publications/eea_report_2006_10. [Accessed 16 October 2015].
EEA-FOEN, 2011. *Landscape Fragmentation in Europe*. Joint EEA-FOEN report, EEA Report No 2/2011, European Environment Agency. [ONLINE] Available at: www.eea.europa.eu/publications/landscape-fragmentation-in-europe. [Accessed 16 October 2015].
European Commission, 2010. *Soil Biodiversity: Functions, Threats and Tools for Policy Makers*. Technical Report 2010 – 049. [ONLINE] Available at: http://ec.europa.eu/environment/archives/soil/pdf/biodiversity_report.pdf. [Accessed 16 October 2015].

Geneletti, D., 2003. Biodiversity impact assessment of roads: an approach based on ecosystem rarity. *Environmental Impact Assessment Review*, 23, pp. 343–365.

Haines-Young, R., 2009. Review: land use and biodiversity relationships. *Land Use Policy*, 26S, pp. S178–S186.

Neil, K. and Wu, J., 2006. Effects of urbanization on plant flowering phenology: a review. *Urban Ecosystems*, 9, pp. 243–257.

Sanderson, F.J., Pople, R.G., Ieronymidou, C., Burfield, I.J., Gregory, R.D., Willis, S.G., Howard, C., Stephens, P.A., Beresford, A.E. and Donald, P.F., 2015. Assessing the performance of EU nature legislation in protecting target bird species in an era of climate change. *Conservation Letters*, 9 (3), pp. 172–180.

Sandström, U.F., 2002. Green Infrastructure planning in urban Sweden. *Planning Practice and Research*, 17 (4), pp. 373–385.

Scalenghe, R. and Ajmone Marsan, F., 2009. The anthropogenic sealing of soils in urban areas. *Review, Landscape and Urban Planning*, 90, pp. 1–10.

Tzoulas, K., Korpela, K., Venn, S., Yli-Pelkonen, V., Kaźmierczak, A., Niemela, J. and James, P., 2007. Promoting ecosystem and human health in urban areas using Green Infrastructure: a literature review. *Landscape and Urban Planning*, 81, pp. 167–178.

Zhao, D., Li, F., Wang, R., Yang, Q. and Ni, H., 2012. Effect of soil sealing on the microbial biomass, N transformation and related enzyme activities at various depths of soils in urban area of Beijing, China. *Journal of Soils and Sediments*, 12, pp. 519–530.

# 10 Impacts of land take and soil sealing on soil carbon

*Klaus Lorenz and Rattan Lal*

## Introduction

Land use and land cover change (LULCC) by urbanization and, in particular, the expansion of urban areas is increasingly affecting the terrestrial carbon (C) stock as the global urban land cover is projected to increase by an area the size of South Africa until 2030 (Seto *et al.*, 2012). Currently, larger than previously thought areas in Europe are already covered by settlement structures including cities, villages, and groups of houses along rivers, roads, and rail tracks or spread into the arable countryside (Figure 10.1). The processes of land take or land consumption interconnected with urban expansion can be defined as an increase of settlement areas over time (European Commission Staff Working Document, 2012). Land take includes the development of scattered settlements in rural areas, the expansion of urban areas around an urban nucleus (including urban sprawl), and the conversion of land within an urban area (densification). By conversion of open into built-up areas, some part of the land take will result in soil sealing by buildings, roads, and parking lots because gardens, urban parks and other green spaces are not covered by an impervious surface (European Commission Staff Working Document, 2012). Otherwise, land take can also be defined as the increase of artificial surfaces (e.g., housing areas; green urban areas; industrial, commercial and transport units; road and rail networks) over time (European Commission, DG Environment, 2011). Soil sealing means the permanent covering of an area of land and its soil by completely or partly impermeable artificial material (e.g., asphalt, concrete), for example, through buildings and roads (European Commission Staff Working Document, 2012). Soil sealing causes the loss of soil and some of its biological functions including C sequestration and loss of biodiversity, either directly or indirectly, due to fragmentation of the landscape (European Commission, DG Environment, 2011).

Both land take and soil sealing seem to be inevitable as most social and economic activities depend on the construction, maintenance and existence of sealed areas and developed land. New housing, business locations and road infrastructure, in particular, are mostly realized on undeveloped land outside or at the border of existing settlements, usually resulting in new

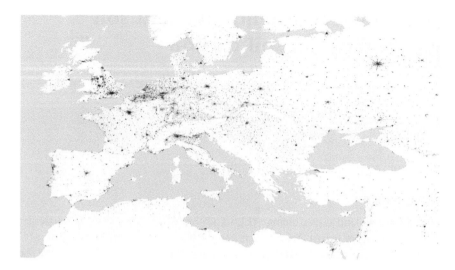

*Figure 10.1* Map of urban structures in Europe in unparalleled precision based on
data acquired by radar satellites (source: DLR, www.dlr.de/eoc/en/
desktopdefault.aspx/tabid-9630/#gallery/24123, accessed 24 March 2016)

soil sealing (European Commission, DG Environment, 2011). The global
extent of soil sealing is further increasing as globalization increasingly influences processes of land change. Land-use changes, in particular, are strongly
influenced by globalized flows of commodities, information, capital and
people, and are increasingly driven by factors in distant markets, often associated with the growing urban consumer class in emerging markets (Lambin
and Meyfroidt, 2011).

Land take and soil sealing for settlements can directly alter biomass C,
and soil C that is comprised of soil inorganic carbon (SIC) and soil organic
carbon (SOC) (Figure 10.2). Land use changes for urbanization occur commonly at the expense of agricultural land as many cities and settlements were
founded in agricultural areas on coastal plains and in river valleys (Hooke
*et al.*, 2012). However, there is little direct quantitative evidence of how,
for example, sealing affects soil C storage (Scalenghe and Marsan, 2009).
Thus, data on urban soil C are urgently needed for an integrated understanding of the processes of urbanization and the impacts of urban areas on C
flows (Romero-Lankao *et al.*, 2014). Human activities associated with land
take such as additions of natural and technogenic materials, and physical soil
disturbance by excavation, export and mixing, and soil sealing alter directly
the soil C balance (Lorenz and Lal, 2009). Indirectly, urban soil C may be
affected by changes in environmental conditions of urban compared to those
of pre-urban environment such as the heat island effect. In the following section are discussed some examples of the effects of land take and soil sealing on
soil C for urban areas (Table 10.1).

*Table 10.1* Maximum relative changes (%) in soil organic and inorganic carbon stocks (Mg C ha$^{-1}$) by land take compared to natural soils, and by soil sealing compared to unsealed soils

| Soil depth (cm) | Process | Soil organic carbon stock (Mg C ha$^{-1}$) | Change (%) | Soil inorganic carbon stock (Mg C ha$^{-1}$) | Change (%) |
|---|---|---|---|---|---|
| 0–10 | Land take | 11.0 → 4.5 | −59 | 0.18 → 0.62 | +344 |
|  |  | 9.7 → 24.5 | +253 |  |  |
|  | Soil sealing | 8.8 → 3.6 | −59 |  |  |
| 0–20 | Land take | 31.3 → 39.3 | +126 | 13.0 → 11.2 | −14 |
|  |  |  |  | 4.2 → 6.4 | +152 |
|  | Soil sealing | 45.2 → 23.5 | −52 |  |  |
| 0–100 | Land take | 172 → 59 | −66 | 0 → 79 |  |
|  |  | 320 → 650 | +203 | 0 → 348 |  |
|  | Soil sealing | 47.5 → 46.0 | −3 |  |  |
|  |  | 127.3 → 234.0 | +184 |  |  |

Sources: see text.

## Effects of land take on soil organic carbon stock

The effects of land take on the SOC stock depends on the balance between the initial SOC stock of the land converted for settlement area, and the C input into urban soils after land take derived directly from plant photosynthesis, organic amendments, and additions during construction of man-made soils (Figure 10.2; Lorenz and Lal, 2015). Thus, depending on the pre-urban soil replaced or disturbed, urban soils may have higher, similar or lower SOC stocks (Lorenz and Lal, 2009).

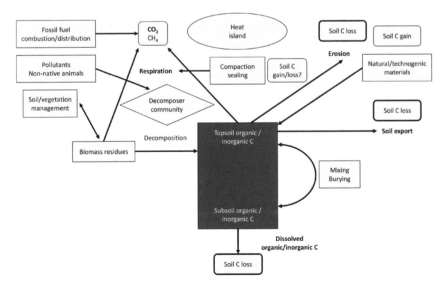

*Figure 10.2* Soil carbon losses (bold arrows) by land take and soil sealing processes in urban areas (source: modified from Lorenz and Lal, 2009)

Hao *et al.* (2013) modeled that the conversion of grassland to urban green land resulted in an increase in SOC stocks to 30 cm depth from 49.0 to 91.7 Mg C ha$^{-1}$ in Tianjin Binhai New Area, China. While the low grassland SOC stock was explained by high soil salinity resulting in low plant C inputs, continuous fertilizer and (human) manure applications at urban green land contributed directly to higher SOC stocks due to higher organic matter (OM) inputs and indirectly due to more productive plant growth (Hao *et al.*, 2013). In Kaifeng city, China, the normal sequence of urban soils was disturbed by construction activities whereas suburban soils were characterized by natural soil sequence (Sun *et al.*, 2010). Frequent fertilizing, watering and scarification enhanced plant growth in urban green spaces and this, together with OM inputs, contributed to SOC accumulation in urban soils. Specifically, urban soils had 2.53-fold more SOC than suburban soils at 0–10 cm depth (24.5 Mg C ha$^{-1}$ vs. 9.7 Mg C ha$^{-1}$). To 100 cm depth, urban soils had 1.56 times more SOC (99.7 Mg C ha$^{-1}$) than suburban soils (63.9 Mg C ha$^{-1}$). The intense human activities altered also the vertical distribution of SOC with 60 percent of the SOC stocks to 100 cm depth in industrial, recreational and traffic districts stored in 0–30 cm depth. In comparison, suburban soils stored only 40 percent of the 100 cm SOC stocks to 30 cm depth (Sun *et al.*, 2010). The SOC stocks to 20 cm depth of urban soils in Shanghai, China, were 1.26 times higher than those of soils in the countryside (39.3 Mg C ha$^{-1}$ vs. 31.3 Mg C ha$^{-1}$) (Xu *et al.*, 2012). However, SOC stocks at 160–180 cm depth were comparable among urban and countryside soils due to limited human influence (15.5 and 15.8 Mg C ha$^{-1}$, respectively). The SOC stocks to 30 cm depth across disturbed village land uses in China, i.e., constructed (mostly housing and roads) and disturbed (mostly unused land around buildings and roads) were 23.6 and 25.5 Mg C ha$^{-1}$, respectively (Jiao *et al.*, 2010). In contrast, ornamental and paddy (*Oryza sativa* L.) land uses had SOC stocks of 33.6 and 33.8 Mg C ha$^{-1}$, respectively, to 30 cm depth. Thus, human residence and not just agricultural practice was an important control on SOC stocks across village landscapes in China (Jiao *et al.*, 2010).

Greenspace soils in Chuncheon, Korea, stored less SOC to 60 cm depth at urban compared to natural lands (24.8 vs. 31.6 Mg C ha$^{-1}$) (Jo, 2002). The lower SOC storage for urban lands may have been the result of sparse tree plantings compared to natural lands, and less composting.

The SOC stocks in urban soils of Moscow and Serebryanye Prudy, Russia, were comparable with or exceeded the SOC stocks in the natural background soils (Vasenev *et al.*, 2013). Specifically, the SOC stock in the topsoil horizons and cultural layer of Moscow was 50 percent higher than that in the zonal soddy-podzolic soil (70–90 Mg C ha$^{-1}$). Further, the SOC stocks to 150 cm depth were 810 Mg C ha$^{-1}$ for Serebryanye Prudy and 610 Mg C ha$^{-1}$ for natural soils in the forest-steppe zone. Thus, the regional C budgets calculated without due account for urban soils may be underestimated (Vasenev *et al.*, 2013). For example, mean SOC stocks at 0–10 cm and 10–150 cm depths for non-urban, conventional and urban-specific maps in Moscow region were 30 and 147 Mg C ha$^{-1}$, 39 and 335 Mg C ha$^{-1}$, and 31 and 156 Mg C ha$^{-1}$, respectively (Vasenev *et al.*, 2014). Total SOC stocks based on the map including urban areas were considerably larger than for those excluding them, with 90 percent of all SOC stored at 10–150 cm depth.

Urban soils in Leicester, UK, stored on average significantly more SOC to 100 cm depth than their counterparts in regional arable lands, i.e., 202 vs. 143 Mg C ha$^{-1}$ (Edmondson *et al.*, 2012). Specifically, SOC storage to 21 cm depth in green spaces was 99 Mg C ha$^{-1}$ compared with 86 Mg C ha$^{-1}$ in pasture and 73 Mg C ha$^{-1}$ in arable lands. The addition of peat, composts, and mulches, and cultivation of trees and shrubs contributed to greater SOC stocks in urban greenspaces compared to those of agricultural soils in the region (Edmondson *et al.*, 2014). Very large SOC stocks were monitored in 55 urban soil profiles in the north east of England, in a region with a history of coal burning and heavy industry (Edmondson *et al.*, 2015). To 100 cm depth, urban SOC stocks ranged between 320 and 650 Mg C ha$^{-1}$. In comparison, soil under semi-natural vegetation and peat (to 100 cm depth) in the UK stored 320 and 520 Mg SOC ha$^{-1}$, respectively. Further, the urban soils also captured a large proportion of black carbon (BC) particulates emitted within urban areas. Thus, UK urban soils may be highly enriched in both BC and SOC (Edmondson *et al.*, 2015).

Pouyat *et al.* (2002) estimated SOC stocks to 100 cm depth for some cities in the northeast and mid-Atlantic region of the US. Residential lawn areas had nearly the same C density as northeastern forests, and higher density than mid-Atlantic forests (155 Mg C ha$^{-1}$ vs. 162 and 112 Mg C ha$^{-1}$). The high SOC stocks of residential lawns may be explained by high rates of nutrient inputs and water, resulting in increases in below-ground productivity, and also a much longer growing season than forests. On a preliminary basis, Pouyat *et al.* (2002) estimated that urban soils in the US stored on average 82 Mg SOC ha$^{-1}$ to 100 cm depth but the net changes in SOC stocks by land take and soil sealing compared to non-urban soils were uncertain.

The SOC stocks to 15 cm depth of well-maintained lawns in Fort Collins, CO, USA, were higher than those of native shortgrass steppe soils (4.8 vs. 2.9 Mg C ha$^{-1}$) (Kaye *et al.*, 2005). However, at 15–30 cm depth, SOC stocks were not significantly different with 2.2 Mg SOC ha$^{-1}$ stored in urban soils and 1.9 Mg SOC ha$^{-1}$ stored in native soils, respectively. Thus, urbanization of arid and semiarid ecosystems may increase SOC stocks only at shallow soil depths.

Estimates for urban SOC stocks to 100 cm depth for Boston, MA, and Syracuse, NY, USA, were 59 and 71 Mg C ha$^{-1}$ in comparison with 172 Mg C ha$^{-1}$ for both cities before urban development (Pouyat *et al.*, 2006). This large difference may result from the high SOC stocks in soils under native forest in the settlement area of both cities, and may have even be higher as wetlands were not considered. In contrast, cities in warmer and/or drier climates had slightly higher SOC stocks for post- than pre-urban measurements. For example, SOC stocks to 100 cm depth for Chicago, IL and Oakland, CA, USA, were estimated at 55 and 59 Mg C ha$^{-1}$, respectively (Pouyat *et al.*, 2006). In contrast, native soils in the settlement areas of Chicago and Oakland had SOC stocks to 100 cm depth of 52 and 57 Mg C ha$^{-1}$, respectively. Pouyat et al. (2006) concluded that there is potential for substantial losses of SOC by urban land take in temperate regions. However, in more arid climates, urban conversions have the potential to increase SOC storage. Nonetheless, only a limited number of data on urban SOC stocks are available for US cities to support this conclusion.

Mean surface (0–10 cm) SOC stocks in Phoenix, AZ, USA, were lower in desert and xeric yards (4.5–5.0 Mg C ha$^{-1}$) than in mesic yards or agroecosystems (7.5–11.0 Mg C ha$^{-1}$) (Kaye et al., 2008). At 10–30 cm depth, mean SOC stock was higher in agricultural soils (10.2 Mg C ha$^{-1}$) than in urban soils (5.3–7.3 Mg C ha$^{-1}$). Thus, SOC stocks to 30 cm depth were not always higher in urban compared to desert areas (Kaye et al., 2008).

The SOC stocks to 90 cm depth of natural forests near Apalachicola, FL, USA, were similar to those under urban land use (Nagy et al., 2014). Specifically, natural forests, urban forests and urban lawns stored (Mg C ha$^{-1}$) on average 73, 107, and 159, respectively, at 0–90 cm depth. However, soils under urban lawns had significantly higher SOC stocks at 0–7.5 cm and 7.5–30 cm depths compared to those under natural forests (14 and 35 vs. 10 and 14 Mg C ha$^{-1}$, respectively). Thus, increases in urban SOC storage are possible with continued urbanization if lawns are incorporated into built-up areas (Nagy et al., 2014).

Despite differences in water and N inputs, and vegetation shifts, SOC stocks to 10 cm in Boston, MA, USA, did not differ among urban and nonurban areas, and ranged from 34 to 44 Mg C ha$^{-1}$ (Raciti et al., 2012b). Otherwise, SOC stocks to 100 cm depth under residential lawns in Baltimore, MD, USA, were higher than those of forested reference sites (69.5 vs. 54.4 Mg C ha$^{-1}$) (Raciti et al., 2011). Lawns on former agricultural land had higher SOC stocks than those on former forest land. In particular, about 0.8 Mg C ha$^{-1}$ accumulated annually in lawn soils built on former agricultural land. Thus, lawn soils in residential areas on former agricultural lands in Baltimore have a large C sink capacity to sequester SOC to 100 cm depth (Raciti et al., 2011).

The SOC stocks to 15 cm depth in undisturbed urban forest soils adjacent to urban interstates in Louisville, KY, USA, were lower than those filled with local or imported material (46.0 Mg C ha$^{-1}$ vs. 54.7 and 79.0 Mg C ha$^{-1}$, respectively) (Trammell et al., 2011). However, SOC stocks were the lowest where A and B horizons had been removed and sub-soils exposed (38.1 Mg C ha$^{-1}$). Trammell et al. (2011) concluded that alterations of soil profiles during highway construction, influx of unknown material post-construction and vegetation management contributed to differences in SOC stocks between the undisturbed and disturbed forest soils adjacent to urban interstates.

Urban soils in Baltimore stored 71.1 Mg SOC ha$^{-1}$ to 100 cm depth or about 35 percent less than native soils but 24 percent more SOC than cultivated soils, respectively (Pouyat et al., 2009). Further, residential turf grass soils stored more SOC to 100 cm depth than rural forest soils (110 vs. 67 Mg C ha$^{-1}$). Otherwise, SOC stocks of two urban forest remnants were not different from those of turf grass soils. However, in turf grass soils only 50 percent of SOC to 100 cm depth accumulated in 0–20 cm depth compared to 70 percent accumulating to 20 cm depth in urban forest remnants. In conclusion, residential turf grass soils in Baltimore have the capacity to sequester large amounts of SOC (Pouyat *et al.*, 2009).

In conclusion, land take has highly variable effects on SOC stocks depending on the pre-urban conditions, and the soil and land-use management practices after conversion to settlement area (Table 10.1). Compared to natural soils, soils affected by land take may have 66 percent lower SOC stocks to 100 cm depth. Otherwise, SOC stocks in 0–10 cm depth may be 253 percent of those in natural soils. However, data for many cities and urban regions are missing (Lorenz and Lal, 2015).

## Sealing and soil organic carbon stock

The sealing of soils by impervious materials is detrimental to SOC storage as exchanges of C, energy, water, and gases between urban soils and the surrounding environment are restricted. The negative effects on SOC originate, in particular, from partial or total loss of soil and its SOC stock, and the partial or total loss of plant cover and its soil C input. However, little specific research is available that describes the effects of sealing on soil properties (Scalenghe and Marsan, 2009). Nevertheless, soils beneath sealed surfaces are part of the urban ecosystems and their properties must be studied (Kida and Kahawigashi, 2015). Some examples of studies on SOC stocks under sealed surfaces are given in the following section.

Artificial sealing of soils in Nanjing City, China, resulted in 40.9 percent and 45.5 percent lower water-soluble organic C and SOC contents, respectively, to 20 cm depth (Wei *et al.*, 2014). Further, the SOC density to 20 cm depth for impervious soils was lower than those for open soils (23.5 vs. 45.2 Mg C ha$^{-1}$). Thus, sealing of urban soils in Nanjing City resulted in a decrease of the SOC sink and degradation of soil fertility (Wei *et al.*, 2014). The SOC stocks to 100 cm depth under impervious cover in China's urban areas were estimated to be highly variable, ranging between 46.0 Mg C ha$^{-1}$ in Xinjiang, northwest China, and 234.0 Mg C ha$^{-1}$ in Heilongjiang, northeast China (Zhao *et al.*, 2013). In comparison, soils under green space to 100 cm depth contained 47.5 Mg C ha$^{-1}$ in Xinjiang and 127.3 Mg C ha$^{-1}$ in Heilongjiang. However, data were uncertain and no statistical analyses were performed. Thus, the effects of sealing on SOC stocks in China's urban areas are unclear (Zhao *et al.*, 2013). Sealed areas across village landscapes in China had SOC stocks of 22.9 Mg C ha$^{-1}$ to 30 cm depth (Jiao *et al.*, 2010). In contrast, soils under vegetation cover tended to have higher SOC stocks with values ranging between 24.4 and 28.8 Mg C ha$^{-1}$. However, differences were statistically significant only in the subtropical hilly region with 32.1 Mg SOC ha$^{-1}$ stored to 30 cm depth under annual land cover and 16.7 Mg SOC ha$^{-1}$ in sealed soils (Jiao *et al.*, 2010).

Kida and Kahawigashi (2015) studied the effects of asphalt sealing on mineral soils beneath the roads in Tokyo, Japan, but without comparing differences by statistical methods. The average TOC concentration in top mineral soils (34.1 g C kg$^{-1}$) was lower than those of surface soils of urban parks covered with turf grass (55.2 g C kg$^{-1}$) or tree plantations (52.9 g C kg$^{-1}$). Otherwise, the TOC in subsoils (~35 g C kg$^{-1}$) did not differ among three land uses (i.e., road, urban park, and tree plantations). However, TOC content in mineral soils beneath asphalt pavement was lower than that in an A-horizon of soil under natural ecosystems. The decline of TOC concentration reflected truncation of surface soil rich in OM by land deformation and pavement construction (Kida and Kahawigashi, 2015).

Piotrowska-Długosz and Charzyński (2015) studied the impact of the degree of sealing on soil properties in Toruń, Poland. Samples from partially sealed and impervious completely covered soils were collected from topmost horizons that survived the process of pavement construction from a depth of 15–25 or 10–20 cm, depending on the thickness of the technic hard rock. Unsealed, vegetated reference sites were sampled to the same depth. The SOC stocks calculated from organic C content and bulk density for 10 cm increments at partially sealed and impervious covered soils were 3.6 and 4.4 Mg C ha$^{-1}$, respectively, in comparison with 8.8 Mg C ha$^{-1}$ at the reference sites. Thus, soil sealing reduced SOC storage at the soil depths studied (Piotrowska-Długosz and Charzyński, 2015).

Soils beneath impervious surfaces in Leicester, UK, stored considerable amounts of SOC to 100 cm depth at a city-wide scale (Edmondson et al., 2012). However, SOC storage beneath impervious surfaces was limited by the depth of excavation for the capping surface. Specifically, SOC stocks in 40–100 cm depth beneath roads and other load bearing surfaces were 67 Mg C ha$^{-1}$ compared with 24 and 135 Mg C ha$^{-1}$ in 45–60 cm and 15–100 cm depth beneath pavements and footpaths, respectively. Soil processes probably remain active potentially accumulating SOC beneath the patches of impervious surface (Edmondson et al., 2012).

The average SOC stock to 15 cm depth under impervious cover for some neighborhoods in New York City, NY, USA, was 22.9 Mg C ha$^{-1}$, and it was significantly lower than those for urban open areas (56.7 Mg C ha$^{-1}$) (Raciti et al., 2012a). Thus, SOC stocks under impervious surfaces cannot be neglected in the assessment of urban SOC storage. However, the fate of SOC lost or depleted from areas now covered by impervious surfaces must also be understood (Raciti et al., 2012a).

In conclusion, in topsoil horizons soil sealing causes losses of SOC stocks relative to those of natural soils, i.e., up to 59 percent of the SOC stock of natural soil may be lost to 10 cm depth (Table 10.1). More uncertain are the effects on SOC at deeper depths as quantitative evidence is scanty. Nevertheless, the SOC stocks of sealed soils cannot be neglected in studies dealing with stock and flux of urban soils.

## Effects of land take on soil inorganic carbon

Soils affected by land take may contain SIC originating from carbonate-bearing soil parent material (Lehmann and Stahr, 2007). Especially in arid and semi-arid regions, the SIC stock may be up to 10 times larger than the SOC stock (Eswaran *et al.*, 2000). In addition, demolition waste (particularly cement and concrete) may contribute to urban SIC storage (Washbourne *et al.*, 2012). The coarse fraction (> 2 mm) of urban soils may substantially contain SIC in the form of demolition waste, limestone, and chalk fragments (Rawlins *et al.*, 2011). Thus, land take for settlements may affect the stocks of SIC aside those of SOC. However, the assessment of changes in SIC stocks by land take is hampered as national and regional databases seldom include the SIC data (Rawlins *et al.*, 2011). Thus, only a few examples are given in the following section.

The SIC stock to 20 cm depth for urban areas in the Jiangsu Province, China, was 1.52 times those of soils in countryside under agricultural ecosystems (6.4 vs. 4.2 Mg C ha$^{-1}$) (Xu and Liu, 2013). The increase in urban SIC stocks may be caused by soil deposition of groundwater containing carbonates and bicarbonates. In contrast, irrigation with alkaline groundwater may not result in increased SIC storage as was the case in Shanghai, China (Xu *et al.*, 2012). Specifically, SIC stocks at 0–20 cm and 160–180 cm depths were 11.2 and 12.4 Mg C ha$^{-1}$, respectively, for urban soils of Shanghai. In countryside soils, 13.0 and 15.8 Mg SIC ha$^{-1}$ were stored at 0–20 and 160–180 cm depths, respectively (Xu *et al.*, 2012).

Input of calcareous building rubble and limestone gravel together with carbonate-rich soil parent material may be the reasons for higher SIC stocks in some urban soils of Stuttgart, Germany, compared to those for adjacent agricultural and forest soils (Stahr *et al.*, 2003). Specifically, SIC stocks to 30 cm depth within Stuttgart ranged between 12 and 82 Mg C ha$^{-1}$. In comparison, rural forest soils contained no carbonates to 30 cm depth while rural agricultural soils stored 7 Mg SIC ha$^{-1}$ to the same depth. At 30–100 cm depth, urban SIC stocks ranged between 67 and 266 Mg C ha$^{-1}$. Again, rural forest soils were carbonate-free while rural agricultural soils contained 2 Mg SIC ha$^{-1}$ at 30–100 cm depth (Stahr *et al.*, 2003).

The SIC stocks of lawns in Fort Collins, CO, USA did not differ from those in soils under adjacent native ecosystems with values ranging between 0.19 and 1.26 Mg C ha$^{-1}$ in 0–15 cm, and between 1.99 and 12.14 Mg C ha$^{-1}$ in 15–30 cm depth (Kaye *et al.*, 2005). In contrast, irrigation water saturated with $CaCO_3$ contributed to higher SIC stocks at 0–10 cm and 10–30 cm depths in urban (i.e., 0.45–0.62 and 0.98–1.04 Mg C ha$^{-1}$, respectively) compared to desert soils in Phoenix, AZ, USA (Kaye *et al.*, 2008). Specifically, desert soils contained 0.18 and 0.64 Mg C ha$^{-1}$ at 0–10 cm and 10–30 cm depths, respectively.

In conclusion, land take has highly variable effects on SIC stocks (Table 10.1). However, strong increases in SIC stocks were observed in urban areas where calcareous demolition waste was buried in urban soils.

## Effects of soil sealing on soil inorganic carbon

Research data are scanty regarding the effects of soil sealing on urban SIC stocks compared to those of unsealed non-urban soils. The pH values and soil Ca contents are often but not always (e.g., Piotrowska-Długosz and Charzyński, 2015) higher in sealed urban soils compared with those in urban open soils (Morgenroth *et al.*, 2013; Wei *et al.*, 2014; Kida and Kahawigashi, 2015). The dissolution of calcareous materials in cement and concrete used for construction of soil sealing may elevate Ca concentration of urban soil, and increase soil pH due to a strong carbonate reaction (Burghardt, 1994). Thus, soil sealing may increase SIC stocks. For example, correlation between inorganic C and CaO content under asphalt in Tokyo indicated that inorganic C exists as $CaCO_3$ in pavement materials and top mineral soils (Kida and Kahawigashi, 2015). In conclusion, soils beneath sealed surfaces may potentially have higher SIC stocks than those in unsealed, non-urban soils, but data are rather scanty.

## Conclusions

Data on urban soil C are urgently needed for an enhanced and integrated understanding on the processes of land take and soil sealing on C flows. While land take has variable effects on SOC stocks, covering soil with impervious layers by sealing generally results in a loss of topsoil SOC stocks, probably because C-rich surface soil is removed during sealing construction and C inputs are altered subsequently. In contrast, SIC stocks may strongly increase by land take as this process is often associated with additions of calcareous materials. However, many more cities and urban areas must be studied for a reliable global assessment on the effects of land take and soil sealing on soil C.

## References

Burghardt, W. (1994) 'Soils in urban and industrial environments', *Journal of Plant Nutrition and Soil Science*. 157. pp. 205–214.

Edmondson, J. L., Davies, Z. G., Mchugh, N., Gaston, K. J. and Leake, J. R. (2012) 'Organic carbon hidden in urban ecosystems'. *Scientific Reports*. 2. 963. doi: 10.1038/srep00963.

Edmondson, J. L., Davies, Z. G., Mccormack, S. A., Gaston, K. J. and Leake, J. R. (2014) 'Land-cover effects on soil organic carbon stocks in a European city', *Science of the Total Environment*. 472. pp. 444–453.

Edmondson, J. L., Stott, I., Potter, J., Lopez-Capel, E., Manning, D. A. C., Gaston, K. J. and Leake, J. R. (2015) 'Black carbon contribution to organic carbon stocks in urban soil'. *Environmental Science & Technology*. 49. pp. 8339–8346.

Eswaran, H., Reich, P. F. and Kimble, J. M. (2000) 'Global carbon stocks'. In: *Global climate change and pedogenic carbonates*. LAL, R., KIMBLE, J. M., ESWARAN, H. and STEWART, B. A. (eds). Boca Raton, FL: CRC Press, pp. 15–25.

European Commission, DG Environment. (2011) *Overview of best practices for limiting soil sealing or mitigating its effects in EU-27*. Brussels: European Communities.

European Commission Staff Working Document. (2012) *Guidelines on best practice to limit, mitigate or compensate soil sealing*. Brussels: European Union.

Hao, C., Smith, J., Zhang, J., Mwng, W. and Li, H. (2013) 'Simulation of soil carbon changes due to land use change in urban areas in China'. *Frontiers of Environmental Science & Engineering.* 7 (2). pp. 255–266.

Hooke, R., Martín-Duque, J. F. and Pedraza, J. (2012) 'Land transformations by humans: a review'. *GSA Today.* 12. pp. 4–10.

Jiao, J.-G., Yang, L.-Z., Wu, J.-X., Wang, H.-Q., Li, H.-X. and Ellis, E. C. (2010) 'Land use and soil organic carbon in China's village landscapes'. *Pedosphere.* 20 (1). pp. 1–14.

Jo, H. K. (2002) 'Impacts of urban greenspace on offsetting carbon emissions from middle Korea'. *Journal of Environmental Management.* 64. pp. 115–26.

Kaye, J. P, Mcculley, R. L. and Burke, I. (2005) 'Carbon fluxes, nitrogen cycling, and soil microbial communities in adjacent urban, native and agricultural ecosystems'. *Global Change Biology.* 11. pp. 575–587.

Kaye, J. P., Majudmar, A., Gries, C., Buyantuyev, A., Grimm, N. B., Hope, D., Jenerette, G. D., Zhu, W. X. and Baker, L. (2008) 'Hierarchical Bayesian scaling of soil properties across urban, agricultural, and desert ecosystems'. *Ecological Applications.* 18 (1). pp. 132–145.

Kida, K. and Kawahigashi, M. (2015) 'Influence of asphalt pavement construction processes on urban soil formation in Tokyo'. *Soil Science and Plant Nutrition.* 61. pp. 135–146.

Lambin, E. F. and Meyfroidt, P. (2011) 'Global land use change, economic globalization, and the looming land scarcity'. *Proceedings of the National Academy of Science, USA.* 108 (9). pp. 3465–3472.

Lehmann, A. and Stahr, K. (2007) 'Nature and significance of anthropogenic urban soils'. *Journal of Soils and Sediments.* 7. pp. 247–296.

Lorenz, K. and Lal, R. (2009) 'Biogeochemical C and N cycles in urban soils'. *Environment International.* 35. pp. 1–8.

Lorenz, K. and Lal, R. (2015) 'Managing soil carbon stocks to enhance the resilience of urban ecosystems'. *Carbon Management.* 6 (1–2). pp. 35–50.

Morgenroth, J., Buchan, G. and Scharenbroch, B. C. (2013) 'Belowground effects of porous pavements: soil moisture and chemical properties'. *Ecological Engineering.* 51. pp. 221–228.

Nagy, R. C., Lockaby, B. G., Zipperer, W. C. and Marzen, L. J. (2014) 'A comparison of carbon and nitrogen stocks among land uses/covers in coastal Florida'. *Urban Ecosystems.* 17 (1). pp. 255–276.

Piotrowska-Długosz, A. and Charzyński, P. (2015) 'The impact of the soil sealing degree on microbial biomass, enzymatic activity, and physicochemical properties in the Ekranic Technosols of Toruń (Poland)'. *Journal of Soils and Sediments.* 15. pp. 47–59.

Pouyat, R., Groffman, P., Yesilonis, I. and Hernandez, L. (2002) 'Soil carbon pools and fluxes in urban ecosystems'. *Environmental Pollution.* 116. pp. S107–118.

Pouyat, R. V., Yesilonis, I. D. and Nowak, D. J. (2006) 'Carbon storage by urban soils in the United States'. *Journal of Environmental Quality.* 35. pp. 1566–1575.

Pouyat, R. C., Yesilonis, I. D. and Golubiewski, N. E. (2009) 'A comparison of soil organic carbon stocks between residential turf grass and native soil'. *Urban Ecosystems.* 12. pp. 45–62.

Raciti, S. M., Groffman, P. M., Jenkins, J. C., Pouyat, R. V., Fahey, T. J., Pickett, S. T. A. and Cadenasso, M. L. (2011) 'Accumulation of carbon and nitrogen in residential soils with different land-use histories'. *Ecosystems.* 14. pp. 287–297.

Raciti, S. M., Hutyra, L. R. and Finzi, A. C. (2012a) 'Depleted soil carbon and nitrogen pools beneath impervious surfaces'. *Environmental Pollution.* 164. pp. 258–261.

Raciti, S. M., Hutyra, L. R., Rao, P. and Finzi, A. C. (2012b) 'Inconsistent definitions of "urban" result in different conclusions about the size of urban carbon and nitrogen stocks'. *Ecological Applications.* 22 (3). pp. 1015–1035.

Rawlins, B. G., Henrys, P., Breward, N., Robinson, D. A., Keith, A. M. and Garcia-Bajoet, M. (2011) 'The importance of inorganic carbon in soil carbon databases and stock estimates: a case study from England'. *Soil Use and Management*. 27. pp. 312–320.

Romero-Lankao, P., Gurney, K., Seto, K., Chester, M., Duren, R. M., Hughes, S., Hutyra, L. R., Marcotullio, P., Baker, L., Grimm, N. B., Kennedy, C., Larson, E., Pincetl, S., Runfola, D., Sanchez, L., Shrestha, G., Feddema, J., Sarzynski, A., Sperling, J. and Stokes, E. (2014) 'A critical knowledge pathway to low-carbon, sustainable futures: integrated understanding of urbanization, urban areas and carbon'. *Earth's Future*. 2. pp. 515–532.

Scalenghe, R. and Marsan, F. A. (2009) 'The anthropogenic sealing of soils in urban areas'. *Landscape and Urban Planning*. 90. pp. 1–10.

Seto, K. C., Güneralp, B. and Hutyra, L. R. (2012) 'Global forecasts of urban expansion to 2030 and direct impacts on biodiversity and carbon pools'. *Proceedings of the National Academy of Science, USA*. 109 (40). pp. 16083–16088.

Stahr, K., Stasch, D. and Beck, O. (2003) 'Entwicklung von Bewertungssystemen für Bodenressourcen in Ballungsräumen'. BWPLUS-Projekt BWC 99001 (www.fachdokumente.lubw.baden-wuerttemberg.de/servlet/is/40148/?COMMAND=DisplayBericht&FIS=203&OBJECT=40148&MODE=METADATA, accessed 24 March 2016).

Sun, Y., Ma, J. and Li, C. (2010) 'Content and densities of soil organic carbon in urban soil in different function districts of Kaifeng'. *Journal of Geographical Sciences*. 20 (1). pp. 148–156.

Trammell, T. L. E., Schneid, B. P. and Carreiro, M. M. (2011) 'Forest soils adjacent to urban interstates: soil physical and chemical properties, heavy metals, disturbance legacies, and relationships with woody vegetation'. *Urban Ecosystems*. 14. pp. 525–552.

Vasenev, V. I., Prokof'eva, T. V. and Makarov, O. A. (2013) 'The development of approaches to assess the soil organic carbon pools in megapolises and small settlements'. *Eurasian Soil Science*. 46 (6). pp. 685–696.

Vasenev, V. I., Stoorvogel, J. J., Vasenev, I. I. and Valentini, R. (2014) 'How to map soil organic carbon stocks in highly urbanized regions?'. *Geoderma*. 226–227. pp. 103–115.

Washbourne, C. L., Renforth, P. and Manning, D. A. C. (2012) 'Investigating carbonate formation in urban soils as a method for capture and storage of atmospheric carbon'. *Science of the Total Environment*. 431. pp. 166–175.

Wei, Z.-Q., Wu, S.-H., Zhou, S.-L., Li, J.-T. and Zhao, Q.-G. (2014) 'Soil organic carbon transformation and related properties in urban soil under impervious surfaces'. *Pedosphere*. 24 (1). pp. 56–64.

Xu, N. and Liu, H. (2013) 'Spatial distribution of soil inorganic carbon in urbanized territories'. *Advanced Materials Research*. 726–731. pp. 188–193.

Xu, N., Liu, H., Wei, F. and Zhu, Y. (2012) 'Urban expanding pattern and soil organic, inorganic distribution in Shanghai, China'. *Environmental Earth Sciences*. 66. pp. 1233–1238.

Zhao, S., Zhu, C., Zhou, D., Huang, D. and Werner, J. (2013) 'Organic carbon storage in China's urban areas'. *PLoS ONE*. 8 (8): e71975.

# 11 Urban sprawl, soil sealing and impacts on local climate

*Luigi Perini, Andrea Colantoni, Gianluca Renzi and Luca Salvati*

## Introduction

Urban dispersion has a considerable impact on ecosystems and ecological resources, which provide social and environmental benefits simply by existing and functioning (Angel *et al.*, 2005). The environmental impact of urban sprawl and the consequent increase in the soil imperviousness rate spans all the geographical scales. An unintended consequence of soil sealing driven by low-density suburban growth is a high resource consumption rate leading to greater environmental damage compared to a compact development pattern (Couch *et al.*, 2007). While an immediate consequence of growing rates of combustion processes of fossil fuels (due to higher consumption rates of low-density urban centers) is air pollution, the carbon dioxide in vehicular emissions and power stations is a major greenhouse gas linked to global warming. Long-term effects of fossil fuel combustion are subjected to a certain degree of uncertainties. Nevertheless, according to the Intergovernmental Panel on Climate Change (IPCC), there is a general agreement that human activities are significantly contributing to the rise in greenhouse gases (GHG) in the atmosphere, which are believed to be responsible for climate changes. If the rationale that urban sprawl leads to higher energy consumption and land use per capita is accepted, then its role in contributing to climate changes must be considered. The present contribution is intended to explore some of the potential effects that urban expansion has on heat balance and climate at the urban scale. We initially show some basic concepts for the study of urban climate. Subsequently, we describe the potential effects of the urban growth on the rise in temperature and precipitation extremes along the urban–rural gradient. Finally, we discusses the need to use methods for analyzing weather and climate specific to the urban climate and to prepare adaptation strategies to the urban climate change.

## The urban heat island

The climatic conditions of the city are significantly different from other populated areas, in particular due to the so-called effect "urban heat island" (UHI), which configures the urban environment as "bioclimatic island" in which specific weather events occur (Oke, 1982). The UHI is related to the increase in temperature of urban areas compared to their rural surroundings. The temperature difference is usually larger at night than during the day, and it is particularly evident when winds are weak, although it is observed during both summer and winter. The UHI is caused mainly by two factors. First, dark surfaces such as roadways and rooftops efficiently absorb heat from sunlight and reradiate it as thermal infrared radiation. Second, urban areas are relatively devoid of vegetation, especially trees, that would provide shade and reduce air temperature through the process of evapotranspiration. As cities sprawl outward, the heat island effect expands, in both geographic extent and intensity. This is especially true if the pattern of development features extensive tree cutting and road construction.

UHI structural factors include the percentage of albedo expressed by urban surfaces, the thermal capacity of the coating materials of soil and surfaces, shape, orientation and ventilation of buildings and, finally, the reduction of evaporating surfaces. These factors create a sort of heat dome of 150–200 meters that—in particular during the winter and in the night—determines a thermal inversion at higher elevations (Figure 11.1). Additional factors include the production of heat from air conditioning systems, vehicle traffic, industry and even the metabolic activities of the inhabitants (Oke, 1982; Arnfield, 2003; Salvati and Forino, 2014).

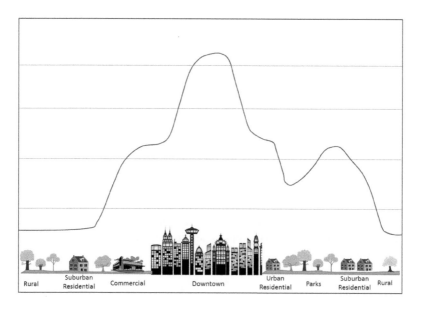

*Figure 11.1* Urban heat island profile

## The heat balance in urban areas

The radiant energy emitted by the sun that reaches the surface of the Earth consists of short wave electromagnetic radiation. Part of this energy is absorbed and then re-emitted as long-wave radiation (infrared or thermal) in the atmosphere. The air is then heated mainly by the emissions of the soil and not directly from the sun. In the case of a natural surface the heat balance is given by:

$$Q + H + E + G = 0$$

where Q is the global net radiation, H and E are respectively the sensible heat and the latent heat absorbed or transmitted by air and soil, G is the heat transferred by thermal soil conductivity. However, in urban areas the heat balance is more complex due to the presence of urban structures. The formula cited above should be integrated as follows by introducing two factors: Qp, indicating the exchange of heat with the road surfaces and buildings, and Qf, the heat generated by the anthropogenic burning of fossil fuels. The thermal balance in a urban area is:

$$Q + Qp + H + E + Qp = 0$$

Under field conditions the equation should be even more complex if considering the tribute of eventual thermal advection.

Industrial areas near urban cities can significantly modify the thermal balance: hot emissions can cause heat accumulation up to four times greater than in non-industrial areas; while the particulate in the polluted air can hinder the incoming solar radiation (10–20 percent less than rural areas) and produce a cooling effect (Bonan, 2008). Domestic heating in the winter and air conditioning in the summer contribute to heating. Moreover, some of the construction materials have high thermal conductivity. Temperature differentials between the exterior and the interior of urban buildings create a heat flow that runs through the thickness of walls from a surface to another (from the outside towards the inside and/or vice versa). Urban areas therefore cool slowly during the night in respect to non-urban areas. Combustion processes in, for example, transport, conditioning devices and industrial machines, produce greenhouse gas emissions released into the atmosphere possibly altering the radiation thermal exchange with the earth's surface by changing the final heat balance. Another factor affecting urban climate is the high concentration of aerosols, tiny particles suspended in the atmosphere, resulting mainly from industrial and car emissions. In addition to damages on human health, they impact both the propagation and absorption of solar radiation, affecting the "transparency" of the air. In other words, they influence the physical processes of condensation of atmospheric moisture, as potential condensation nuclei promoting the formation of smog and mists.

The growing demand for mobility implies growing emissions. For example, in 2005 transport emissions accounted for 20 percent of greenhouse gas emissions in the European Union (EU-25), while road transport was responsible for 93 percent of total emissions in the transport sector with about 900 million tons of $CO_2$. In the period 1990–2002, the number of kilometers of road passenger in the EU-25 increased by 26 percent, while the number of cars increased by 35 percent, with about 40 cars per 100 inhabitants in the EU-15. In the same period the number of tonnes of goods per kilometer also increased by 36 percent, while $CO_2$ emissions from road transport increased by 18 percent (István, 2010).

## Profiling urban climate

Urban areas have a similar structure to the natural canyons, in terms of absorption of solar radiation, surface temperature, evaporation rates, storage/heat radiation and direction and intensity of the wind. The amount of solar radiation received by an urban canyon depends on the height of the buildings and the orientation of the road. In an urban canyon, as in the natural one, the so-called phenomenon of "trapping" of solar energy is quite common. Due to the wall-to-wall reflection within the canyon, this phenomenon contributes to the increase of the fraction of energy absorbed by land surfaces. As a general rule, about 60 percent of the net radiation is released in the atmosphere in the form of sensible heat, 30 percent is stored in the surface of roads and buildings and 10 percent is used for the evaporation of green areas, streams or wetlands (Spronken Smith *et al.*, 2006). The temperature ranges are closely connected to the surface and the shape of the buildings, land cover, the presence of vegetation and man-made radiation sources (Giridharan *et al.*, 2004; Jonsson, 2004; Unger, 2004; Johnson and Wilson, 2009).

Profiling urban climate regimes may benefit from the comparative analysis of weather variables in strictly urban and neighboring rural sites (Hawkins *et al.*, 2004; Sakakibara and Owa, 2005). A survey carried out on the basis of the criteria recommended by the World Meteorological Organization (WMO) on gauging stations located along the urban gradient in Rome and Milan (Italy) shows different patterns between maximum and minimum temperatures. In Rome, the differences between the values recorded inside and outside the city are reduced slightly in the warmer months, persisting throughout the year. In Milan, the values are instead more correlated with a similar pattern and significant differences only in a deficiency months (Figure 11.2). The fact is that the minimum temperatures are the result of thermal conditions expressed by the atmospheric layer close to the ground, while the maximum temperatures, depending generally on convection heating and consequent mixing of the air mass above the soil, are representative of the thermal conditions of the troposphere (Beltrano and Perini, 1997). These outcomes can therefore confirm that the difference between urban and rural areas during the day is low, increasing gradually during the evening and night.

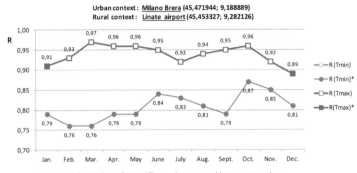

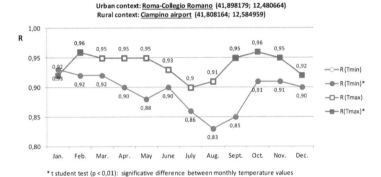

*Figure 11.2* Correlation coefficient of mean monthly temperatures (Tmin and Tmax) between urban and rural environmental contexts in Milan (a) and Rome (b), Italy

## Climate and urban form

Urban climate regimes can be seen as the product of a "cultural mediation" between the diversity of approaches in designing and planning cities in terms of materials and structures. Traditional and compact urban forms may improve specific micro-climate conditions. The Physiological Equivalent Temperature (PET) index, which combines temperature, humidity and wind conditions, was proposed to assess the variations of the thermal conditions (and the related "thermal comfort") in the cities, according to selected architectural parameters and considering the season and hour of the day (Matzarakis *et al.*, 2007). In a neighborhood with compact settlements and east–west orientation in Fez (Morocco), the threshold of well-being (PET = 33°C) is abundantly exceeded at street level for most of the day. Under porches, in particular those to the north side of the road, the thermal comfort is always instead at better levels (Ahmed, 2003; Johansson, 2006). Urban elements may thus be effective in the mitigation of environmental conditions. Conversely, the geometry of the buildings can result in extreme weather, including storms of considerable intensity (Ntelekos *et al.*, 2007).

A detailed analysis of the urban climate should benefit from a classification of different types of settlement along the urban–rural gradient. A standard classification was proposed based on the Local Climatic Zones (LCZ) (Bechtel *et al.*, 2015). The LCZ are intended as homogeneous areas whose characteristics influence the thermal properties as the fraction of built-up area, the aspect ratio of buildings, the sky view factor, the height of the elements that constitute the "roughness" of the surface, the heat flux of anthropogenic origin and the surface of heat radiation (Stewart and Oke, 2009). The scaling factor is also important as the representativeness of meteorological stations varies based on the instrument adopted, the conditions around the station and the surface geometry (Oke, 2004). The standard measurement of temperature in gauging stations is generally less reliable in built-up areas in respect to the open field. The spatial dimension of a LCZ varies depending on the measurement conditions imposed by the site. Table 11.1 shows an example of LCZ classification.

*Table 11.1* An example of Local Climatic Zones for the analysis of urban contexts

| Built types | Description | Land cover types | Description |
|---|---|---|---|
| Compact high-rise | Context of tall buildings (tens of floors). Land cover mainly paved. Presence of concrete, steel, stone and glass building materials. Deficiency or absence of trees. | Dense trees | Rich landscape of vegetation. Forest area, plant cultivation or urban park. Mostly permeable land cover. |
| Compact midrise | Context of midrise buildings (3–9 floors). Land cover mainly paved. Presence of concrete, tile, brick and stone building materials. Deficiency of trees. | Sparse trees | Landscape vegetation dispersed. Forest area, plant cultivation or urban park. Mostly permeable land cover. |
| Compact low-rise | Context of low-rise buildings (1–3 floors). Land cover mainly paved. Presence of concrete, tile, brick and stone building materials. Deficiency of trees. | Bush and scrub | Presence of bush and scrub. Agricultural or natural scrubland areas. Mostly permeable land cover. |
| Open high-rise | Open collocation of tall buildings (tens of floors). Presence of concrete, steel, stone and glass building materials. Plenty of permeable land cover (low plants and sparse trees). | Low plant | Shapeless landscape of grass or herbaceous crops with deficiency or absence of trees. Natural grassland, agricultural area or urban park. |

| Open midrise | Open collocation of midrise buildings (3–9 floors). Presence of concrete, steel, stone and glass building materials. Plenty of permeable land cover (low plants and sparse trees). | Bare rock or paves | Shapeless landscape of rock or paved cover with deficiency or absence of trees. Natural desert or urban transportation. |
|---|---|---|---|
| Open low-rise | Open collocation of low-rise buildings (1–3 floors). Presence of concrete, brick, wood, stone and tile building materials. Plenty of permeable land cover (low plants and sparse trees). | Bare soil or sand | Shapeless landscape of soil or sand cover with deficiency or absence of trees. Natural desert or agricultural area. |
| Lightweight low-rise | Dense mix of buildings with one floor. Land cover mainly hard-packed. Lightweight construction materials. Deficiency of trees. | Water | Water bodies. |
| Large low-rise | Open collocation of low-rise buildings (1–3 floors). Land cover mainly hard-packed. Presence of concrete, steel, stone and metal building materials. Deficiency of trees. | | |
| Sparsely built | Dispersed collocation of small-medium buildings in natural context. Abundance of permeable land cover. | | |
| Heavy industry | Low-rise and midrise industrial buildings. Land cover mainly paved. Presence of concrete, metal and steel building materials. Deficiency of trees. | | |

*Variable land cover proprieties*

Land cover variables that considerably change with synoptic weather conditions, agricultural practices and/or seasonal cycles.

| Bare trees | Bare deciduous trees. Increased factor of the sky view and reduced albedo. |
|---|---|
| Snow cover | Snow cover ( > 10 cm in depth), low admittance and high albedo |
| Dry ground | Parched land, low admittance, large Bowen ratio and high albedo. |
| Wet ground | Waterlogged ground, high admittance, small Bowen ratio and reduced albedo. |

## Monitoring climate and planning sprawl in urban areas

Urban climatic regimes have characteristics that justify a specific approach for permanent monitoring and adaptation strategies. For example, it is not possible to assess climate variables strictly according to the measurement criteria recommended by WMO. New technologies are applied to the analysis of the effects of urban areas on the formation of clouds, precipitation and the storms. Satellite remote sensing (Schumacher and Houze, 2000), LIDAR (Zhou *et al.*,

2004) and Doppler radar (Russo *et al.*, 2005) allow a detailed analysis of the rainfall spatial variability at a disaggregated geographical scale. Using such methodologies, Souch and Grimmond (2006) confirm that urbanization has effects on precipitation by increasing the hygroscopic nuclei of condensation of atmospheric moisture due to the air turbulence caused by the increased "roughness" of the land surface and to convection caused by the properties and different thermal states of the materials (see Lowry, 1998). Average air temperatures are 1–2°C higher in urban areas than in the surrounding rural areas, particularly at night and during summer. Vehicular traffic, the air conditioning of buildings and the quality of material for the covering of land surfaces contribute to heating, while the scarcity of green areas, associated to the lower ventilation, reduces the efficiency of the natural forms of mitigation during extreme events. This implies that the negative effects of climate change can be exacerbated in strictly urban areas and reduced along the urban–rural gradient (Szymanowski, 2005).

Multi-scalar adaptation strategies at both national and local level are also necessary to cope with meso-scale climate changes in metropolitan regions. Specific measures to adapt to climate change at the urban scale are thus necessary (Hallegatte *et al.*, 2011; Hunt and Watkiss, 2011).

One of the major objectives of urban planning is to promote efficient settlement forms that rely less on the consumption of fossil fuels and agricultural/forest land reducing the local-scale impact of climate variations. For example, the European Commission proposed specific policies coping with climate changes in metropolitan regions with the aim to balance the bio-climatic regimes and to affect positively local communities, the activities of policy-making and the dissemination of good behavior in the daily life of the inhabitants (Castan Broto and Bulkeley, 2013). Several European countries have adopted national strategies for adaptation to climate change (Westerhoff *et al.*, 2011). The issue seems to be pressing national authorities in the aftermath of the exceptional 2003 when more than 3,000 deaths were directly related to the repeated heat waves affecting large urban areas only in Italy (Conti *et al.*, 2005). Adaptation strategies at the local level were proposed to include specific measures which adapt the urban structure to the risk of heat waves (MATTM, 2013). Additional actions are targeted (1) to stimulate the use of weather-alert systems, (2) to promote the reduction of energy consumption and the thermal efficiency of public and private structures, (3) to restore green spaces and to promote the re-naturalization of riparian areas and the proper management of urban waterways.

## Conclusions

Cities are complex integrated systems interconnected by infrastructures of transport, energy, water and services. With urban dispersion, peri-urban areas became progressively more vulnerable, especially to the impacts of climate change, such as floods, drought or heat waves, depending on local characteristics such as urban topography, economic structure and socio-spatial organization (Hallegatte and Corfee Morlot, 2011). The present contribution outlines that the main features of the urban climate are not represented by simple biophysical factors, while being dependent on the shape and spatial organization of each city. Urban planning and socioeconomic policies may contain the weakness caused by climate change when addressing place-specific and multifaceted factors integrating the biophysical and socioeconomic dimension. At the same time, monitoring urban climate cannot be reduced to schematic interpretations, while opening up an in-depth discussion on how urban life styles may affect weather conditions at the local scale. Improving the quality of urban life, for example by enlarging and better designing green urban areas, promoting sustainable architecture and renewable energy policies, developing public transport networks and "soft mobility," are actions mitigating land and population vulnerability to climate change and reducing the socioeconomic loss due to extreme weather events.

## References

Ahmed K.S. (2003), "Comfort in urban spaces: defining the boundaries of outdoor thermal comfort for the tropical urban environments", Energy and Buildings, vol. 35, pp. 103–110.

Angel S., Sheppard S.C. and Civco D.L. (2005), "The dynamics of global urban expansion", Department of Transport and Urban Development, The World Bank, (www.williams.edu/Economics/UrbanGrowth/DataEntry.htm, accessed 12 June 2016).

Arnfield A.J. (2003), "Two decades of urban climate research: a review of turbulence, exchanges of energy and water, and the urban heat island", International Journal of Climatology, vol. 23, pp. 1–26.

Bechtel B., Alexander P.J., Böhner J., Ching J., Conrad O., Feddema J., Mills G., See L. and Stewart I. (2015), "Mapping local climate zones for a worldwide database of the form and function of cities", ISPRS International Journal of Geo-Information, vol. 4, pp. 199–219.

Beltrano M.C. and Perini L. (1997), "Comparazione tra le temperature estreme giornaliere urbane ed extraurbane a Roma e Milano", Nimbus, vol. 3/4, pp. 48–51.

Bonan G.B. (2008), Ecological Climatology: Concepts and Applications, Cambridge University Press.

Castan Broto V. and Bulkeley H. (2013), "A survey of urban climate change experiments in 100 cities", Global Environmental Change, vol. 23, pp. 92–102.

Conti S., Meli P., Minelli G., Solimini R., Toccaceli V., Vichi M., Beltrano M.C. and Perini L. (2005), "Epidemiologic study of mortality during Summer 2003 heat wave in Italy", Environmental Research, vol. 98, n. 3, pp. 390–399.

Couch C., Leontidou L. and Petschel-Held G. (2007), Urban Sprawl in Europe: Landscape, Land-use Change and Policy, Blackwell.

Giridharan R., Ganesan S. and Lau S.S.Y. (2004), "Daytime urban heat island effect in high-rise and high-density residential developments in Hong Kong", Energy and Buildings, vol. 36, pp. 525–534.

Hallegatte S. and Corfee-Morlot J. (2011), "Understanding climate change impacts, vulnerability and adaptation at city scale: an introduction", Climatic Change, vol. 104, pp. 1–12.

Hallegatte S., Henriet F. and Corfee-Morlot J. (2011), "The economics of climate change impacts and policy benefits at city scale: a conceptual framework", Climatic Change, vol. 104, pp. 51–87.

Hawkins T.W.B., Stefanov W.L., Bigler W. and Saffell E.M. (2004), "The role of rural variability in urban heat island determination for Phoenix, Arizona", Journal of Applied Meteorology, vol. 43, pp. 476–486.

Hunt A. and Watkiss P. (2011), "Climate change impacts and adaptation in cities: a review of the literature", Climatic Change, vol. 104, pp. 13–49.

István L.B. (2010), "Urban sprawl and climate change: a statistical exploration of cause and effect, with policy options for the EU", Land Use Policy, vol. 27, pp. 283–292.

Johansson E. (2006), "Influence of urban geometry on outdoor thermal comfort in a hot dry climate: a study in Fez, Morocco", Building and Environment, vol. 41, pp. 1326–1338.

Johnson D.P. and Wilson J.S. (2009), "The socio-spatial dynamics of extreme urban heat events: the case of heat-related deaths in Philadelphia", Applied Geography, vol. 29, pp. 419–434.

Jonsson P. (2004), "Vegetation as an urban climate control in the subtropical city of Gaborone, Botswana", International Journal of Climatology, vol. 24, pp. 1307–1322.

Lowry W.P. (1998), "Urban effects on precipitation amount", Progress in Physical Geography, vol. 22, pp. 477–520.

Mattm (2013), Elementi per una Strategia Nazionale di Adattamento ai Cambiamenti Climatici. Documento per la consultazione pubblica.

Matzarakis A., Georgiadis T. and Rossi F. (2007), "Thermal bioclimate analysis for Europe and Italy", Il Nuovo Cimento, vol. C30, pp. 623–632.

Ntelekos A.A., Smith J.A. and Krajewski W.F. (2007), "Climatological analyses of thunderstorms and flash floods in the Baltimore metropolitan region", Journal of Hydrometeorology, vol. 8, n. 1, pp. 88–101.

Oke T.R. (1982), "The energetic bases of the urban heat island", Quarterly Journal of the Royal Meteorological Society, vol. 108, pp. 1–24.

Oke T.R. (2004), "Siting and exposure of meteorological instruments at urban sites", Proceedings of 27th NATO/CCMS International Technical Meeting on Air Pollution Modelling and Its Application, Banff, 25–29 October 2004.

Russo F., Napolitano F. and Gorgucci, E. (2005), "Rainfall monitoring systems over an urban area: the city of Rome", Hydrological Processes, vol. 19, pp. 1007–1019.

Sakakibara Y. and Owa K. (2005), "Urban rural temperature differences in coastal cities: influence of rural sites", International Journal of Climatology, vol. 25, pp. 811–820.

Salvati L. and Forino G. (2014), "A 'laboratory' of landscape degradation: social and economic implications for sustainable development in peri-urban areas", International Journal of Innovation and Sustainable Development, vol. 8, n. 3, pp. 232–249.

Schumacher C. and Houze R.A. (2000), "Comparison of radar data from the TRMM satellite and Kwajalein oceanic validation site", Journal of Applied Meteorology, vol. 39, pp. 2151–2164.

Souch C. and Grimmond S. (2006), "Applied climatology: urban climate", Progress in Physical Geography, vol. 30, n. 2, pp. 270–279.

Spronken-Smith R.A., Kossmann, M. and Zawar-Reza, P. (2006), "Where does all the energy go? Surface energy partitioning in suburban Christchurch under stable wintertime conditions", Theoretical and Applied Climatology, vol. 84, pp. 137–150.

Stewart I.D. and Oke T.R. (2009), "Classifying urban climate field sites by 'local climate zones': the case of Nagano, Japan", Preprints, Seventh International Conference on Urban Climate, Yokohama, Japan, June 29–July 3.

Szymanowski M. (2005), "Interactions between thermal advection in frontal zones and the urban heat island of Wroclaw, Poland", Theoretical and Applied Climatology, vol. 82, pp. 207–224.

Unger J. (2004), "Intra-urban relationship between surface geometry and urban heat island: review and new approach", Climate Research, vol. 27, pp. 253–264.

Westerhoff, L., Keskitalo E.C.H. and Juhola S. (2011), "Capacities across scales: local to national adaptation policy in four European countries", Climate Policy, vol. 11, n. 4, pp. 1071–1085.

Zhou G.Q., Song C., Simmers J. and Cheng, P. (2004), "Urban 3D GIS from LiDAR and digital aerial images", Computers and Geosciences, vol. 30, pp. 345–353.

# 12 Impacts of urban sprawl on landscapes

*Marie Cugny-Seguin*

## Introduction

Landscapes can be seen from many views depending on the phenomenon under consideration. For landscape ecology, focused on the understanding of the interactions between spatial heterogeneity and ecological processes, 'a landscape is an area that is spatially heterogeneous in at least one factor of interest' (Turner *et al.*, 2001; Turner, 2005). Other authors insist on anthropogenic aspects: 'A heterogeneous area comprising interacting ecosystems that are repeated in similar form throughout, including both natural and anthropogenic land cover, across which humans interact with their environment' (Forman and Godron, 1981). For social science, landscape is understood 'as an arena where conflicting interests meet, but also as sites of importance for people's individual and collective memories and identifications' (Tengberg *et al.*, 2012). According the European Landscape Convention, 'landscape means an area, as perceived by people, whose character is the result of the action and interaction of natural and/or human factors' (Committee of Ministers of the Council of Europe, 2000). The convention promotes the integration of landscapes in any policies with possible direct or indirect impacts on landscapes such as cultural, environmental, agricultural, social and economic policies, using a participatory approach. That means to integrate landscape issues into spatial and urban planning policies and to develop strategies and guidelines to create, enhance, protect, restore and manage landscapes. For this contribution we have adopted the definition of the European Landscape Convention.

Therefore, the notion of landscapes comprises not only physical and spatial parameters but also cultural, social, historical, aesthetic and even religious connotations. Landscapes are crucial for the quality of life of people everywhere (in urban areas and in the countryside), the formation of local cultures and the consolidation of the identity of a place. The Millennium Ecosystem Assessment (MA) related to cultural and amenity services stresses that human cultures, knowledge systems, religions, heritage values and social interactions have always been influenced and shaped by the nature of the ecosystems and the ecosystem conditions in which a culture is based (MA, 2005a).

Landscapes are dynamic systems. They are continuously affected by human activities and natural processes and these continual land use changes have a significant effect on ecosystem services supply (Maes *et al.*, 2011). Landscapes evolve because of individual and unrelated actions upon the environment (e.g. actions of inhabitants), local decisions (e.g. urban planning), external factors (e.g. change in the hierarchy of cities due to the globalisation of economy), changes in technology (e.g. mobility by car), change in lifestyle (e.g. the preference for a detached house with private garden), and the action of natural forces (e.g. floods, cyclones).

Since the second half of the twentieth century, with the rise of mobility and commuting, cities have physically expanded around a major urban centre, mainly into the surrounding agricultural areas. This expansion of built-up areas has generated urban sprawl characterised by areas of low density, patchy and scattered development (EEA, 2006). This growth of artificial surfaces associated with the development of linear transportation infrastructure has fragmented the landscapes and generated adverse ecological effects. The fragmentation contributes significantly to the decline and loss of wildlife populations, the increasing endangerment of species in Europe (e.g. through the isolation of populations) and the spread of invasive species (EEA, 2011); it also affects the water regime and the aesthetic and recreational quality of landscapes.

Landscapes are made by humans and reflect changes in society (e.g. culture, values, behaviour, lifestyle) and its relationship to the natural environment (Antrop, 2000a). The speed, frequency and magnitude of changes has varied according to historical periods. During many centuries changes were local and slow (Antrop, 2005). The use of land resulted in a traditional landscape with a recognisable structure and significant aesthetic values that give clear identity to a place. With the economic rationalisation of agriculture and the rapid urbanisation, landscapes of large areas have lost and continue to lose their diversity and territorial identity.

The major challenge of landscape planning and management is to minimise the disturbance effects of human interventions while satisfying the human needs for activities. Land ownership is the main difficulty of landscape management; land is owned by many people who all have their own particular interest. However, landscape is a common good that provides habitats for flora and fauna and is the base for human activities. The detrimental effects on a landscape are seen not only by the citizen, generally the owner, who has decided the change of a landscape element (e.g. to build a house, to cut a hedgerow) but shared by all society (inhabitants, tourists, visitors) and for a long time. In the same way, ecosystems services provided by landscapes benefit the well-being of all people living or visiting the place and not only the citizen who has decided on the transformation.

The following key questions will be analysed in this chapter:

- What are the ecological impacts of landscape fragmentation?
- What are the impacts of urban sprawl on cultural services of landscape degradation?

## The ecological impacts of landscape fragmentation

Landscape fragmentation is the product of the linkage of built-up areas via linear infrastructure, such as roads and railroads. It is the result of transforming large habitat patches into smaller and more isolated fragments of habitat (EEA, 2011). These transformations generate an increase in the amount of patches, and therefore in the amount of patch edges, changes of their shape and their spatial arrangements and the interspersion of anthropogenic and natural land. Landscape fragmentation has an impact on landscape structure, including changes in landscape configuration and heterogeneity[1] (Matthew *et al.*, 2015).

Patches and corridors are the key spatial elements for increasing connectivity and therefore preventing fragmentation (Forman, 2008). The size and the shape of a patch affect how a patch functions on its own or in relation with the other patches; larger patches provide larger habitats and are more effective. The corridors (e.g. linear landscape elements such as hedgerows) between patches determine the opportunities of movement across the landscape.

There is a trade-off between the level of fragmentation and the supply of services. Road density eases the accessibility to specific services, their supply and their exploitation. However, at the same time, the supply of different services are affected by landscape fragmentation and the scale at which fragmentation occurs, in particular when their flows depend on the movement of organisms, matter, energy or people across landscapes (e.g. fresh water provision, water quality, natural hazards). 'Regulations and maintenance services' such as species movement, water-related services (e.g. with the increase of imperviousness, less water infiltrates and run-offs increase) or erosion prevention are particularly impacted. 'Provisional services' such as food or timber production (e.g. due to small land parcels or reduced quality of agricultural products along roads) are also affected as well as 'cultural services' (e.g. aesthetic value of a landscape).

---

### Box 12.1   Landscape fragmentation in Europe

By using the method of 'effective mesh density', it is possible to quantify the degree to which the possibilities for movement of wildlife in the landscape are interrupted by barriers. The effective mesh density values across the 28 investigated European countries cover a large range, from low values in large parts of Scandinavia to very high values in western and central Europe. Many highly fragmented regions are located in Belgium, the Netherlands, Denmark, Germany, France, Poland and the Czech Republic. High fragmentation values are mostly found in the vicinity of large urban areas and along major transportation corridors. The lowest levels of fragmentation are usually associated with mountain ranges or remoteness.

The density of the transportation network and the extent of landscape fragmentation is largely a function of interacting socioeconomic drivers such as population density and geophysical factors such as topography. According the report *Landscape Fragmentation in Europe*, the most relevant variables affecting landscape fragmentation are population density, gross domestic product per capita, volume passenger density and the quantity of goods loaded and unloaded per capita (EEA, 2011).

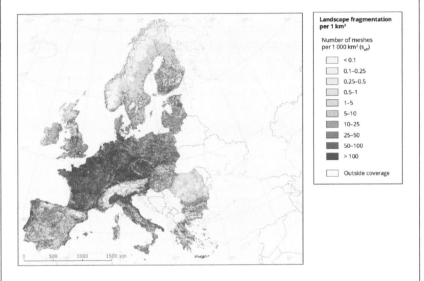

*Figure 12.1* Landscape fragmentation per 1 km² grid in 2009 (source: EEA, 2011)

Note: Landscape fragmentation was calculated using fragmentation geometry FG-B2.

Landscape fragmentation is a major cause of the decline of many wildlife populations. It creates smaller habitat patches that support fewer species, and contain smaller and more vulnerable populations with a reduced genetic variability (Forman and Alexander, 1998; IUCN, 2001). It increases the edge effect that negatively affects the persistence of native species (Dobson et al., 2006; Matthew et al., 2015). It contributes to the destruction of established ecological connections between areas of the landscape (Jaeger et al., 2005) and reduces the ability of plant and animal species to move across landscapes. Roads and traffic reduce their access to the different types of habitat they need during their life cycle (e.g. foraging and breeding habitats), enhance mortality due to collisions with vehicles and generate disturbance and dispersal events. Several examples of the detrimental effect of landscape fragmentation, combined with intensive agricultural practices, exist (e.g. the continuous decline of the brown hare (*Lepus europaeus*) populations in Switzerland (EEA, 2011)).

In addition, physical processes such as radiation, flows of water and wind speed can be changed by removing large pieces of native vegetation. These physical changes affect biological processes such as litter decomposition, nutrient cycling, composition and structure of vegetation, and hydrological regime. The greatest impacts occur mainly at the edge of patches but changes occurring in one patch are accumulated across the landscape and finally have an impact on the entire landscape. For example, in the Western Australian wheat belt massive loss of native vegetation has resulted in a rise of groundwater, bringing stored salt to the surface and finally reducing agriculture productivity (Bennett and Saunders, 2010). The effect of fragmentation of habitats on the biota take many years to be expressed, in particular for long-lived organisms (e.g. trees) that can persist decades before disappearing without replacement.

## The impacts on cultural services

Cultural services are defined as 'the non-material benefits people obtain from ecosystems through spiritual enrichment, cognitive development, reflection, recreation, and aesthetic experiences' (MA, 2005b). Landscapes contain not only physical information but also exhibit the social history and the identity of the place. They show the beliefs, values, shared habits and preferences of the different cultures that have formed the landscapes. From a social point of view, they are the result of the past representations and future expectations of society (Black, 2003).

Cultural aspects (e.g. heritage and aesthetic values) of landscapes are threatened and can be irreversibly lost by uncontrolled urbanisation and the development of transportation networks. Europe and North America experienced a first wave of urbanisation in the course of two centuries (1750–1950) and poorer and emerging countries are currently experiencing a 'second wave' of demographic, economic and urban transitions, much bigger and much faster than the first (UNFPA, 2007). After the 1960s, urban sprawl became a worldwide problem, not only in North America, Western Europe (EEA, 2006) and Japan, but also in some large cities in developing countries. The causes of sprawl can be different according to country and period; for example, in Western countries, urban sprawl is the result of suburbanisation[2] (Mills, 2003) whilst in China, it is mainly due to low-density urbanisation and industrial development at the urban fringe (Zhao, 2010, 2011).

In Europe, cities experienced three major transformations until the twentieth century. First, with the increase of the urban population, at a level incomparable to the previous periods in history, cities became denser and covered a larger space. Second, during the industrial period, urban landscapes were transformed by the need of places for mass production, easy access to raw materials and energy, transportation infrastructure and workers' settlements close to the manufactories. Third, the introduction of new means of transportation (train, electric subway, metro and cars) changed the structure of European cities that became more dispersed, overcrowded and distributed over larger spaces. During the second half of the twentieth century, urban sprawl occurred with the development of road networks, the rise of income, the increase in private car ownership, the preference for single-family houses, the land market attracting people in the periphery and insufficient or non-relevant urban policies.

Then, with the decline of industrial production and the emergence of a new kind of knowledge-based and service-oriented urban economy, European cities entered a new phase of development (Cremaschi and Eckardt, 2011). Post-industrial landscapes, without any cultural background and looking all alike, are being created. Nowadays changes in landscape are more driven by external decisions (e.g. from multinational firms) or global tendency (e.g. economic crisis) rather than local or regional decisions. The growth of the service sector is extremely visible in today's urban landscape. Since the 1980s, large public and private investments have been made in and near the city centres or at the edge of the cities, changing the physical appearance of the cities. Huge office buildings, with an architecture focused on status and prestige (impressive architecture is designed to suggest economic power), have generated new urban landscapes that express a new economic reality. In the post-industrial city, wastelands – remains of the previous industrial transformation – have become strategic places for new urban development (e.g. harbours, industrial brownfields, abandoned rail tracks etc.) and for changing the place (e.g. new landscape of Marseille waterfront, culture-led regeneration of Psiri in Athens).

The different phases of urbanisation have transformed landscapes. According to the country and regional area, different phases of urbanisation have been identified and described to explain the concentric zones of influence around urban centres. For example, five concentric zones of influence have been described for the cities of Western Europe (Antrop, 2000b): the urban core (completely built up area with different periods – Middle Ages, nineteenth century etc.), the inner urban fringe (post-Second World War garden cities with a dense housing pattern), the outer urban fringe (urban landscape characterised by a complex mosaic of land use), the rural commuting zone, with important functional changes due to demographic transition (emergence of exurbs), the depopulating countryside with relics of old landscapes.

With the rise of car-mobility, landscapes have become fragmented by highways and roads, even into the urban fabric. Transport networks connect to destinations (e.g. commercial malls, airports, zones of activities, office parks, allotments) rather than to 'places' that refer to identity. During the last few decades, single-use zoning has produced landscapes characterised by a functional homogeneity. In the same way, the intensification of agriculture and the removal of small landscape elements (e.g. hedgerows, isolated trees) reduce spatial variation. Territories, isolated from the others and with a unique function (e.g. commerce, housing, gated communities), and often with similar architecture, have been created. Between these new developed areas, open spaces can be left for potential urbanisation over years. In this world of 'hypermobility', space and distance are measured in time.

Urban sprawl has major impacts not only on the environment (surface sealing, ecosystem fragmentation, emissions from transport, run-off etc.) but also on the social structure of an area (by spatial segregation, lifestyle changes etc.) and on the economy (via distributed production, land and housing prices, scale issues etc.). Residential segregation that is often combined with fewer services for the population (e.g. poor transportation, health, deprived housing) is happening in most big metropolises everywhere in the world and in different manners (racial groups, ethnicity, religion or income status) according to the cultural and historical context. In some counties, spatial segregation of the poor often occurs within informal settlements characterised by chaotic landscape without formal streets. With the proliferation of gated communities, voluntary segregation has become a new force in both the Northern and Southern hemispheres (e.g. India, USA) because of the demand (e.g. perception of security, new lifestyle) and the supply (better profitability with large-scale internalisation of externalities).

---

**Box 12.2**

Urban sprawl results in discontinuous, scattered urban and low density growth. It creates interstitial open spaces, generates interwoven agricultural enclaves in urbanised areas, and wastes valued productive agricultural land (UN-HABITAT, 2012). The transition between urban and rural

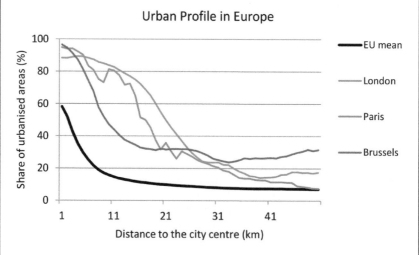

*Figure 12.2* Urban profile in Europe (source: Corilis 2006 (2000 for Greek cities) based on Corine Land Cover version 16)

Source: EEA, 2015.

Note: Above, graph showing the urbanisation pattern from the city centre to a maximum distance of 50 km for the Urban Audit's selection of cities over 50,000 inhabitants. Each line represents the share of urbanised area (per cent) in a 1 km buffer ring from the city centre (centroid of city boundaries as defined by Urban Audit) for selected cities (London, Paris and Brussels) and the mean value at European level (EU28 + Norway and Switzerland).

areas is a continuum. The distinction between urban and rural patterns are more diffuse and fuzzy. In 2010, according the new urban–rural typology, 40.4 per cent of EU-27 population were living in urban areas and 35.4 per cent in intermediate areas.[3] This diffusion of artificial areas into rural landscapes, that have a distinct and recognisable structure, contributes to rapidly changing traditional rural landscapes that have been formed rather slowly by rural lifestyles and therefore have harmoniously integrated natural conditions and cultural patterns (Antrop, 2000a). On the contrary, suburban landscapes are highly dynamic and new landscapes, that a bit chaotic, are created.

Urban landscape is the physical environment, where people live their daily life, work, move, do shopping, have their social interactions etc. The quality of urban life depends on the quality of public spaces (including green open areas) that can be considered as the 'living room' of the city (Burgers, 2000; EEA, 2009); they range from grand central plazas and squares, streets and their pavements to small, local neighbourhood parks. In a context of urban adaptation to climate change, accessible green open spaces, as well as green walls and green roofs, are also crucial in providing health and wellbeing benefits for a city's residents (de Vries *et al.*, 2003; EEA, 2012) and might become an important part of urban landscape in the future. Architecture is also a key component contributing to the quality of the urban environment and the transformation of the urban landscape; new architectural design produces buildings with innovative forms that are radically changing the physical landscape of the cities.

Finally, to ensure their long-term viability, some cities are trying to develop an urban sustainable model that lead cities to manage urban development in a way that minimises the environmental impacts and the land use per capita as well as promotes a mix of land use and proximity. With increasing concerns about climate change, the reign of cars is finishing; walking, cycling and using public transport are becoming ideal models of mobility in urban areas. Compactness and integrated urban development, such as eco-district or eco-city, are becoming mainstream in new urbanised areas where urbanism can be strongly integrated. In Europe, urban sustainability is mainly based on the retrofitting of existing urban infrastructure and building stocks, the conversion of underused or abandoned industrial areas, the conversion of low-density suburban environments into high-density areas and the upgrading of non-sustainable settlements. These changes in urban systems should produce a new cycle of urban landscape.

## Conclusion

The main challenges for landscape management are to integrate landscape ecological needs and to supply ecosystems services. Management landscape has to be focused, at the same time, on the preservation of the quality of landscape for natural resources (such as biodiversity, habitats, connectivity, water cycle) and the delivery of ecosystems services that are crucial for the well-being of society, in particular the delivery of cultural services (e.g. scenic beauty, heritage landscape) that are crucial for the identity of the place and the sense of the community. Understanding the different aspects of services provision, and what features of landscape structure, fragmentation and heterogeneity control those services, can significantly improve the ability to manage landscapes for ecosystem services.

More balanced relationships between humans and nature as well as between rural hinterland and urban areas are needed. The permanent and dynamic changes of urban and peri-urban landscapes provide an opportunity for planning authorities to consider not only the quality of a landscape itself but also the level of existing development and the ability of the landscape to absorb further development without altering its character. That means to understand the supply of ecosystem services at landscape scale and, at the same time, to analyse the potential impacts on the close environment of people, even in the most common landscapes, where a transformation can easily be perceived (e.g. changes in scenic beauty, quality of open public spaces or noise).

Landscape planning and landscape management face major challenges. First, the perception of landscape is subjective and depends on the person who looks at it. Second, elements of a landscape have no absolute value; the change of one significant element can contribute to change the whole landscape and the same element in another geographical context may have a different value (Antrop, 2000a). Third, the changes usually occur in a gradual manner and are not immediately perceived as dramatic and the cumulative impact therefore underestimated.

Participation of stakeholders at the decision-making process is a way not only to know the expectations of people with regard to the landscape features of their surroundings, but also to raise the landscape awareness of the entire society. The perception of the urban landscape depends of how people move through the city. Each city resident develops their own experience of a city according the place they live, work or socialise. The personal understanding of a city by its residents cannot match with the 'real' city because it does not take into account the 'real' scale (e.g. metropolitan area), the degree of complexity (e.g. interactions between activities), the interrelationship between nature and human, the past and the future etc. To contribute effectively to urban sustainability and prosperity, landscape planning has to be based on a shared vision.

Finally, urban sprawl is recognised as a major issue in several countries and strategies limiting land take in order to mitigate the negative effects of market-led development have been developed (e.g. the Federal Sustainable Development Strategy of the German federal government, and strategies for urban containment in China) (EEA, 2016). Many cities have also developed their own objectives in order to achieve compactness (e.g. the Master Plan project for the Paris metropolitan areas) (OECD, 2012) and to limit land take. The implementation of these policies needs strong leadership by local authorities, monitoring progress and proposing regular and transparent reporting. All these land policies can contribute to preserve urban and peri-urban landscapes, but we need to underline that land policies are different from landscape policies.

## Notes

1 The number of habitats is generally higher in a heterogeneous landscape than in simpler landscapes and this affects species richness positively.

2 Suburban: generally of lower density contiguous built-up areas that are attached to inner urban areas and where houses are typically not more than 200 metres apart (*Peri-Urbanisation in Europe: Synthesis Report*, Plurel FP7 project www.plurel.net/ images/Peri_Urbanisation_in_Europe_printversion.pdf, accessed 15 August 2015). It is a patchwork of residential, commercial, municipal, and industrial land uses and related transportation and utility corridors often adjacent to urban centres. See also Australian Government, Department of Infrastructure and Transport (2011).

3 http://ec.europa.eu/eurostat/statistics-explained/index.php/File:Share_of_population_ according_to_the_original_OECD_classification_and_the_new_urban-rural_typology_ new.png, (Nuts3) accessed 18 August 2015.

## References

Antrop, M. (2000a) 'Background Concepts for Integrated Landscape Analysis', *Agriculture, Ecosystems & Environment*, 77(1–2), 17–28. doi:10.1016/S0167-8809(99)00089-4.

Antrop, M. (2000b) 'Changing Patterns in the Urbanized Countryside of Western Europe', *Landscape Ecology*, 15, 257–270.

Antrop, M. (2005) 'Why Landscapes of the Past Are Important for the Future', *Landscape and Urban planning*, 70(1–2), 21–34. doi:10.1016/j.landurbplan.2003.10.002.

Australian Government, Department of Infrastructure and Transport (2011) *Our Cities, Our Future: A National Urban Policy for a Productive Sustainable and Liveable Future*. https://infrastructure.gov.au/infrastructure/pab/files/Our_Cities_National_ Urban_Policy_Paper_2011.pdf, accessed 28 August 2015.

Bennett, Andrew F. and Saunders, Denis A. (2010) 'Habitat Fragmentation and Landscape Change', in N.J. Sodhi and P.R. Ehrlich (eds) *Conservation Biology for All*. Oxford: Oxford University Press, 88–108. doi:10.1093/acprof:oso/9780199554232.003.0006. www.oxfordscholarship.com/view/10.1093/acprof:oso/9780199554232.001.0001/ acprof-9780199554232-chapter-6, accessed 10 August 2015.

Black, I. (2003) '(Re) Reading Architectural Landscapes', in I. Robertson and P. Richards (eds) *Studying Cultural Landscapes*. London: Arnold, 19–46.

Burgers, J. (2000) 'Urban Landscapes: On Public Space in the Post-industrial City', *Journal of Housing and the Built Environment*, 15(2), 45–164.

Committee of Ministers of the Council of Europe (2000) *European Landscape Convention*. www.coe.int/en/web/landscape, accessed 2 January 2017.

Cremaschi, M. and Eckardt, F. (eds) (2011) *Changing Places: Urbanity, Citizenship and Ideology in New European Neighbourhoods*, European Urban Research Series, 3. Amsterdam: Techne Press.

de Vries, S., Verheij, R.A., Groenewegen, P.P. and Spreeuwenberg, P.P. (2003) 'Natural environments – healthy environments?', *Environmental Planning*, 35, 1717–1731.

Dobson, A., Lodge, D., Alder, J., Cumming, G.S., Keymer, J., McGlade, J., Mooney, H., Rusak, J.A., Sala, O., Wolters, V., Wall, D., Winfree, R. and Xenopoulos, M.A. (2006) 'Habitat Loss, Trophic Collapse, and the Decline of Ecosystem Services', *Ecology*, 87, 1915–1924.

EEA (2006) *Urban Sprawl in Europe: The Ignored Challenge*. EEA Report No 10/2006. European Environment Agency.

EEA (2009) *Ensuring Quality of Life in Europe's Cities and Towns*. EEA Reports No 5/2009. European Environment Agency.

EEA (2011) *Landscape Fragmentation in Europe.* EEA Report No 2/2011. European Environment Agency.

EEA (2012) *Urban Adaptation to Climate Change in Europe.* EEA Report No 2/2012. European Environment Agency.

EEA (2015) *Urban System.* SOER 2015. European Environment Agency.

EEA (2016) *Urban Sprawl in Europe.* Joint EEA-FOEN report. Technical report N°11/2016. European Environment Agency.

Forman, R.T. (2008) *Urban Region: Ecology and Planning Beyond the City.* Cambridge: Cambridge University Press.

Forman, R.T. and Alexander, L.E. (1998) 'Roads and Their Major Ecological Effects', *Annual Review of Ecology and Systematics*, 29, 207–231 and C2.

Forman, R.T. and Godron, M. (1981) 'Patches and Structural Components for a Landscape Ecology', *Bioscience*, 31, 733–740.

IUCN (International Union for Conservation of Nature and Natural Resources) (2001) *IUCN Red List Categories.* Gland, Switzerland: IUCN.

Jaeger, J.A.G., Bowman, J., Brennan, J., Fahrig, L., Bert, D., Bouchard, J., Charbonneau, N., Frank, K., Gruber, B. and Tluk von Toschanowitz, K. (2005) 'Predicting When Animal Populations Are at Risk from Roads: An Interactive Model of Road Avoidance Behavior', *Ecological Modelling*, 185, 329–348.

MA [Millennium Ecosystem Assessment] (2005a) *Ecosystems and Human Well-being: Current State and Trends.* Volume 1: *Findings of the Conditions and Trends Working Group*, ed. R. Hassan, R. Scholes and N. Ash. Washington, DC: Island Press.

MA [Millennium Ecosystem Assessment] (2005b) *Ecosystems and Human Well- being: Synthesis.* Washington, DC, Island Press.

Maes, J., Paracchini, M.L. and Zulian, G. (2011) *A European Assessment of the Provision of Ecosystem Services: Towards an Atlas of Ecosystem Services.* Luxembourg: European Commission Joint Research Centre/Institute for Environment and Sustainability.

Matthew, G.E. Mitchell, Suarez-Castro, A.F., Martinez-Harms, M., Maron, M., McAlpine, C., Gaston, K.J., Johansen, K. and Rhodes, J.R. (2015) 'Reframing Landscape Fragmentation's Effects on Ecosystem Services', *Trends in Ecology & Evolution*, 30(4), 190–198.

Mills, E.S. (2003) 'Book Review of Urban Sprawl Causes, Consequences and Policy Responses', *Regional Science and Urban Economics*, 33, 251–252.

OECD (2012) *Compact City Policies: A Comparative Assessment.* Paris: OECD.

Tengberg, A., Fredholm, S., Eliasson, I., Knez, I., Saltzman, K. and Wetterberg, O. (2012) 'Cultural Ecosystem Services Provided by Landscapes: Assessment of Heritage Values and Identity', *Ecosystem Services*, 2, 14–26.

Turner, M.G. (2005) 'Landscape Ecology: What Is the State of the Science?', *Annual Review of Ecology, Evolution, and Systematics*, 36, 319–344.

Turner, M.G., Gardner, R.H. and O'Neill, R.V. (2001) *Landscape Ecology in Theory and Practice: Patterns and Processes.* New York: Springer-Verlag.

UNFPA (2007) *State of World Population 2007*, New York. www.unfpa.org/sites/default/files/pub-pdf/695_filename_sowp2007_eng.pdf, accessed 15 August 2015.

UN-HABITAT (2012) *State of the World's Cities 2012/2013. Prosperity of Cities.* United Nations Human Settlements Programme. Nairobi: UN-HABITAT.

Zhao, P. (2010) 'Sustainable Urban Expansion and Transportation in a Growing Megacity: Consequences of Urban Sprawl for Mobility on the Urban Fringe of Beijing', *Habitat International*, 34, 236–243.

Zhao, P. (2011) 'Managing Urban Growth in a Transforming China: Evidence from Beijing', *Land Use Policy*, 28, 96–108.

# Part III

# Case studies

# 13 Soil consumption monitoring in Italy

*Michele Munafò and Luca Congedo*

## Introduction

In Italy, ISPRA is undertaking several activities related to land cover monitoring to assess soil consumption evolution over the last few decades. In particular, ISPRA developed a soil consumption monitoring network based on the sampling approach and the photo interpretation of very high resolution images; also, in the frame of the Copernicus initiative, ISPRA validated and enhanced the High Resolution Layers (HRLs) of 2012, which are land cover rasters. In addition, a Very High Resolution Layer (VHRL) of built-up was produced for 2012 with a spatial resolution of 5 m.

## Italian Soil Consumption Monitoring Network by ISPRA

In 2005, ISPRA and the National System for Environmental Protection (ARPA/APPA) developed a Soil Consumption Monitoring Network in order to overcome the lack of updated and homogenous data about soil consumption. This system allows for the assessment of soil consumption trends in Italy from the 1950s to today with a stratified sampling that implements the photointerpretation of very high resolution images and topographical maps. The monitoring network is the official benchmark for national soil consumption in the National Statistic Program 2014–2016.

The survey is integrated with cartographic data required for the validation and to ensure coherence with spatial data, in particular with the activities undertaken in the Copernicus framework. This network based on the interpretation of very high resolution data is not affected by the constraints of the minimum mapping unit, allowing for more accurate and reliable estimation. In particular, it is possible to include sparse built-up that individually covers small impervious surfaces for the assessment of small land cover changes. These data allow for the calculation of soil consumption indicators, accuracy assessment and error estimation.

The network has about 180,000 samples – 12,000 belong to the national monitoring network, about 28,000 to the regional network and the remainder to the municipal network. The result is a two-level classification of land cover and soil sealing.

*Table 13.1* Soil consumption in Italy (km$^2$ and percentage over the national surface)

|  | 1950s | 1989 | 1996 | 1998 | 2006 | 2008 | 2013 | 2014 |
|---|---|---|---|---|---|---|---|---|
| Soil consumption | 8,100 (2.7%) | 15,300 (5.1%) | 17,100 (5.7%) | 17,600 (5.8%) | 19,400 (6.4%) | 19,800 (6.6%) | 20,800 (6.9%) | 21,000 (7.0%) |

Source: Munafò *et al.* (2015).

Estimates show that soil consumption in Italy is significantly increasing, although the pace of growth is slowing: between 2008 and 2013 soil consumption involved about 55 ha per day, with a speed of 6–7 m$^2$/s. In particular, the temporal analysis shows that soil consumption increased from 2.7 per cent in the 1950s to 7.0 per cent in 2014, with a difference of 4.3 percentage points. Globally, about 21,000 km$^2$ of soil have been occupied by built-up (see Table 13.1).

Soil consumption is continuously occupying natural and agricultural areas, where impervious surfaces like asphalt and concrete are growing with buildings, roads and infrastructures, often in low-density urban areas.

In 2013, 15 regions reached 5 per cent of soil consumption, with higher values in Lombardia and Veneto (northern Italy), and Campania and Puglia (southern Italy), as illustrated in Table 13.2. It is worth noticing that

*Table 13.2* Percentage of soil consumption (range) in Italian regions

|  | 1950s | 1989 | 1996 | 1998 | 2006 | 2008 | 2013 |
|---|---|---|---|---|---|---|---|
| Piemonte | 2.2–3.9 | 4.4–6.3 | 4.7–6.7 | 4.8–6.8 | 5.0–7.0 | 5.1–7.1 | 5.9–8.2 |
| Valle d'Aosta | 1.1–2.3 | 1.7–3.0 | 1.8–3.1 | 1.8–3.1 | 2.0–3.4 | 2.0–3.4 | 2.2–3.7 |
| Lombardia | 3.9–5.8 | 6.8–9.0 | 7.5–9.9 | 7.7–10.1 | 8.5–11.0 | 8.8–11.3 | 9.6–12.2 |
| Trentino-Alto Adige | 0.9–2.0 | 1.5–2.7 | 1.6–2.8 | 1.6–2.9 | 1.8–3.1 | 1.8–3.1 | 1.8–3.2 |
| Veneto | 3.0–4.8 | 5.0–7.1 | 6.2–8.3 | 6.5–8.7 | 7.7–10.1 | 8.3–10.8 | 8.6–11.1 |
| Friuli-Venezia Giulia | 2.2–3.8 | 4.4–6.3 | 5.0–7.0 | 5.1–7.1 | 5.5–7.5 | 5.6–7.7 | 5.8–7.9 |
| Liguria | 2.0–3.5 | 4.2–6.1 | 5.0–7.0 | 5.2–7.2 | 5.6–7.7 | 5.6–7.7 | 5.9–8.0 |
| Emilia-Romagna | 1.8–3.0 | 5.7–7.7 | 6.4–8.4 | 6.6–8.7 | 6.7–8.8 | 6.8–8.8 | 6.9–8.9 |
| Toscana | 1.6–3.0 | 3.7–5.5 | 4.5–6.4 | 4.5–6.5 | 5.1–7.2 | 5.2–7.2 | 5.3–7.4 |
| Umbria | 1.1–2.3 | 2.6–4.2 | 3.1–4.8 | 3.2–4.9 | 4.2–6.2 | 4.2–6.2 | 4.3–6.3 |
| Marche | 1.9–3.5 | 3.9–5.8 | 4.6–6.6 | 4.8–6.8 | 5.1–7.3 | 5.3–7.4 | 5.7–7.9 |
| Lazio | 1.3–2.4 | 4.5–6.3 | 5.5–7.4 | 5.9–7.9 | 6.1–8.0 | 6.1–8.1 | 6.4–8.4 |
| Abruzzo | 1.0–2.2 | 2.7–4.3 | 3.2–4.9 | 3.3–5.0 | 3.6–5.5 | 4.0–5.8 | 4.2–6.1 |
| Molise | 1.3–2.7 | 2.2–3.7 | 2.4–4.0 | 2.5–4.1 | 2.7–4.3 | 2.8–4.5 | 3.0–4.7 |
| Campania | 3.5–5.4 | 6.0–8.2 | 6.5–8.7 | 6.6–8.8 | 7.2–9.5 | 7.5–9.8 | 7.8–10.2 |
| Puglia | 2.6–4.3 | 5.3–7.2 | 6.0–8.0 | 6.3–8.4 | 7.1–9.3 | 7.3–9.6 | 7.4–9.7 |
| Basilicata | 1.5–3.0 | 2.2–3.7 | 2.6–4.1 | 2.7–4.3 | 3.3–5.1 | 3.4–5.2 | 3.6–5.3 |
| Calabria | 1.6–3.1 | 3.1–4.8 | 3.4–5.2 | 3.4–5.2 | 3.9–5.7 | 4.3–6.1 | 4.5–6.4 |
| Sicilia | 1.4–2.8 | 4.5–6.5 | 4.9–6.9 | 5.0–7.0 | 5.5–7.7 | 5.5–7.7 | 5.8–7.9 |
| Sardegna | 1.1–2.3 | 2.0–3.3 | 2.3–3.7 | 2.4–3.8 | 3.2–4.8 | 3.3–5.0 | 3.4–5.0 |

Source: Munafò *et al.* (2015).

estimation is provided with a confidence interval of 95 per cent, depending on the regions, the characteristics of the monitoring network and the estimation error.

## Copernicus High Resolution Layers

Copernicus is a European initiative aimed at monitoring the environment through several products and services. The High Resolution Layers (HRLs) are rasters from 2012 that monitor the land cover of European countries with a high level of detail (i.e. 20 m – higher resolution than Corine Land Cover). In particular the following issues are monitored: soil imperviousness, forest, grassland, wetland and surface water.

The production of HRLs was multi-step: the first phase was performed by different service providers that processed the IMAGE2012 dataset, which is composed of remote sensing images such as RapidEye; subsequently, intermediate HRLs were validated and enhanced, using regional and local cartography and ancillary data.

It is worth noting that Copernicus services and data are provided free of charge to users. The intermediate HRLs were produced using a semi-automatic approach, and in Italy were validated and enhanced by ISPRA.

The Degree of Imperviousness is specifically designed to monitor soil consumption, in particular providing a percentage of soil sealing per pixel. The spatial resolution of HRLs is particularly useful at the regional level for assessing urban sprawl, defined as unplanned, low-density urban expansion, characterized by a mix of land uses on the urban fringe (European Environmental Agency, 2006).

## High Resolution Land Cover Map of Italy

ISPRA developed and distributed a High Resolution Land Cover Map of Italy (20 m spatial resolution), which is the result of the integration of HRLs 2012. The Degree of Imperviousness was reclassified in order to obtain a binary map, where imperviousness values greater than 29 per cent were considered built-up (Maucha *et al.*, 2011).

Figure 13.1 shows the map with the following land cover classes:

- Built-up
- Broadleaved forest
- Coniferous forest
- Grassland
- Wetland
- Permanent Water Bodies
- Other
- Unclassified.

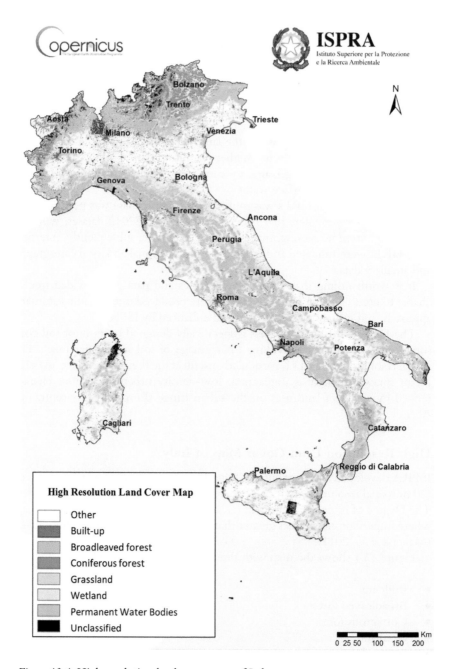

*Figure 13.1* High resolution land cover map of Italy

## Very High Resolution Layer of built-up

In Italy, ISPRA has developed a Very High Resolution Layer (VHRL) that identifies built-up areas with a spatial resolution of 5 m (see Figure 13.2).

*Figure 13.2* Very high resolution layer of built-up area

Similar to Copernicus HRL, VHRL production was based on the semi-automatic classification of satellite images (i.e. RapidEye acquired in 2012) and the integration of local ancillary data such as OpenStreetMap, in order to identify the built-up.

It is worth pointing out that this VHRL and the HRL Degree of Imperviousness are different in terms of spatial resolution (5 m and 20 m),

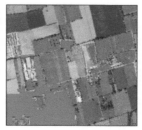

*Figure 13.3* Comparison of resolutions for orthoimagery (left), HRL (centre) and VHRL (right)

classification system (binary map and percentage of imperviousness) and class definition: railway lines, dump sites and mines are excluded from the Copernicus HRL, but included in the ISPRA VHRL; in fact, the inclusion of these features in HRLs is still debated, which for ISPRA should be considered soil consumption. Figure 13.3 compares VHRL and HRL, showing that the VHRL outperforms the HRL especially for the identification of streets and small buildings.

## Urbanization pattern

Understanding urbanization structures and patterns is a requirement for defining effective policies for limiting soil consumption and fostering sustainability governance. HRLs allowed for the assessment of urban dynamics through the calculation of landscape metrics that are measures describing the characteristics of landscape patches regarding the structure, function and changes thereof, initially developed for ecological studies (McGarigal and Marks, 1995). Spatial metrics are useful for assessing the physical characteristics and patterns of landscape, in particular for studying land cover change in urban areas (Huang *et al.*, 2009).

For the Italian case study, the following landscape metrics were calculated using the HRL Degree of Imperviousness (Munafò *et al.*, 2015):

- Largest Class Patch Index (LCPI): percentage of landscape occupied by the largest patch, indicating landscape compactness;
- Residual Mean Patch Size (RMPS): the mean patch area excluding the largest patch, providing the dimension of urban sprawl around the city centre;
- Edge Density (ED): the perimeter of urban areas dividing the area thereof, describing urban fragmentation;
- Urban Sprawl Index: ratio of high-density and low-density areas, describing the variation of urban density related to urban sprawl.

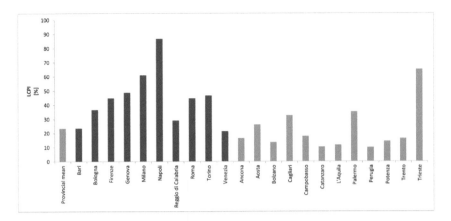

*Figure 13.4* LCPI at the provincial level (source: Munafò *et al.*, 2015)

LCPI, a compactness indicator, has higher values in cities with a large urban centre, lower values where urban sprawl is predominant. Results at the provincial level are shown in Figure 13.4, where Napoli, Milano and Trieste have the highest value.

The RMPS is highly influenced by the study scale, and it provides the dimension (in hectares) for sprawl around cities; high RMPS values imply polycentric cities while low values mean fragmentation of the periphery not connected to the city centre. The results at the provincial level are shown in Figure 13.5, where Milano has the highest value due to the presence of compact areas around the city. In order to understand this phenomenon it is necessary to combine this with evaluation of the other metrics, especially LCPI.

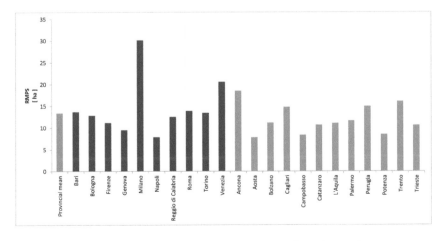

*Figure 13.5* RMPS at the provincial level (source: Munafò *et al.*, 2015)

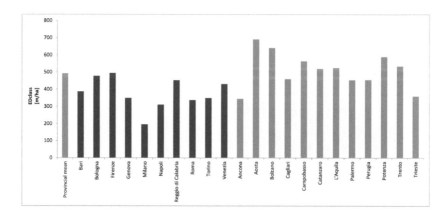

*Figure 13.6* ED at the provincial level (source: Munafò *et al.*, 2015)

In order to assess urban sprawl, the ED was calculated as it is related to the morphological characteristics of urban boundaries, which are influenced also by altitude and slope. In particular, higher values mean irregularity of urban boundary, while low values are related to compact shapes with regular boundaries. The ED results are shown in Figure 13.6, where large urban areas such as Milano and Napoli have lower values.

The Urban Sprawl Index describes urban dispersion and fragmentation, as it is discontinuous areas divided by the total area; low values mean compactness while higher values represent sprawling cities (European Environmental Agency, 2006; ESPON, 2011). Milano and Napoli again have lower values, because of their characteristic compactness (see Figure 13.7).

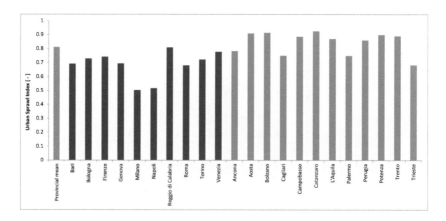

*Figure 13.7* Urban Sprawl Index at the provincial level (source: Munafò *et al.*, 2015)

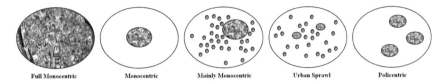

*Figure 13.8* Classes of urban development

The above results refer to the provincial level, but it is worth noting that analysis was performed also at the municipal level and described in ISPRA (2013).

Consequently, urban areas were grouped into five classes (Figure 13.8):

- 'Full Monocentric': municipalities with compact urban development, also beyond the municipal boundaries;
- 'Monocentric': municipalities with compact urban development, only within the municipal boundaries;
- 'Mainly Monocentric': municipalities with a centre tending to sprawl in the periphery;
- 'Urban Sprawl': municipalities affected by fragmentation without a city centre;
- 'Policentric': municipalities with several small centres.

The results of the landscape metrics calculated for provincial capitals are shown in Figure 13.9, where classes of urban development are defined.

Municipalities that have sprawling features such as Mainly Monocentric and Urban Sprawl classes are affected by the worst risk caused by the negative effects of urban fragmentation. Also, greater attention is required for Full Monocentric cities, such as Milano and Torino, which exceed municipal boundaries.

Most Italian cities are Mainly Monocentric, such as Campobasso and Reggio Emilia, although several cities are Monocentric, such as Firenze, Genova and Bologna. Polycentric cities (e.g. Venezia, Bari, Taranto) are less numerous, and the shape thereof is influenced by the morphology of the ground, coast line and growth of industrial areas or infrastructures.

There are also several Urban Sprawl cities, characterized by the interspersion of urban features in natural and agricultural areas, such as Trapani, Latina and Ferrara.

The results of this analysis provide an important step forward in understanding landscape dynamics and urban shapes that are crucial for environmental sustainability.

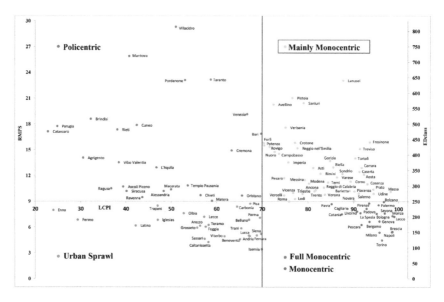

*Figure 13.9* Distribution of provincial capitals according to the landscape metrics and classes of urban development (source: Munafò *et al.*, 2015)

## Urban fragmentation

Urban fragmentation and configuration was analysed through landscape metrics calculated using the High Resolution Land Cover Map, in order to assess spatial configuration and heterogeneity.

Table 13.3 shows the landscape metrics calculated for the Italian landscape at the provincial level. Metrics with a high degree of correlation have been excluded from the calculation, in order to avoid redundant information in the analysis (Bogaert, 2005).

*Table 13.3* Landscape metrics calculated at the provincial level

| Indicator | Description |
| --- | --- |
| MPA (*Mean Patch Area*) | Average of the area of individual patches for each class, tends to increase with the increasing homogeneity of the landscape |
| PD (*Patch Density*) | Number of patches dividing the landscape area, high values mean landscape fragmentation |
| PLADJ (*Percentage of Like Adjacencies*) | Percentage of adjacencies between pixels belonging to different classes, high values mean landscape heterogeneity |
| SHDI (*Shannon Diversity Index*) | Indicator combines class abundance and landscape homogeneity, describing the landscape diversity |
| MSI (*Mean Shape Index*) | Indicator describing patch shape: 1 indicates regular shapes (e.g. circle) and the number tends to increase with shape irregularity and complexity |

These indicators are designed to characterize the degree of homogeneity (MPA) and complexity (MSI) of the landscape, the heterogeneity and diversity of forms present (PLADJ, SHDI) and the fragmentation of landscape units (PD). The combination of these indicators enables the assessment of the Italian landscape according to the classification system provided by the High Resolution Land Cover Map. Figures 13.10, 13.11, 13.12, 13.13 and 13.14 show the calculated metrics.

Results show that Italian landscape is generally not homogenous with a high level of fragmentation (i.e. MPA < 15 per cent and PLADJ > 90 per cent), also with irregular shapes (i.e. MSI between 1.2 and 1.5). The high variability of MPA and SHDI describe the diversity of landscape, where some provinces

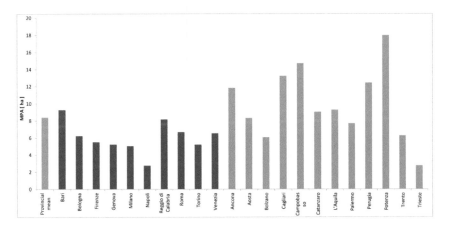

*Figure 13.10* Mean Patch Area calculated at the provincial level (source: Munafò *et al.*, 2015)

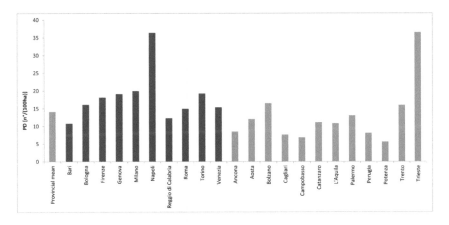

*Figure 13.11* Patch Density calculated at the provincial level (source: Munafò *et al.*, 2015)

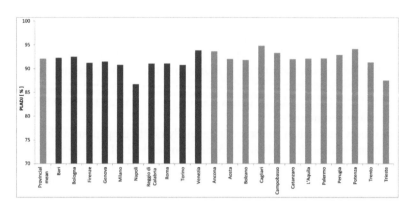

*Figure 13.12* Percentage of Like Adjacencies calculated at the provincial level (source: Munafò *et al.*, 2015)

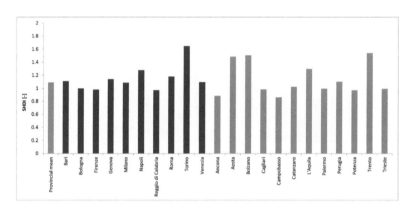

*Figure 13.13* Shannon Diversity Index calculated at the provincial level (source: Munafò *et al.*, 2015)

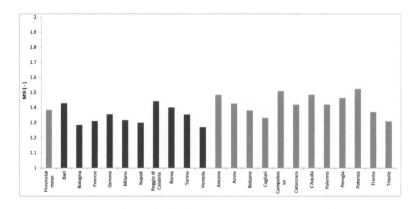

*Figure 13.14* Mean Shape Index calculated at the provincial level (source: Munafò *et al.*, 2015)

such as Ancona and Cagliari are highly homogenous (as shown by PD and SHDI values). Other provinces are more fragmented as shown by higher PD values and lower MPA values; in particular, the high SHDI of Napoli is characteristic of landscape variability.

It is worth noting the influence of local morphology on these metrics – for example, Potenza and Campobasso have high MSI values due to the mountain areas, and present homogenous characteristics (as shown by low MPA and SHDI values).

An overall analysis of metropolitan areas reveals a trend of landscape metrics around average values, which can be explained by the predominance of the built-up class (having more regular shape and less fragmentation) over the natural and semi-natural areas.

## Soil consumption

The VHRL of 2012 allows for the accurate estimation of soil consumption at the local level, and thus ISPRA (Munafò *et al.*, 2015) calculated soil consumption for all Italian municipalities. Nevertheless, soil consumption estimates are generally lower (i.e. about 1 per cent) than actual soil consumption (i.e. sampling method) due to the cartographic method that tends to omit very small or narrow surfaces (e.g. small roads).

At the municipal level, Rome has the highest soil consumption surface (about 30,000 ha) while the provincial capitals have very high values (Milano, Torino and Napoli with values higher than 4,000 ha). However, several non-capital cities have high values of soil consumption (e.g. Marsala in Sicily).

*Table 13.4* Soil consumption (%) at the municipal level for the top 20 municipalities, 2012

|  | *Municipality* | *Province* | *Soil consumption [%]* |
|---|---|---|---|
| 1 | Casavatore | Napoli | 85.4 |
| 2 | Arzano | Napoli | 78.9 |
| 3 | Melito di Napoli | Napoli | 76.0 |
| 4 | Cardito | Napoli | 67.9 |
| 5 | Frattaminore | Napoli | 66.9 |
| 6 | Torre Annunziata | Napoli | 65.2 |
| 7 | Lissone | Monza e Brianza | 64.0 |
| 8 | Casoria | Napoli | 63.1 |
| 9 | Portici | Napoli | 62.3 |
| 10 | San Giorgio a Cremano | Napoli | 60.1 |
| 11 | Aversa | Caserta | 60.0 |
| 12 | Mugnano di Napoli | Napoli | 59.1 |
| 13 | Lallio | Bergamo | 59.1 |
| 14 | Frattamaggiore | Napoli | 59.1 |
| 15 | Curti | Caserta | 59.0 |
| 16 | Sant'Antimo | Napoli | 58.1 |
| 17 | Fiera di Primiero | Trento | 57.9 |
| 18 | Torino | Torino | 57.6 |
| 19 | Napoli | Napoli | 57.0 |
| 20 | Sesto San Giovanni | Milano | 56.8 |

Table 13.4 contains the highest percentages of soil consumption at the municipal level; it is worth noting that several municipalities belonging to the Province of Napoli and Caserta, Milano and Torino have values higher than 50 per cent. Urban development in these municipalities appears to be strongly influenced by the economic power of capital cities.

## References

Bogaert J. (2005) 'Metriche del paesaggio: definizioni e utilizzo', *Estimo e Territorio* 9: 8–14.

ESPON (2011) *ESPON Climate: Climate Change and Territorial Effects on Regions and Local Economies. Final Report Annex 4: Case Study Mediterranean Coast of Spain,* Technical report. Dortmund: ESPON and IRPUD ESPON.

European Environmental Agency (2006) *Urban Sprawl in Europe: The Ignored Challenge.* Copenhagen: EEA/OPOCE.

Huang, S.-L., Wang, S.-H. and Budd, W.W. (2009) 'Sprawl in Taipei's peri–urban zone: responses to spatial planning and implications for adapting global environmental change', *Landscape and Urban Planning* 90(1–2): 20–32.

ISPRA (2013) 'Il monitoraggio del consumo di suolo in Italia', *Ideambiente* 62: 20–31. www.isprambiente.gov.it/files/ideambiente/ideambiente_62.pdf, accessed 21 June 2016.

McGarigal, K. and Marks, M. (1995) 'FRAGSTATS: spatial pattern analysis program for quantifying landscape structure', General Technical Report PNW-GTR-351. Portland, OR: Pacific Northwest Research Station.

Maucha, G., Büttner, G. and Kosztra, B. (2011) 'European validation of GMES FTS soil sealing enhancement', Data 31st EARSeL Symposium and 35th General Assembly 2011, EARSeL, 223–238.

Munafò, M., Assennato, F., Congedo, L., Luti, T., Marinosci, I., Monti, G., Riitano, N., Sallustio, L., Strollo, A., Tombolini, I. and Marchetti, M. (2015) *Il consumo di suolo in Italia,* Edizione 2015. Rapporti 218/2015. Rome: ISPRA.

# 14 Urban land expansion and its impacts on cultivated land in the Pearl River Delta, China

*Xiaoqing Song and Zhifeng Wu*

## Introduction

Urban land expansion has been a key driving factor for multifunctional rural land loss. Moreover, cultivated land loss in the process of urban–rural interaction has posed major challenges for food security worldwide (Foley *et al.*, 2005). Since the reform and opening up in 1978, intensive land use change involving urban land expansion and cultivated land use conversion have been witnessed in China (Deng *et al.*, 2015; Liu *et al.*, 2014; Wang *et al.*, 2012). For example, small towns grew into megacities, especially in the coastal regions such as the Pearl River Delta, which led to the loss of a large amount of highly productive cultivated land (Seto *et al.*, 2002; Altrock and Schnoon, 2014). In the meantime, a vast amount of additional cultivated land with lower productivity was converted from forest and grassland in northeastern and western China (Yan *et al.*, 2009). To ensure food and ecological securities, the central government has implemented a series of land use policies for guiding the smart growth of urban land. The proportion of urban population, however, amounted to 54.77 percent of the total population in 2014, according to the *China Statistical Yearbook* (National Bureau of Statistics of China, 2015). It is projected that urban land expansion will continue, which will inevitably lead to massive cultivated land use change in the future. Thus it is urgent to seek solutions for coordinating sustainable urbanization, food security, and ecosystem services.

The Pearl River Delta, located in the subtropical monsoon zone of south Asia, is one of the greatest urban agglomerations in China. Cultivated land in this delta is diverse with high productivity. After the reform and opening up, urbanization in this delta proceeded rapidly, forced by an export-oriented economic model, which successfully promoted conversion from manufacturing to the service industry (Altrock and Schnoon, 2014). It is acknowledged that urbanization in this delta has transformed into the post-urbanization stage with the most developed economy and the highest proportion of urban population in China. Thus, analysis, from the perspective of urbanization transition, of urban land expansion and its impacts on cultivated land use change in this delta is of significance for policy making with regard to coordinating sustainable urbanization, food security, and ecosystem services.

## Data and methodology

The land use dataset used in this study was produced using the ERDAS IMAGINE software based on Landsat TM images, with a ground resolution of 30 m. The time dimension covers the years 1980, 1990, 2000, 2005, and 2010. Urban land and cultivated land were categorized according to land use classification system in China employing both unsupervised classification and supervised classification methods.

Urbanization transition refers to any change in urbanization from one state to another. From the perspective of urbanization transition, land area (Area) and number of land parcels (NP) were used as key indicators for analyzing change in urban land expansion and cultivated land use change. Specifically, urban land expansion changed from experiencing increase to undergoing a decrease while the number of urban land parcels went from increasing to decreasing, indicating a transition in urban land use. Cultivated land loss also changed from experiencing increase to undergoing decrease while the number of cultivated land parcels went from increasing to decreasing, indicating a transition in cultivated land use, too. Urban land parcels were categorized into five levels using 50 hm², 100 hm², 500 hm², and 1,000 hm², considering the average size of urban landscape in the Pearl River Delta. Cultivated land parcels were categorized into six levels using 1 hm², 5 hm², 10 hm², 50 hm², and 100 hm², considering the average size of cultivated landscape in the Pearl River Delta. Then, changes of Area and NP among the different size levels were analyzed. Additionally, cultivated land use structure changes were analyzed using the ratio of dry farmland area to paddy fields area.

## Results

### Change in urban land expansion in 1980–2010

Table 14.1 shows that the total area of urban land increased by 81.22 percent from 1980 to 2010. Urban land expansion, however, slowed down after 2005. Specifically, the annual expansion of urban land increased from 5,050.33 hm² to 15,925.70 hm² from 1980–1990 to 2000–2005. In 2005–2010, the annual expansion of urban land decreased to 11,776.40 hm². Although the total number of urban land parcels increased by 3.64 percent before 2000, it decreased by 8.15 percent in 1980–2010. Annual growth of the average area of urban land parcels increased from 0.14 hm² to 0.74 hm² from 1980–1990 to 2000–2005. However, it decreased to 0.54 hm² in 2005–2010.

Among the different size levels, urban land parcels of more than 1,000 hm² had the largest expansion with the expansion area of 276,540.42 hm² in 1980–2010. Meanwhile, urban land parcels smaller than 50 hm² and between 50 hm² and 100 hm², however, had much less expansion with the expansion area of 1,778.66 hm² and 2,106 hm², respectively. The number of urban land parcels smaller than 50 hm² decreased by 2,909. The number of urban land parcels between 100 and 500 hm² had the largest increase (60), followed by the number of urban land parcels bigger than 1,000 hm² (30).

*Table 14.1* Change in urban land parcels with different sizes in 1980–2010, Pearl
River Delta (unit: hm$^2$)

|  |  | 1980 | 1990 | 2000 | 2005 | 2010 |
|---|---|---|---|---|---|---|
| Total | Area | 376,377.44 | 426,880.78 | 543,569.51 | 623,198.03 | 682,080.05 |
|  | NP | 33,982 | 34,222 | 35,221 | 32,548 | 31,211 |
| <50 hm$^2$ | Area | 136,141.10 | 139,325.57 | 149,593.63 | 140,415.80 | 137,919.77 |
|  | NP | 33,333 | 33,547 | 34,426 | 31,797 | 30,424 |
| 50–100 hm$^2$ | Area | 23,589.88 | 24,681.32 | 27,515.44 | 23,508.53 | 25,695.88 |
|  | NP | 337 | 351 | 394 | 338 | 367 |
| 100–500 hm$^2$ | Area | 52,366.78 | 52,439.37 | 65,470.93 | 63,738.72 | 65,766.10 |
|  | NP | 263 | 265 | 324 | 316 | 323 |
| 500–1,000 hm$^2$ | Area | 17,111.86 | 19,855.68 | 19,281.29 | 29,914.53 | 28,990.07 |
|  | NP | 24 | 28 | 30 | 44 | 42 |
| ≥1,000 hm$^2$ | Area | 147,167.81 | 190,578.84 | 281,708.22 | 365,620.45 | 423,708.24 |
|  | NP | 25 | 31 | 47 | 53 | 55 |

### Change in cultivated land loss in 1980–2010

Table 14.2 shows that the total area of cultivated land decreased by 29.08
percent from 1980 to 2010. Cultivated land loss, however, slowed down after
2005. Specifically, the annual loss of cultivated land increased from 8,190.20
hm$^2$ to 18,203.25 hm$^2$ from 1980–1990 to 1990–2000. From 2000–2005
to 2005–2010, however, the annual loss of cultivated land decreased from
13,586.85 hm$^2$ to 8,842.94 hm$^2$. The annual loss of total number of cultivated
land parcels increased from 494 to 715 from 1980–1990 to 1990–2000. In
2000–2005 and 2005–2010, the annual loss of total number of cultivated land
parcels was 1,196 and 1,823 respectively. The average area of cultivated land
decreased from 12.97 hm$^2$ to 11.75 hm$^2$. From 2000 to 2010, however, it
increased from 11.78 hm$^2$ to 12.65 hm$^2$.

Among the different size levels, the annual loss of area of cultivated land parcels less than 1 hm$^2$, between 1 hm$^2$ and 5 hm$^2$, and between 5 hm$^2$ and 10 hm$^2$
increased by 310.61 hm$^2$, 1,900.98 hm$^2$, and 698.18 hm$^2$ from 1980–1990
to 2005–2010, respectively. The annual loss of area of cultivated land parcels between 10 hm$^2$ and 50 hm$^2$, between 50 hm$^2$ and 100 hm$^2$, and bigger
than 100 hm$^2$, increased by 2,559.83 hm$^2$, 1,828.93 hm$^2$, and 5,447.14 hm$^2$
from 1980–1990 to 1990–2000. From 1990–2000 to 2005–2010, however,
the three annual losses above decreased by 2,511.92 hm$^2$, 2,358.16 hm$^2$, and
7,222.85 hm$^2$, respectively. The annual loss of number of cultivated land parcels smaller than 1 hm$^2$ and between 1 hm$^2$ and 5 hm$^2$ increased by 1,155 and
3,747 from 1980–1990 to 2005–2010, respectively. The annual loss of number
of cultivated land parcels between 5 hm$^2$ and 10 hm$^2$, between 10 hm$^2$ and
50 hm$^2$, between 50 hm$^2$ and 100 hm$^2$, and bigger than 100 hm$^2$, increased
by 26, 509, 106, and 83 from 1980–1990 to 1990–2000, respectively. From
1990–2000 to 2005–2010, however, the four annual losses above decreased by
382, 1622, 363, and 266, respectively.

*Table 14.2* Change in cultivated land with different sizes in 1980–2010, Pearl River Delta (unit: hm²)

|  |  | 1980 | 1990 | 2000 | 2005 | 2010 |
|---|---|---|---|---|---|---|
| Total | Area | 129,3486.66 | 1,211,584.63 | 1,029,552.08 | 961,617.62 | 917,402.93 |
|  | NP | 99,712 | 94,774 | 87,622 | 81,642 | 72,526 |
| 0–1 hm² | Area | 14,038.73 | 13,761.37 | 13,365.79 | 12,145.50 | 10,453.78 |
|  | NP | 20,502 | 19,426 | 17,246 | 15,622 | 13,391 |
| 1–5 hm² | Area | 112,356.66 | 108,341.44 | 105,544.91 | 99,460.37 | 87,947.88 |
|  | NP | 46,739 | 45,476 | 45,012 | 42,245 | 37,235 |
| 5–10 hm² | Area | 95,851.98 | 88,313.72 | 77,903.49 | 73,514.15 | 66,254.14 |
|  | NP | 13,575 | 12,515 | 11,099 | 10,471 | 9,437 |
| 10–50 hm² | Area | 313,369.30 | 286,680.83 | 234,394.06 | 219,772.85 | 206,189.08 |
|  | NP | 14,926 | 13,665 | 11,291 | 10,582 | 9,830 |
| 50–100 hm² | Area | 151,268.38 | 140,879.14 | 112,200.54 | 101,868.95 | 99,320.44 |
|  | NP | 2,189 | 2,036 | 1,626 | 1,476 | 1,429 |
| ≥100 hm² | Area | 606,601.60 | 573,608.12 | 486,143.29 | 454,855.80 | 447,237.61 |
|  | NP | 1,781 | 1,656 | 1,348 | 1,246 | 1,204 |

### Change in cultivated land use structure in 1980–2010

Table 14.3 shows that the ratio of total area of dry farmland to total area of paddy fields increased by 0.69 from 1980 to 2005, without the slightest decrease in 2000–2005. In 1980–1990 and 1990–2000, growth of the ratio was 0.09 and 0.24 respectively. Growth of the ratio increased by 0.35 in 2000–2010.

Among the different size levels, the ratios of area of dry farmland to area of paddy fields between 10 and 50 hm² and bigger than 100 hm² increased by 0.73 and 0.70 in 1980–2010, respectively. Meanwhile, the ratios of area of dry farmland to area of paddy fields smaller than 1 hm² and between 1 and 5 hm² increased by only 0.17 and 0.40, respectively. Moreover, in 2000–2010, the ratio of area of dry farmland to area of paddy fields bigger than 100 hm² increased by 0.40, which was the greatest increment among the six size levels.

*Table 14.3* Change in the ratio of area of dry farmland to area of paddy fields with different sizes in 1980–2010, Pearl River Delta (unit: hm²)

|  | 1980 | 1990 | 2000 | 2005 | 2010 |
|---|---|---|---|---|---|
| Total | 0.79 | 0.88 | 1.12 | 1.09 | 1.48 |
| 0–1 hm² | 1.08 | 1.10 | 1.11 | 1.11 | 1.25 |
| 1–5 hm² | 1.12 | 1.20 | 1.28 | 1.27 | 1.52 |
| 5–10 hm² | 0.93 | 1.07 | 1.28 | 1.29 | 1.60 |
| 10–50 hm² | 0.86 | 0.98 | 1.28 | 1.33 | 1.59 |
| 50–100 hm² | 0.81 | 0.92 | 1.27 | 1.25 | 1.31 |
| ≥100 hm² | 0.67 | 0.75 | 0.97 | 0.90 | 1.37 |

# Discussion

## *Urban land use transition in the Pearl River Delta*

In the process of urban land expansion in 1980–2010, urban land use transition occurred as an annual expansion of urban land, and the annual growth of the average area of urban land parcels turned from increase to decrease after 2005. This transition in essence mainly resulted from the increasing land value. Specifically, as urban expansion and economic growth proceeded, land scarcity became the key factor in rising land value, which in turn restrained the expansion of urban land especially smaller urban land parcels. Meanwhile, more expansion was allocated to the larger urban land parcels to achieve economies of scale and to raise land use efficiency, such as urban land parcels bigger than 1,000 $hm^2$.

## *Cultivated land use transition in the Pearl River Delta*

Results of change in cultivated land loss show that annual loss of cultivated land turned from increase to decrease after 2000. Meanwhile, the average area of cultivated land turned from decrease to increase. This implies that cultivated land use transition occurred in 2000. This transition mainly resulted from the growing value of cultivated land and the strict land use control policies for cultivated land protection (Lichtenberg and Ding, 2008), e.g., the dynamic balance policy in which the total amount of cultivated land converted was compensated for by reclaimed land with equivalent qualities (Li *et al.*, 2009). Moreover, cultivated land parcels with the largest size had the most remarkable transition, and vice versa. Thus large-scale land management with the advantage of economies of scale contributes to hindering cultivated land conversion.

## *Implications for coordinating sustainable urbanization, food security, and ecosystem services*

Urban land use transition mainly resulting from the increasing land value in the Pearl River Delta hints at a sustainable urbanization model for other regions in China. Moreover, cultivated land use transition presents an opportunity for alleviating the contradiction between urbanization and food security in China. However, urban land use transition and cultivated land use transition are not determinative and not a certainty. Both of these transitions interact with and are forced by economic and institutional development (Figure 14.1).

Additionally, the growing ratio of dry farmland area to paddy fields area was forced by the farmers' desire for more income. Specifically, more and more paddy fields with rice farming were converted to dry farmland for the large-scale management of cash crops such as vegetables, fruits, and flowers, especially at the level of large cultivated land parcels in 1980–2010. This conversion,

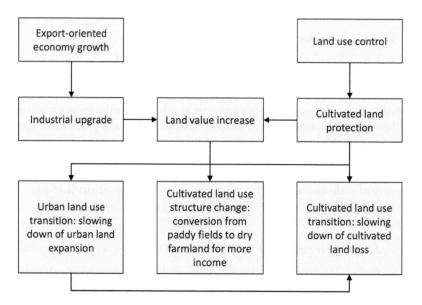

*Figure 14.1* Mechanisms of urban land use transition and cultivated land use transition in the Pearl River Delta, China

however, has posed major challenges for sustaining ecosystem services. First, paddy fields could provide more climate regulating services and soil formation services than dry farmland. Second, chemical input intensity on dry farmland was much higher than in paddy fields, which contributed to massive biodiversity loss in farmland. For example, according to the *Data Compilation of the National Agricultural Costs and Returns*, nitrogen fertilizer used for vegetables in dry farmland is 330.15 kg/hm², which is 189.90 kg/hm² more than that in paddy fields for rice farming (Department of Price of National Development and Reform Commission, 2013). Thus, more attention should be paid to ecosystem services maintenance to guide the smart conversion from paddy fields to dry farmland in the process of cultivated land use transition.

## References

Altrock U. and Schnoon S. (2014) *Maturing Megacities: The Pearl River Delta in Progressive Transformation*. Dordrecht: Springer.

Deng, X., Huang, J., Rozelle, S., Zhang, J. and Li, Z. (2015) 'Impact of urbanization on cultivated land changes in China', *Land Use Policy*, 45, 1–7.

Department of Price of National Development and Reform Commission (2013) *Data Compilation of the National Agricultural Costs and Returns*. Beijing: China Statistics Press (in Chinese).

Foley, J. A., DeFries, R., Asner, G. P., Barford, C., Bonan, G., Carpenter, S. R., Chapin, F. S., Coe, M. T., Daily, G. C., Gibbs, H. K., Helkowski, J. H., Holloway, T., Howard, E. A., Kucharik, C. J., Monfreda, C., Patz, J. A., Prentice, I. C., Ramankutty, N. and Snyder, P. K. (2005) 'Global consequences of land use', *Science*, 309(5734), 570–574.

Li, W., Feng, T. T. and Hao, J. M. (2009) 'The evolving concepts of land administration in China: cultivated land protection perspective', *Land Use Policy*, 26, 262–272.

Lichtenberg, E. and Ding, C. R. (2008) 'Assessing farmland protection policy in China', *Land Use Policy*, 2008, 25, 59–68.

Liu, J., Kuang, W., Zhang, Z., Xu, X., Qin, Y., Ning, J., Zhou, W., Zhang, S., Li, R., Yan, C., Wu, S., Shi, X., Jiang, N., Yu, D., Pan, X. and Chi, W. (2014) 'Spatiotemporal characteristics, patterns, and causes of land-use changes in China since the late 1980s', *Journal of Geographical Sciences*, 24(2), 195–210.

National Bureau of Statistics of China (2015) *China Statistical Yearbook 2015*. Beijing: China Statistical Press (in Chinese).

Seto K. C., Kaufmann R. K. and Woodcock C. E. (2002) 'Monitoring land use change in the Pearl River Delta, China', in *Linking People, Place, and Policy*, ed. Walsh, S. J. and Crews-Meyer, K. A. Dordrecht: Springer, 69–90.

Wang, L., Li, C., Ying, Q., Cheng, X., Wang, X., Li, X., Hu, L., Liang, L., Yu, L., Huang, H. and Gong, P. (2012) 'China's urban expansion from 1990 to 2010 determined with satellite remote sensing', *Chinese Science Bulletin*, 57(22), 2802–2812.

Yan, H., Liu, J., Huang, H. Q., Tao, B. and Cao, M. (2009) 'Assessing the consequence of land use change on agricultural productivity in China', *Global and Planetary Change*, 67(1), 13–19.

# 15 Urbanization in Latin America with a particular emphasis on Mexico

*René R. Colditz, María Isabel Cruz López,*
*Adrian Guillermo Aguilar Martínez, José Manuel*
*Dávila Rosas and Rainer A. Ressl*

## Introduction

Most Latin American cities were established by the sixteenth century. These urbanized core areas were symbols of territorial possession and centers from which the surrounding countryside could be administered and exploited. In Central American and Andean countries they are often located away from the coast and build upon pre-colonial settlements. The most important towns in colonial Latin America were political and cultural centers, for example, Mexico City, Lima and Buenos Aires, which were also capitals of viceroyalties. Other economically important cities were mining centers (Taxco, Potosi) and major ports along the coast of the Caribbean sea and Atlantic ocean such as Cartagena, Santo Domingo, Rio de Janeiro and Montevideo. The political and administrative centers of the past remain the major urban centers in Latin America today.

By the mid-twentieth century industrialized countries such as Argentina, Brazil and Mexico achieved rapid growth with manufacturing activities concentrated in the capitals and chief ports. This pattern of concentration became most prominent after World War II and it had an impact on urban growth, migration and regional development strategies. Away from these manufacturing cores, export-processing industries created economic enclaves in intermediate cities and peripheral zones. Urban primacy became a distinctive geographic feature in most of Latin America (Aguilar and Vieyra 2008).

In the early 1980s, Latin America adopted a free-market economic model. Opening-up national economies led to increasing deindustrialization, deteriorated labor conditions, growth of the informal sector and an increase in urban poverty. This, in turn, shifted growth from large metropolitan areas towards middle-sized urban centers that became more competitive in the global economy, such as border towns like Tijuana in Mexico, export-oriented manufacturing poles such as Medellin in Colombia or Ciudad Juarez in Mexico, and tourist centers like Cancun, Panama City or Rio de Janeiro (Aguilar and Vieyra 2008).

## Data sets

In the following sections regional definitions from the United Nations (UN 2014a) were adapted in the following way: Latin America was defined as all land from Mexico to Tierra del Fuego including all Caribbean islands. This area was subdivided into three regions: Central America (Mexico to Panama), Caribbean (islands of Greater and Lesser Antilles) and South America (the remainder). For population analysis statistical data of the World Urbanization Prospects 2014 were employed with population data from 1950 to 2050 and urban agglomerations (more than 300,000 inhabitants) from 1950 to 2030 (UN 2014b), excluding countries with an area smaller than 5,000 km$^2$. The National Institute for Statistics and Geography (INEGI) provides population census data for Mexico for 1950, 1960, 1970 and 1990 to 2010 at five-year intervals (INEGI 2014). Urban areas were defined as localities with more than 2,500 inhabitants, but city analysis only focused on agglomerations with more than 15,000 people.

Defense Meteorological Satellite Program – Operational Linescan System (DMSP-OLS, Elvidge *et al.* 1997) images of annual average stable lights from 1992 to 2009 were cross-calibrated (Elvidge *et al.* 2009) and employed for defining urban areas uniformly in space and time using threshold value DN>=55 or 87 percent of the data range (Imhoff *et al.* 1997, Small *et al.* 2005). For countries and states with more than 200 urban pixels linear least-square regression was used for trends estimation and F-test for statistical significance analysis of the regression model.

Land take was analyzed using a land cover map of Latin America derived from 500 m Moderate Resolution Imaging Spectroradiometer (MODIS) images for the year 2008 (Blanco *et al.* 2013). For the country of Mexico and local analysis of Cancun and Merida a 250 m MODIS-based land cover time series (2005–2011) was employed (Colditz *et al.* 2012, Colditz *et al.* 2014a, Colditz *et al.* 2014b). INEGI vegetation data (1970, 2012) were used for studying the urban expansion of Mexico City (INEGI-INE 1999, INEGI 2013).

## Urbanization in Latin America, its regions and countries

In 2015, the estimated population of Latin America was 630 million or 8.6 percent of the world's population (Table 15.1; UN 2014b). Over the course of time from 1950 to 2050 the total population growth rate declined more rapidly than global numbers. The reason for slower population growth in Latin America is the stable low rate of mortality and decreasing rate of fertility which puts most countries in stage 3 out of 4 of the demographic transition model (Pacione 2009). Latin America comprises a total area of 2,055 million ha or 15.1 percent of the global land surface excluding Antarctica (Table 15.1). In 2015, population density in Latin America was only 30.7 people/km$^2$ (53.8 people/km$^2$ for the world) with notable regional disparities.

*Table 15.1* Area, total population and urban population (selected years) for the
World, Latin America and its regions and Mexico which were used to
calculate population density and urban proportions

| | Area [million ha] | Total population [millions] | | | | Urban population [millions] | | | |
|---|---|---|---|---|---|---|---|---|---|
| | | 1950 | 2000 | 2015 | 2050 | 1950 | 2000 | 2015 | 2050 |
| World | 13,616 | 2,525.8 | 6,127.7 | 7,324.8 | 9,550.9 | 746.5 | 2,856.1 | 3,957.3 | 6,338.6 |
| Latin America | 2,055 | 167.9 | 526.3 | 630.1 | 781.6 | 69.3 | 396.3 | 502.8 | 673.6 |
| Caribbean | 23 | 17.1 | 38.4 | 43.1 | 47.6 | 6.2 | 23.5 | 30.3 | 38.4 |
| Central America | 248 | 38.3 | 139.6 | 171.9 | 228.8 | 15.0 | 96.1 | 126.9 | 187.2 |
| South America | 1,783 | 112.5 | 348.2 | 415.1 | 505.1 | 48.1 | 276.6 | 345.6 | 448.0 |
| Mexico | 196 | 28.3 | 103.9 | 125.2 | 156.1 | 12.1 | 77.6 | 99.2 | 134.8 |

Source: UN (2014b).

Note: Area for the world excludes Antarctica.

The Caribbean, by far the smallest region (1.1 percent), also hosts the smallest population proportion of 6.8 percent, but the population density of 184.2 people/km$^2$ is the highest among global regions (UN 2014b). Central America, with 12.1 percent of the land surface and 27.3 percent of the population, shows an intermediate density of 69.3 people/km$^2$. South America is the largest region but population density is low (23.3 people/km$^2$), also because of large, nearly uninhabited areas like the Amazon, which puts it among the sparsely populated regions of the world.

In 2015, in Latin America 502.8 million people, that is 79.8 percent, live in urban areas (Table 15.1). This puts it in second place with a slightly lower urban population proportion than North America (81.6 percent) and well above the global average (54.0 percent, UN 2014b). While there is a relatively linear increase in global urban population by approximately 0.37 percent per year, Figure 15.1A shows for Latin America an increase of, on average, 0.69 percent until 2000 and since then 0.21 percent. It should be noted that the growth of urban population was above the growth of total population; hence there is a steady decline in rural population proportion and for most countries also a decrease in absolute numbers due to rural-to-urban migration. Regional disparities can be noted in Figure 15.1B, e.g. Guatemala, Guyana, Honduras, Nicaragua and Paraguay show lower than average Latin American percentages of urban population in 2015, while Argentina and Uruguay are well above average. Most countries show increasing trends in urban population proportion (Figure 15.1C) with Brazil, Costa Rica, the Dominican Republic, Haiti, Honduras and Puerto Rico clearly above Latin American and global trends.

The percent urban area, estimated from DMSP between 1992 and 2009 (Elvidge *et al.* 1997, 2009), replicates the above-described pattern of population density (Figure 15.1D). While the world shows a nearly zero trend over 18 years, Latin America and regional tendencies are all positive but not

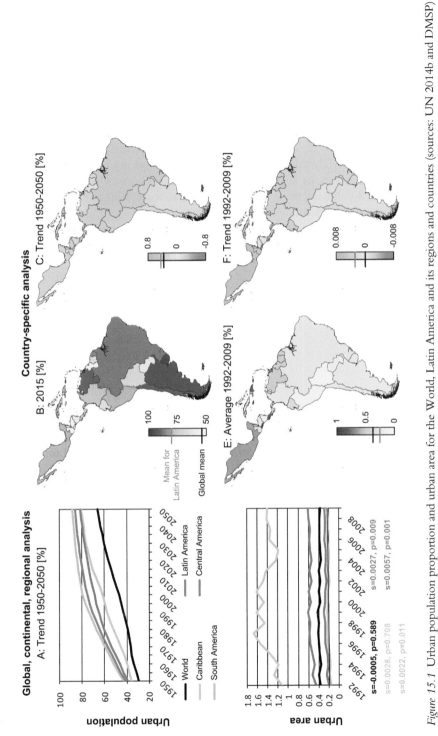

*Figure 15.1* Urban population proportion and urban area for the World, Latin America and its regions and countries (sources: UN 2014b and DMSP)

Note: s. . .linear trend, p. . .significance of F-test for linear regression model. Lines in scale bars indicate global and Latin American average. Grey indicates countries not analyzed due to too few urban area pixels.

always significant (p>5 percent). Percent urban area in Bolivia, Nicaragua and Peru is clearly below the Latin American average, while Costa Rica, Dominican Republic, El Salvador, Jamaica, Mexico and Puerto Rico are above (Figure 15.1E). Trends also vary widely (Figure 15.1F) with Guatemala, Mexico and Trinidad and Tobago showing significant, above-average trends of urban area growth. All negative trends, e.g. for Colombia, Costa Rica, El Salvador, Jamaica, Puerto Rico, Uruguay and Venezuela, were not significant (p>0.05). These countries indicate a particular tendency to urban densification, e.g. constructing higher buildings or reducing individual space with smaller apartments to accommodate the growing urban population in nearly the same area. This development may be fostered in countries with a small national territory or large cities in mountainous areas which, due to construction in floodplains or steep slopes increases susceptibility to natural hazards such as mudslides and inundations.

Land transformation due to urban growth is difficult to analyze due to lacking long time series of spatially-explicit land cover information. A simple attempt was undertaken using a MODIS-based land cover map of Latin America for the year 2008 (Blanco *et al.* 2013) and assuming uniform urban growth of 2 km around each urban agglomeration. The proportion of land cover classes potentially transformed to urban was summarized for each country in pie charts (Figure 15.2A). Notable is the high proportion of cropland loss in many countries. Large reductions of forested land are shown for Brazil, Colombia, Ecuador, Paraguay, Trinidad and Tobago and Venezuela. Bolivia, Chile and Peru also depict a transformation of high elevation barren land to urban areas. However, land transformation is a local process and depends on local actors and the dominating land cover in this region.

The majority of the world's urban population lives in centers smaller than 300,000 inhabitants, but this proportion is declining as more people agglomerate in large cities (above 1 million) and megacities (above 10 million) (UN 2014b). Latin America is no exception; the urban population proportion of bigger settlement categories of Argentina, Chile, Colombia, Mexico and Peru have already or will soon surpass the group with fewer than 300,000 inhabitants.

In 2015, there are 205 cities with more than 300,000 people in Latin America: 83 with 300,000–500,000, 55 with 500,000–1 million, and 59 with 1–5 million (UN 2014b). There are 4 cities with 5–10 million habitants: Lima (9.8 m), Bogota (9.7 m), Santiago de Chile (6.5 m) and Belo Horizonte (5.7 m). Out of the 29 global megacities with more than 10 million people Latin America hosts four: Sao Paulo (21.0 m), Mexico City (20.9 m), Buenos Aires (15.1 m) and Rio de Janeiro (12.9 m); by 2030 Bogota and Lima are expected to join this group. Figure 15.2B shows the location of all 205 cities with the diameter indicating the proportion of urban population residing in those centers relative to the total urban population of each country. In the Caribbean and small Central American countries as well as Paraguay and Uruguay most

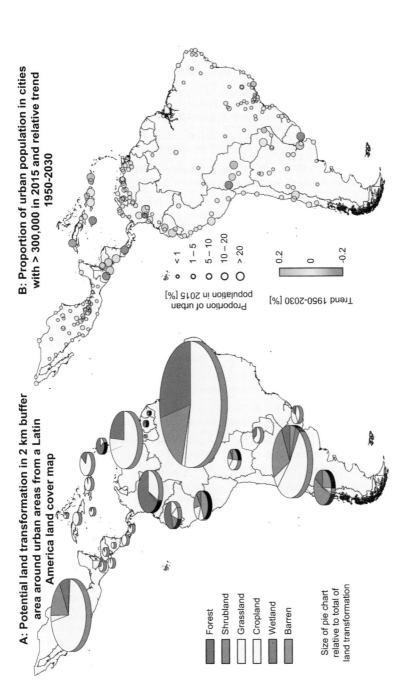

*Figure 15.2* A: Potential land cover change in 2 km buffer zone around urban areas from a MODIS–based land cover map of 2008 (source: Blanco et al. 2013). B: Urban population proportion for cities with more than 300,000 inhabitants in 2015 (urban population of city in relation to urban population of each country) (source: UN 2014b)

of the urban population concentrates in one city, usually the capital. In other counties, despite a higher number of urban centers, one agglomeration clearly dominates with 20–40 percent of the total urban population, e.g. Argentina, Chile, Colombia, Cuba, Ecuador, Mexico and Peru. This concentration reflects the above-mentioned centralized political and economic development of most Latin American countries. Brazil and Venezuela form a group in which, despite large cities, urban population proportion is not concentrated in only one major center, e.g. only 12.1 percent of the Brazilian urban population resides in Sao Paolo and 10.5 percent in Caracas. A singular case is Bolivia with three major centers: Cochabamba, La Paz and Santa Cruz. In both, Brazil and Bolivia the government moved to another city which led to notable proportional decreasing trends in Rio de Janeiro and La Paz (colors in Figure 15.2B indicate trends in urban population proportion). Other notable decreasing trends of urban population proportion are noted for Buenos Aires, Caracas, Montevideo and Quito, nevertheless, all cites have gained population in absolute numbers. However, there are also large urban centers with relative increases, e.g. Bogota, Lima, Santiago de Chile, San Juan (Puerto Rico), Santa Cruz (Bolivia) and Ciudad de Este (Panama).

## Urbanization in Mexico at the national and state level

In 2015, Mexico was home to 1.7 percent of the global and almost 20 percent of Latin Americas population (Table 15.1). The country multiplied its population almost five times between 1950 and 2015 but population growth is slowing down. In 2015, the urban population is almost 80 percent and is expected to reach 86 percent in 2050. In terms of area the country makes up almost 1.5 percent of the global land surface and nearly 10 percent of Latin America. The population density of 63.8 people/km$^2$ in 2015 is above Latin American and global numbers. With respect to urban population, urban area and city development in general, Mexico is a representative example for Latin America.

Figure 15.3 indicates urban population proportion (INEGI 2014) and urban area from DMSP (Elvidge *et al.* 1997, 2009) at the state level (for names see Figure 15.3B). While high urban population proportion in the center of the country is due to the highly centralized system around Mexico City, concentration in the northern states (Figure 15.3A) is due to water scarcity. Touristic development is the reason for above-average urban population in Quintana Roo. Chiapas and Oaxaca are the only states with a higher rural than urban population. Figure 15.3B shows a positive trend for all states, except the Federal District with a nearly zero growth of urban population (at a level of 99.5 percent of urban population in 2010). In fact, the capital has spread into the surrounding State of Mexico and larger export-oriented industries have settled in a wider realm in the states of Queretaro, Puebla and Tlaxcala. Migration due to employment in tourism and relocating elderly, partly foreign residents has caused urban population growth in Baja California Sur and

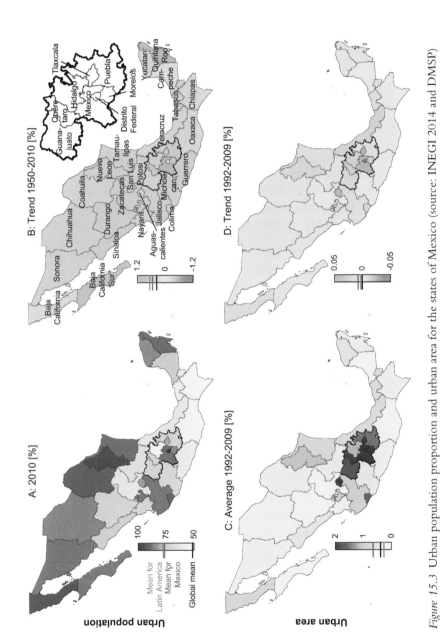

*Figure 15.3* Urban population proportion and urban area for the states of Mexico (source: INEGI 2014 and DMSP)

Note: For state names see 15.3B. Lines in scale bars indicate the average for Mexico, Latin America and the world.

Quintana Roo. Also, there is a national migration pattern towards the northern states and in particular border towns to the United States due to employment in local export-oriented manufacturing industries and the eventual goal of working in the United States. Over time regional disparities at the state-level have increased as rural states such as Campeche, Veracruz, Oaxaca and Chiapas depict lower than average trends.

In contrast to the average of 0.62 percent urban area on the national level (Figure 15.3C), 60.5 percent of the Federal District is urban, 9.7 percent in State of Mexico, 5.9 percent in Morelos, 5.4 percent in Tlaxcala, 1.9 percent in Queretaro and 1.5 percent in Puebla (data from DMSP). The urban area proportion below the national average in the north and south indicates for the former that few people live in larger agglomerations due to limiting environmental factors and for the latter a generally higher population in small settlements dispersed over the state territory. The trends in Figure 15.3D indicate the expected pattern with substantial urban growth around the Federal District, while the district itself shows non-significant (p>5 percent) negative tendencies. Notable are growing urban areas in Baja California, Jalisco, Nuevo Leon, Quintana Roo, Tamaulipas and Yucatan. A particular case is Aguascalientes, a small but highly industrial state with significant urban growth.

A spatially-explicit change product based on 250 m MODIS data from 2005–2011 was employed for estimating land take due to growth of urban areas (Colditz *et al.* 2014a, Colditz *et al.* 2014b). The total annual change varies between 0.08 and 0.11 percent of which 2 to 4 percent were urban changes. The bar totals in Figure 15.4 depict gain and loss of class urban for each bi-annual comparison and the colors indicate class-specific from-to change. The smallest urban expansion occurred between 2005 and 2006 and highest between 2008 and 2009. Even though almost 30 percent of the national territory is forested land, 37 percent shrubland and nearly 9 percent grassland (Colditz *et al.* 2012), few of these semi-natural areas were transformed to urban. In all years the majority of area transformed to urban was managed cropland (20 percent of the total national territory). Transformations from water to urban is a result of spatially unconstrained change detection and unlikely in reality.

In the period 1990–2010 the number of urban centers with at least 15,000 inhabitants increased from 312 to 384 (Figure 15.5). In 2010, there were 11 cities with more than 1 million inhabitants, which can be distinguished in two groups, cities with 2.5 million or more (Guadalajara, Mexico City, Monterrey and Puebla), with growth rates below the national average of 2 percent and the remainder (Ciudad Juarez, Leon, Queretaro, San Luis Potosi, Tijuana, Toluca, Torreon), which are more dynamic and with a growth rate mostly above 3 percent. Notable is the historically low growth rate of Mexico City with 0.9 percent between 2000 and 2010 (Aguilar and Graizbord 2014).

A second important aspect is the metropolization process that is affecting mostly the bigger cities. Whereas in 1990 there were 37 metropolitan zones in the country with 31.5 million people living in them, by 2010 there were 50 of these zones with 63.8 million inhabitants. These metropolitan centers have

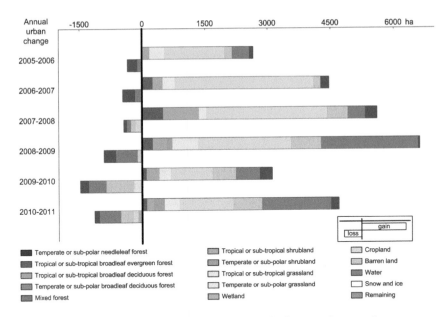

*Figure 15.4* Gain and loss of urban area in hectares for bi-annual comparisons (2005–2011) of MODIS-based land cover maps (source: Colditz *et al.*, 2014a)

Note: Colors indicate class-specific from-to changes.

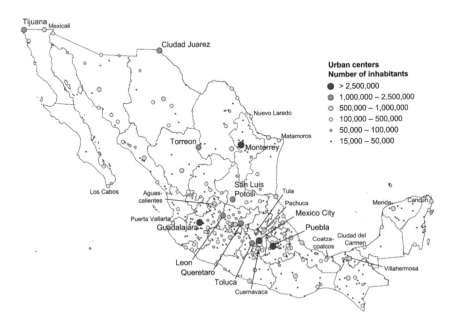

*Figure 15.5* Location and population of urban areas of Mexico in 2010

emerged as the nodes of higher hierarchy in the urban system because they concentrate 56.8 percent of total population and generate approximately 75 percent of the national gross domestic product. Although they have a favorable influence in their respective regions to impulse socioeconomic development, they also demand a high quantity of land for the excessive expansion of built-up areas often with a notable peri-urbanization process accompanied by lower densities and disperse urbanization (Aguilar 2014). This corresponds to the above-noted loss of managed agricultural areas and indirectly moves the frontiers reducing also natural land and its supporting, provisioning and regulating ecosystem services such as forests, water retention and purification, biodiversity etc. In addition, a great proportion of population living in peripheral areas constitute informal settlements in precarious conditions that contribute to environmental damage. In general terms, in the last 30 years, during which urban population in all urban centers has doubled, the expansion of their built-up areas has, on average, multiplied by a factor of seven (SEDESOL 2011).

Mid-sized cities with a population between 100,000 and 1 million have multiplied in different regions and are now the nodes of the urban deconcentration process. In the last 20 years (1990–2010) this number increased from 55 to 84 and its inhabitants almost doubled passing from 17.6 to 30.30 million people. They are now important centers for productive activities such as oil exploitation (Ciudad del Carmen, Coatzacoalcos, Villahermosa) or new export-oriented manufacturing centers (Aguascalientes, Leon, Queretaro, San Luis Potosi), thriving border towns with significant service for the US market (Matamoros, Mexicali, Nuevo Laredo, Tijuana) or touristic centers which are also the preferred destiny of foreign migrants (Cancun, Los Cabos, Puerto Vallarta).

## Moving the frontiers—the expansion of Mexico City

Administratively, Mexico City consists of 16 boroughs (delegations) which form the Ciudad de México (before January 29, 2016, Federal District), but has grown beyond those borders into the surrounding State of Mexico. In 2005/2006 the Metropolitan Area of the Valley of Mexico (MAVM) was established which today consists of all 16 boroughs, 59 municipalities of the State of Mexico and one of the state of Hidalgo.

In 1325, the settlement was founded on islands in the Texcoco lake as the capital of the empire of the Mexica from which also originates its name. The early urban growth of Mexico City is related to historic processes such as the arrival of the Spanish conquerors in 1521 and establishment of the vice-royalty of New Spain in 1535, the independence from Spain in 1810 and the Mexican revolution in 1910. The concentration of political and economic power in one place has shaped Mexico City over the centuries and is still relevant for business decisions today.

Figure 15.6A shows all municipalities of the MAVM from its historic core to the most recent expansions. Between the end of the Mexican revolution in 1929 and the first decade of the twenty-first century several authors found seven phases of urban growth and associated them to models of concentric rings (Negrete *et al.* 1993, Delgado 1988, SEDESOL CONAPO INEGI 2012). Starting with the expansion from its core in Cuauhtémoc into the boroughs of Miguel Hidalgo, Venustiano Carranza and Benito Juarez from 1930 to 1950, the second phase (1950–1970) followed due to significant industrial development, extending in all cardinal directions and for the first time including four municipalities from the State of Mexico. The third phase (1970–1986) incorporated four boroughs in the south, six municipalities to the north and two to the east. By 1990, among others, the last borough of the

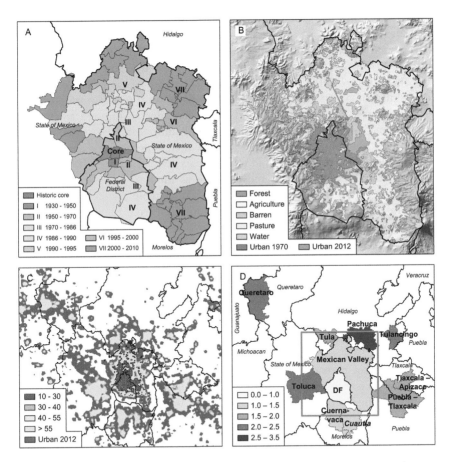

*Figure 15.6* Urban expansion of Mexico City. A: Seven phases of urban growth between 1930 and 2010. B: Land take due to urban expansion between 1970 and 2012. C: Stable lights (2009) in the larger Mexico City area and light contamination beyond the urban area extent. D: Megalopolis Mexico City and growth rates in percent between 2000 and 2010

Note: The red box indicates the subset for Figures 15.6A and 15.6B.

Federal District (Milpa Alta) was included. During this time the growth of Mexico City was fostered by the growing cities outside the Mexican Valley (Toluca, Cuernavaca, Puebla and Pachuca). The fifth phase (1990–1995) includes four large municipalities in the north including one from Hidalgo and during the sixth (1995–2000) there is an expansion to the northeast and division of previously included municipalities in the east and north. During the last phase (2000–2010) 24 municipalities were added. Today, more than 20 million people (18 percent of the total population) live in the MAVM with an area of only 7,800 km² (0.4 percent of the national territory). This corresponds to 2,564 people per square kilometer, the highest population concentration in the country (INEGI 2010).

Figure 15.6B shows the state of the urban area in 1970 in red, in red cross-hairs the built-up area in 2012 and in different colors the land use and land cover, such as grassland, agriculture, forest and shrubland (INEGI-INE 1999, INEGI 2013). Over the four decades the city grew by 235 percent from 610 km² in 1970 to 1,440 km² in 2012. Over 90 percent of the land take affected agricultural area and grassland in relatively flat terrain to the north and east. Approximately 8 percent of the expansion has affected forested areas at the western and southern edge of the valley, mainly during the last two decades. In those areas steep slopes limit the growth by increasing construction costs and loss of valuable environments and related ecosystem services.

In particular water is a scarce resource as the city requires a supply of 35.2 m³/s with a current deficit of 3 m³/s of potable water (FCEA 2015). Statistics of the National Water Commission (CONAGUA) indicate that the natural water supply of the Mexican Valley for all land uses (urban, agriculture, industry) is far from being self-sustainable. 54 percent of the water is therefore supplied by other hydrologic regions several hundred kilometers away and 46 percent by internal sources such as the Rio Magdalena and ground water (GDF 2008). All forms of water supply cause serious and large-scale secondary effects on the environment. Energy is needed to pump water from lower hydrologic regions into the city. In addition there are conflicts in water use and amount of water extraction with local municipalities and other large cities. The over-exploration of the local aquifer causes significant subsidence as large parts of the city were constructed on sediments of the former Texcoco lake. For instance, the city center has subsided by approximately 10 m over the last 60 years (SACMEX 2012). Water infiltration is achieved in conservation areas but expansion of the impervious cover has reduced its surface area and percolation to the subsiding areas is slow. Wastewater removal is another issue, for which drainage systems and retention bodies were constructed to limit ground water contamination. Wastewater treatment is just in its initial stages.

Air contamination is another pressing environmental problem in Mexico City, in particular due to its topographic location in a valley and frequent formation of atmospheric inversions. High aerosol ozone concentrations affect many citizens by respiratory and skin sickness and cause high societal costs for health care. The main cause for emissions is transportation (45 percent), followed by industry (21 percent), housing (20 percent) and 14 percent by others (GDF 2015). The list of environmental issues caused by urban agglomerations can be continued, e.g. soil reduction, contamination and erosion with increasing risks during earthquakes, biodiversity loss and changes in species behaviors in surrounding areas, nutrient loss, fuel consumption including fossil fuel, high water and energy use, waste dumping, treatment and recycling, increased velocity of contagious diseases, etc.

A less studied issue is nighttime light contamination and its effect on surrounding areas. Figure 15.6C shows data of the DMSP sensor for the year 2009 in tones from brown to yellow and black cross hairs marking the urban area of Mexico City in 2012. Considering values of 55 and higher in yellow the area is 17 percent larger than urban mapped from official sources (INEGI 2013) and includes several natural areas in the southern mountainous region. Disturbance by artificial light not only affects the behavior of nocturnal species but the functioning of ecosystems in general (López Acosta *et al.* 2009, Meyer and Sullivan 2013, Gaston *et al.* 2013).

The continuing expansion of Mexico City today needs to be seen in a larger context. Although there is a physical separation by high mountain ranges, the pass elevation to the south, east and west entrance is above 3,000 m, the city is highly interconnected with small and large surrounding cities (Figure 15.6C), which by themselves often form metropolitan areas. Altogether they form the megalopolis Mexico City, and colors in Figure 15.6D indicate their growth between 2000 and 2010. The higher positive trend of surrounding cities in comparison to the core area, also known as polarization reversal (Aguilar and Rodríguez 1995), is in line with our previous analysis at state level using DMSP. The attractive location in a large megalopolis with short connections to business partners and political stakeholders, space for larger industrial plants and proximity to recreational facilities, but still with the option to take advantage of the cultural urban life in the core area and not suffering from all the negative issues, is the main reason for increased growth in the periphery, either by new businesses or relocation. It needs to be seen if at some point urban areas completely connect even across physical barriers. First tendencies can be noted, e.g. between Mexico City and Pachuca, Toluca and Tula (Figure 15.6C).

## Selected drivers for urbanization

### *Tourism—Cancun*

The state of Quintana Roo on the Yucatan peninsula is internationally recognized for its beautiful white beaches with excellent offshore reef snorkeling and diving opportunities along the Mexican Caribbean coast, known as the Riviera Maya and Costa Maya. The tourist development started in the late 1960s, primarily as a result of a government-initiated study to develop a counterbalance to the Mexican Riviera on the Pacific coast and also to compete with resort destinations on several Caribbean islands (Collins 1979). In 1968, the Tourism Infrastructure Promotion Fund (INFRATUR, later FONATUR) was created and together with the Bank of Mexico they promoted six sites for major tourist developments, among those Cancun. At that time the fisherman's village of Cancun accounted for 120 inhabitants and increased rapidly during the different development stages in the 1970s, 1980s and 1990s to a population of 628,306 in 2010 (INEGI 2010). Cancun has undergone a radical transformation, becoming the most important tourist resort of the country. Already in 1990, Cancun accounted for 110 hotels, with more than 17,000 rooms and received approximately 1.5 million visitors annually (FONATUR 2001). Successively the entire Caribbean coast south of Cancun to the border of Belize has been developed with significant impact on coastal ecosystems and the environment.

The urbanization trends in Cancun and surrounding areas are dramatic with increased dynamics since 1990. The spatial changes of these trends can be easily detected with satellite remote sensing. For instance, analysis of 250 m MODIS land cover maps between 2005 and 2010 (Colditz *et al.* 2014b) revealed an increase of 2,500 ha of urban area around Cancun, that is 500 ha per year, which corresponds to public statistics reporting an annual growth rate of 616 ha or 3.1 percent for the Benito Juarez municipality to which Cancun belongs (Veloz Avilés 2011). Urban growth spatially occurs around the airport and along the western part of the city with new commercial and housing sections to accommodate the increasing population, but fewer hotel developments were detected along the coastal strip during this period in comparison to previous times.

### Commerce—Merida

The rise of Merida, capital of the state of Yucatan, started in the late nineteenth century as one of the centers of sisal (*henequén*) cultivation, a fibrous plant for twine and rope production, also known as the "green gold" (Duch Colell 1998).

Until the beginning of the twentieth century agriculture was the primary activity in the area around Merida. Over the course of the twentieth century Merida gained importance as the peninsula's center of commerce, in particular based on many assembly plants that were established since 1980 and tourism. All these factors progressively attracted the rural population on the Yucatan peninsula migrating to and working in Merida, which resulted also in a strong increase of accompanying service industries. In 2010, 78.7 percent of the economically active population of Merida was occupied in the tertiary sector, 20.3 percent in the secondary and only 1 percent in the primary sector (SEDESOL 2013).

According to the 2010 census, the population of Merida reached 777,615 inhabitants (INEGI 2010). High birth rates and a continuous rural-to-urban migration result in constant urban sprawl, which is also expressed by the high population density of 938 people/km$^2$ in the municipality (SEDESOL 2013). These demographic and socioeconomic changes in the region have caused pressure on the city of Merida. The resulting spatial consequences of regional migration can be documented with satellite imagery. For instance, there are significant expansions and new developments of urban area between 2005 and 2010 in Merida, which can even be noted in coarse resolution MODIS satellite data (Colditz *et al.* 2014a). The "Fraccionamiento Las Américas" was constructed for an expected population of 20,000 habitants, which was mapped as a newly constructed urban area of 225 ha in the land cover map of 2010. The largest new city section, Ciudad Caucel west of Merida, encompasses an area of 875 ha and is designed for 30,000 new houses and an expected population of 100,000 new residents. In fact, Ciudad Caucel was found to be the largest single patch of newly constructed urban surface between 2005 and 2010 in entire Mexico. Annual images of this site indicate that most of the construction occurred in the years 2007 and 2009.

### International migration—Tijuana

Over the last 110 years Tijuana has transformed from a ranch with 224 inhabitants in 1900 to one of the 10 most important cities in Mexico. In 2010 the population was more than 1.5 million (INEGI 2010), and with San Diego it forms the largest binational conurbation in the world. Located directly along the border to the state of California the city attracts large groups of legal and illegal emigrants. Most migrants are from the Federal District, Jalisco, Michoacán, Oaxaca and Sinaloa of which most work in factories, often awaiting permission of entry to the United States. There is also a significant number of immigrants, mainly from China, Central American and Andean countries and ultimately from the United States due to lower living costs. Daily or weekly commuters as well as visitors from all over the world make Tijuana–San Diego the busiest land-border passage in the world with more than 300,000 daily crossings.

The growth of Tijuana was always linked to the political and economic situation of the United States. For instance, during the years of prohibition (1919–1933) in the United States, Tijuana offered the respective services of alcohol retail and consumption in bars and night clubs, which caused a population increase by 1,000 percent. During the Great Depression, foreigners working in the United States were forced to return to their country. Approximately 400,000 Mexicans returned, of which many stayed in Tijuana, mostly in precarious conditions hoping for permission of reentry. During World War II the United States required manual laborers in the agricultural areas and implemented programs such as Braceros (1942–1964). Thousands of migrants arrived in Tijuana and those who were not allowed to enter the United States frequently stayed in this city. Another boost occurred upon the end of Braceros when returning workers often remained close to the border.

This rapid population growth is also reflected in the growth of the urban area. By 1950 the urban area was approximately 1,450 ha (Padilla 1985), by 1973 it had expanded to 6,620 ha, which coincides with the major population growth during the 1960s. In 1993 the urban area had grown to 16,830 ha (Bocco and Sánchez 1996) and reached 24,240 ha in 2010 (INEGI 2010). It is estimated that by 2030 the city of Tijuana will have 2.8 million inhabitants (IMPLAN 2010) which poses substantial challenges to urban planners to provide the urgently needed public services.

## Perspectives

The urban population of Latin America is expected to reach 673.6 million in 2050, which is 86.1 percent of its total population (UN 2014b). The further growth of already overpopulated cities will have additional consequences on the environment and requires innovative solutions for already existing social issues. Successful sustainable urbanization requires competent, responsive and accountable governments charged with the management of cities and urban expansion. Attention not only has to go to big cities, but also to new urban forms that have acquired importance in recent years like peri-urbanization, rural–urban transition zones, and intermediate and small cities, that suffer similar problems to those of a big metropolis.

Water availability and wastewater treatment, increasing needs of energy, higher demands on agricultural lands also in distant regions to feed the rural and urban population, transportation of commodities and people, and air and water contamination are worrisome environmental concerns as many large cities and megacities already seem on the "verge of collapse." In addition the existing social conflicts will intensify, such as the gap between the rich and poor living together on very limited space, a still too-small middle class, lack of education, violence and crime, corruption and lack of law enforcement or even impunity. Therefore, a change in policies is needed, away from the highly centralized political and economic power present in most Latin American countries and towards a more balanced distribution across several cities in various regions. These policies can help responding to the challenges of providing urban infrastructure and basic social services for the urban poor, and mitigating the negative environmental impacts associated with large and rapidly growing urban agglomerations.

There are indicators that the growth of megacities larger than 10 million people attracts fewer businesses and population growth is slower in comparison to cities with 1–5 million inhabitants. However, there is a risk that those cities undergo unplanned and uncontrolled growth which soon will expose them to similar negative environmental issues and social conflicts. Urbanization will continue to be the distinctive geographical feature for Latin America at an even faster pace affecting all levels of society and environment.

# References

Aguilar, A.G., (2014) 'El reparto poblacional en el territorio. Tendencias recientes y desafíos futuros', In: Ávila J.L., Hernández Bringas H. and Narro Robles J. (Eds.) *Cambio Demográfico y Desarrollo en México*, UNAM, Mexico.

Aguilar, A.G. and Graizbord B., (2014) 'La distribución espacial de la población, 1990–2010: Cambios recientes y perspectivas diferentes', In: Rabell Romero, C. (Ed.) *Los Mexicanos. Un Balance del Cambio Demografico*, Fondo de Cultura Economica, Mexico.

Aguilar, A.G and Rodríguez, F., (1995) 'Tendencias de desconcentración urbana en México', In: Aguilar A.G., Castro, L.J. and Juárez, A. (Eds.) *El desarrollo urbano de México a fines del siglo XX*, Instituto de Estudios Urbanos de Nuevo León y Sociedad Mexicana de Demografía, Mexico.

Aguilar, A.G. and Vieyra, A., (2008) 'Urbanization, migration, and employment in Latin America: A review of trends', In: Jackiewicz, E. and Bosco, F. (Eds.) *Placing Latin America: Contemporary Themes in Human Geography*, Rowman and Littlefield, USA.

Blanco, P.D., Colditz, R.R., López Saldaña, G., Hardtke, L.A., Llamas, R.M., Mari, N.A., de los Angeles Fischer, M., Caride, C., Aceñolaza, P.G., del Valle, H.F., Lillo-Saavedra, M., Coronato, F.R., Opazo, S.A., Morelli, F., Anaya, J.A., Sione, W.F., Zamboni, P. and Barrena Arroyo, V., (2013) 'A land cover map of Latin America and the Caribbean in the framework of the SERENA Project', *Remote Sensing of Environment*, 132, 13–31.

Bocco, G. and Sánchez, R., (1996) 'Cuantificación del crecimiento de la mancha urbana usando percepción remota y sistemas de información geográfica. El caso de la ciudad de Tijuana (BC), México (1973–1993)', *Boletín de Investigaciones Geográficas*, 4, 123–129.

Colditz, R.R., López Saldaña, G., Maeda, P., Argumedo Espinoza, J., Meneses Tovar, C., Victoria Hernández, A., Ornelas de la Anda, J.-L., Zermeño Benítez, C., Cruz López, I. and Ressl, R., (2012) 'Generation and analysis of the 2005 land cover map for Mexico using 250m MODIS data', *Remote Sensing of Environment*, 123, 541–552.

Colditz, R.R. Pouliot, D., Llamas, R.M., Homer, C., Latifovic, R., Ressl, R.A., Meneses Tovar, C., Victoria Hernández, A. and Richardson, K., (2014a) 'Detection of North American land cover change between 2005 and 2010 with 250m MODIS data', *Photogrammetric Engineering & Remote Sensing*, 80 (10), 918–924.

Colditz, R.R., Llamas, R.M. and Ressl, R.A., (2014b) 'Detecting change areas in Mexico between 2005 and 2010 using 250m MODIS images', *IEEE Journal on Selected Topics in Applied Earth Observation and Remote Sensing*, 7 (8), 3358–3372.

Collins, C.O., (1979) 'Site and situation strategy in tourism planning: A Mexican case study', *Annals of Tourism Research*, 6, 351–366.

Delgado, J., (1988) 'El patrón de la ocupación territorial de la ciudad de México al año 2000', In: Terrazas, O. and Preciat, E. (Eds.) *Estructura territorial de la Ciudad de México*, Plaza y Valdés Editores, Mexico.

Duch Colell, J., (1998) *Yucatán en el tiempo*, Inversiones Cares, Mérida.

Elvidge, C.D., Baugh, K.E., Kihn, E.A., Kroehl, H.W. and Davis, E.R., (1997) 'Mapping city lights with nighttime data from DMSP Operational Linescan System', *Photogrammetric Engineering & Remote Sensing*, 63, 727–734.

Elvidge, C.D., Ziskin, D., Baugh, K.E., Tuttle, B.T., Ghosh, T., Pack, D.W., Erwin, E.H. and Zhizhin, M., (2009) 'A fifteen year record of global natural gas flaring derived from satellite data', *Energies*, 2, 595–622.

FCEA (2015) 'Agua en México', *Fondo para la Comunicación y la Educación Ambiental*. www.agua.org.mx/h2o/index.php?option=com_content&view=section&id=6&It emid=300004, accessed April 9, 2015.

FONATUR (2001) *Fondo Nacional de Fomento al Turismo*, Costa Maya. www.fonatur. gob.mx, accessed May 29, 2015.

Gaston, K., Bennie, J., Davies, T. and Hopkins, J., (2013) 'The ecological impacts of nighttime light pollution: A mechanistic appraisal', *Biological Reviews*, 88, 912–927.

GDF (2008) 'Fuentes de abastecimiento', *Transparencia D.F.* www.transparencia medioambiente.df.gob.mx/index.php?option=com_content&view=article&id= 86%3Afuentes-de-abastecimiento&catid=57%3Aimpactos-en-la-vida-cotidiana& Itemid=415, accessed February 24, 2015.

GDF (2015) 'Sistema de monitoreo atmosférico', *Gobierno del Distrito Federal y SEDEMA*. www.aire.df.gob.mx/, accessed April 7, 2015.

Imhoff, M.L., Lawrence, W.T., Stutzer, D.C. and Elvidge, C.D, (1997) 'A technique for using composite DMSP/OLS "City Lights" satellite data to map urban area', *Remote Sensing of Environment*, 61, 361–370.

IMPLAN (2010) 'Actualización del programa de desarrollo urbano del centro de población de Tijuana', B.C. (PDUCP T 2010-2030), Reporte del desarrollo.

INEGI (2010) 'Censo de población y vivienda', Instituto Nacional de Estadística y Geografía, Mexico.

INEGI (2013) 'Conjunto de datos vectoriales de uso del suelo y vegetación, serie V (capa unión), escala 1:250,000', Instituto Nacional de Estadística y Geografía, Mexico.

INEGI (2014) 'Datos Socioeconómicos', Instituto Nacional de Estadística y Geografía, Mexico.

INEGI-INE (1999) 'Datos Vectoriales de la Carta de Uso de Suelo y Vegetación, Serie I, Escala 1:250,000', Instituto Nacional de Estadística, Geografía e Informática, Instituto Nacional de Ecología – Dirección de Ordenamiento Ecológico, Mexico.

López Acosta, J.C., Lira Noriega, A., Cruz, I. and Dirzo, R., (2009) 'Proliferación de luces nocturnas: un indicador de actividad antrópica en México', In: CONABIO, *Capital Natural de México*, Volume II: *Estado de conservación y tendencias de cambio*.

Meyer, L.A. and Sullivan, S.M.P., (2013) 'Bright lights, big city: Influences of ecological light pollution on reciprocal stream-riparian invertebrate fluxes', *Ecological Application*, 23 (6), 1322–1330.

Negrete, M.E., Graizbord, B. and Ruíz, C., (1993) *Población espacio y medio ambiente en la zona metropolitana de la Ciudad de México*, Colegio de México, Mexico.

Pacione, M., (2009) *Urban Geography: A Global Perspective*, third edition, Routledge, London and New York.

Padilla Corona, A., (1985) 'Desarrollo Urbano', In: Piñera Ramírez, D. (Ed.) *Historia de Tijuana, Semblanza General*, Centro de Investigaciones Históricas UNAM-UABC, XI Ayuntamiento de Tijuana.

SACMEX (2012) 'El gran reto del agua en la Ciudad de México, pasado, presente y prospectivas de solución para una de las ciudades más complejas del mundo', Sistema de Aguas de la Ciudad de México, Mexico, D.F.

SEDESOL (2011) 'La expansión de las ciudades 1980–2010', Secretaría de Desarrollo Social, Mexico.

SEDESOL (2013) 'Unidad de microrregiones. Cedulas de Información Municipal (SCIM)', Secretaría de Desarrollo Social, Mexico.

SEDESOL CONAPO INEGI (2012) 'Delimitación de las zonas metropolitanas de México 2010', Consejo Nacional de Población, Mexico, D.F. www.conapo.gob.mx/es/CONAPO/Zonas_metropolitanas_2010, accessed November 4, 2014.

Small, C., Pozzi, F. and Elvidge, C.D., (2005) 'Spatial analysis of global urban extent from DMSP-OLS night lights', *Remote Sensing of Environment*, 96, 277–291.

UN (2014a) 'World urbanization prospects: The 2014 revision. Classification of countries by major area and region of the world and income group', United Nations, Department of Economic and Social Affairs.

UN (2014b) 'World urbanization prospects: The 2014 revision. Highlights', United Nations, Department of Economic and Social Affairs, Population Division.

Veloz Avilés, C.A., (2011) 'La planeación urbana en la ciudad de Cancún, el siguiente paso', Tercer congreso internacional de arquitectura y ambiente, Mexico D.F., October 17–19, 2011.

# 16 Monitoring built-up areas in Dar es Salaam using free images

*Michele Munafò and Luca Congedo*

This case study presents the activity of land cover monitoring of Dar es Salaam (Tanzania), in the frame of the ACC Dar Project (Adapting to Climate Change in Coastal Dar es Salaam, www.planning4adaptation.eu).

Dar es Salaam is located in eastern Tanzania on the Indian Ocean coast, covering an area of about 1,800 km². The ACC Dar project aimed at improving the effectiveness of municipal initiatives in Dar es Salaam to support coastal peri-urban dwellers, who are partially or totally dependent on natural resources, in adapting to climate change impacts. In particular, one of the ACC Dar activities was the development of a methodology for land cover monitoring in order to understand land cover change drivers in the Dar es Salaam region, with special attention to peri-urban development within the coastal plain.

During the last few decades, Dar es Salaam has grown rapidly because of unplanned settlement development and a regulatory framework characterized by long administrative procedures to make land available (Kironde, 2006).

The ACC Dar project had the specific goals of: developing a methodology for the monitoring of urban sprawl, analysing urban development and land cover change, investigating the relationships between urban sprawl and population growth. The methodology used was designed especially to fulfil these requirements and involved: the use of free or very low-cost remote sensing images; the availability of images for past years; the use of semi-automatic classification to reduce the time and cost of land cover mapping; preprocessing and processing phases that were achievable with open-source software.

Landsat images were used because of their medium spectral resolution (i.e. 7 bands), although images have coarse spatial resolution (i.e. 30 m), and a large image archive for the past few decades is available; images are provided for free by the USGS, therefore allowing for the affordable classification of land cover, especially impervious surfaces (Fan *et al.*, 2007).

The use of a supervised, semi-automatic, maximum likelihood algorithm allowed for the classification of each image pixel based on spectral similarity with spectral signatures, assuming a multivariate normal distribution of the classes' probability (Richards and Jia, 2006; Song *et al.*, 2001). Furthermore, the use of vegetation indices (i.e. Normalized Difference Vegetation Index and

Enhanced Vegetation Index) and knowledge-based classification improved the identification of urban areas (Congedo and Munafò, 2012).

During the data processing, the following land cover classes were identified:

- Continuously Built-up: a high-density urbanized class
- Discontinuously Built-up: a low-density urbanized class, characterized by a mix of urban and vegetation or soil pixels
- Full Vegetation: a high density vegetation class
- Mostly Vegetation: a vegetation class with medium density
- Soil: bare soil surfaces
- Water: surface water.

The image processing was performed with commercial software; however, a free open-source program was produced within the ACC Dar project – Semi-Automatic Classification Plugin for QGIS – which allows for the semi-automatic classification of multispectral images and can satisfactorily replace commercial software for land cover classifications (Congedo and Munafò, 2014).

The land cover classification results for years 2002, 2004, 2007, 2009 and 2011 are listed in Table 16.1.

The Continuously Built-up and Discontinuously Built-up classes have increased during the past few years; in particular, the fast rate of increase from 2007 on is noticeable. The growth of the Discontinuously Built-up class is related to urban sprawl, which has occurred along Dar es Salaam's main roads. Over the years, some of the low-density areas have changed to Continuously Built-up areas.

It is worth noting that the 30 m Landsat spatial resolution is a constraint in land cover monitoring, because a pixel in the image could contain a mixture of cover types, causing a mixed spectral signature (i.e. mixed pixel) depending on composition and kinds of materials on the ground (Small, 2006).

The validation of land cover classifications was performed through the calculation of error matrices and accuracy statistics for every land cover class. For this purpose, about 500 sample units were selected randomly, and photo interpreted using high-resolution images (i.e. images freely available from Google Earth software, developed by Google). A field survey was performed for 100 of these samples in order to improve the photo interpretation process.

*Table 16.1* Land cover classification results (in hectares)

|                          | *2002*  | *2004*  | *2007*  | *2009*  | *2011*  |
|--------------------------|---------|---------|---------|---------|---------|
| Continuously Built-up    | 8,415   | 10,025  | 10,447  | 12,370  | 14,808  |
| Discontinuously Built-up | 8,098   | 9,134   | 12,509  | 17,318  | 23,678  |
| Soil                     | 102,079 | 95,732  | 76,011  | 57,385  | 66,791  |
| Water                    | 193     | 276     | 304     | 7       | 199     |
| Full Vegetation          | 14,887  | 13,172  | 14,905  | 26,751  | 18,195  |
| Mostly Vegetation        | 35,164  | 40,631  | 54,798  | 55,144  | 45,313  |

Furthermore, fuzzy error matrices were calculated (i.e. considering the presence of secondary classes) in order to improve the assessment of mixed classes; mixed pixels frequent appear in Landsat images because of pixel size. The fuzzy error matrix for the land cover classification of 2011 is shown in Table 16.2.

Table 16.3 shows the accuracy statistics calculated for each land cover class. Considering the secondary class of the fuzzy error matrix, the accuracy statistics improve significantly for the built-up classes, as shown in Table 16.4.

The statistics of the fuzzy error matrix for the urban Continuous Built-up class are considerably improved; the user's accuracy of the Discontinuous Built-up class has also particularly improved, while the producer's accuracy of the same class is slightly better (66.9 per cent considering only the primary class,

*Table 16.2* Fuzzy error matrix calculated for land cover classification based on Landsat images of 2011

| | Reference data | | | | | |
| --- | --- | --- | --- | --- | --- | --- |
| | Continuous Built-up | Discontinuous Built-up | Full Vegetation | Most Vegetation | Soil | Total |
| Continuous Built-up | 93 | (4)(2) | (0)(0) | (0)(0) | (0)(0) | 99 |
| Discontinuous Built-up | (1)(2) | 81 | (0)(0) | (3)(1) | (3)(0) | 91 |
| Full Vegetation | (0)(0) | (0)(0) | 29 | (5)(7) | (0)(3) | 44 |
| Most Vegetation | (0)(0) | (0)(0) | (3)(4) | 56 | (6)(16) | 85 |
| Soil | (0)(5) | (2)(32) | (0)(3) | (2)(55) | 47 | 146 |
| Total | 101 | 121 | 39 | 129 | 75 | 465 |

*Table 16.3* The accuracies of user and producer

| Class | User's accuracy [%] | Producer's accuracy [%] |
| --- | --- | --- |
| Continuous Built-up | 93.9 | 92.1 |
| Discontinuous Built-up | 89.0 | 66.9 |
| Full Vegetation | 65.9 | 74.4 |
| Most Vegetation | 65.9 | 43.4 |
| Soil | 32.2 | 62.7 |

*Table 16.4* The accuracies of user and producer

| Class | User's accuracy [%] | Producer's accuracy [%] |
| --- | --- | --- |
| Continuous Built-up | 98.0 | 93.1 |
| Discontinuous Built-up | 96.7 | 71.9 |
| Full Vegetation | 77.3 | 82.1 |
| Most Vegetation | 76.5 | 51.2 |
| Soil | 34.9 | 74.7 |

71.9 per cent considering the secondary class). The field survey has confirmed the reliability of the photo interpretation, and allowed for the creation of a photographic database.

The accuracy results show that land cover monitoring can be affordable and reliable; however, the class Discontinuous Built-up is more affected by errors due to the coarse pixel size, which is larger than small buildings and caused mixed spectral signatures. In particular, the class Discontinuous Built-up is related to urban sprawl areas where a single pixel is covered by impervious and pervious surfaces, therefore the definition of a spectrally distinct land cover class is difficult.

## References

Congedo, L. and Munafò, M. (2012) *Development of a Methodology for Land Cover Classification in Dar es Salaam using Landsat Imagery*. Technical report. Rome: Sapienza University, ACC Dar Project Sapienza University.

Congedo, L. and Munafò, M. (2014) 'Urban Sprawl as a Factor of Vulnerability to Climate Change: Monitoring Land Cover Change in Dar es Salaam', in *Climate Change Vulnerability in Southern African Cities*, edited by S. Macchi and M. Tiepolo. Cham, Switzerland: Springer, 73–88.

Fan, F., Weng, Q. and Wang, Y. (2007) 'Land Use and Land Cover Change in Guangzhou, China, from 1998 to 2003, Based on Landsat TM /ETM+ Imagery', *Sensors*, 7, 1323–1342.

Kironde, J.M.L. (2006) 'The Regulatory Framework, Unplanned Development and Urban Poverty: Findings from Dar es Salaam, Tanzania', *Land Use Policy*, 23(4), 460–472.

Richards, J.A. and Jia, X. (2006) *Remote Sensing Digital Image Analysis: An Introduction*. Berlin: Springer.

Small C. (2006) 'Comparative Analysis of Urban Reflectance and Surface Temperature', *Remote Sensing of Environment*, 104(2), 168–189.

Song, C., Woodcock, C.E., Seto, K.C., Lenney, M.P. and Macomber, S.A. (2001) 'Classification and Change Detection Using Landsat TM Data: When and How to Correct Atmospheric Effects?', *Remote Sensing of Environment*, 75(2), 230–244.

# Part IV

# Policy and good practices

# 17 The European approach

Limitation, mitigation and compensation

*Gundula Prokop and Stefano Salata*

## Introduction

In the mid-2000s, growing awareness of the risk caused by an uncontrolled growth of urbanization forced the European Commission to introduce the Thematic Strategy for Soil Protection (COM(2006) 231) that was a binding measure for European Union (EU) member states to limit the process of soil consumption due to land take caused by urbanization.

Even if at first the proposal (2006) was intended in a few years to become a binding measure (Soil Framework Directive), the binding aspect wasn't passed for political reasons. In March 2010, in the Environment Council, a minority of member states blocked the progress of the proposal for reasons of subsidiarity, excessive cost and administrative burden (COM(2012) 46 final). This first, fundamental proposal was blocked at the Council's table, and a clear political message about the competence on land use management was launched: member states don't want to be forced by the EU to adhere to communitarian legislation regarding land use constraints or quantitative thresholds for the containment of urban expansion.

In the absence of communitarian legislation on soil, the European Commission approved a number of measures focused on soil-related issues, demonstrating that the Soil Framework Directive was not a unique and systematic approach to protecting and monitoring soil from the risk of degradation. Independent of the legal aspect, knowledge about the status and the quality of soils remains fragmented between member states, and soil protection is not undertaken in an effective and coherent way in all of them (COM(2012) 46 final).

Furthermore, the political orientation of the European Commission seems to be focused on the application of a guidelines document, which is not as binding as a Soil Framework Directive, but which focuses on common targets for limiting, mitigating and compensating soil sealing. Even if the application of the guidelines document at the national scale is not compulsory, the general impression is that land use management has drawn greater attention at the European level, and the enforcement of land use monitoring requires the major use of technical instruments to present, assess and control the trends of land use change in Europe.

Nevertheless, the absence of European legislation on soil still demonstrates the lack of a common agreement on a field that has many related aspects (e.g. the real estate market). Overcoming this position is now fundamental: some recent initiatives aim to introduce new strategies to define the risk of uncontrolled growth and to clarify that land take is a matter not only of 'quantity' of soil that is no longer available for other uses (e.g. food production), but also of the 'quality' of citizens' lives in the urban environment.

## Soil Thematic Strategy: an important key driver for awareness

Even if it has been argued how the originally planned way to achieve common soil legislation has failed, it has to be considered that the proposal has acted as a key driver for the development of a huge number of studies, indicators and databases on land use variation and soil related issues, including conferences, debates, scientific and non-scientific publications on soil, climate change and biodiversity. Consequently a working group on Awareness Raising and Education in the context of the European Soil Bureau Network (ESBN) and a European Network for Soil Awareness (ENSA) (COM(2012) 46 final) have been established. Thus the proposal at least acted to promote studies and technical reports: around 25 research projects have been started, including those listed below (for a deeper understanding of the mission and final results of each project see the specific websites):

- RAMSOIL, www.ramsoil.eu/UK/
- ENVASSO, http://eusoils.jrc.ec.europa.eu/projects/envasso/
- SOILSERVICE, www.lu.se/soil-ecology-group/research/soilservice
- LUCAS, http://eusoils.jrc.ec.europa.eu/projects/Lucas/
- BIOSOIL, http://forest.jrc.ec.europa.eu/contracts/biosoil.

All projects are integrated by specific policies focused on the sustainable use of the soil:

- Common Agriculture Policy (CAP): focused on promoting good agricultural and environmental conditions, limiting erosion, improving organic matter and avoiding compaction.
- Industrial Installation: focused on saving soil quality from potential negative impacts of future industrial plants.
- Cohesion Policy: focused on the rehabilitation of industrial sites and contaminated land.
- State Aids for the remediation of soil contamination.

In this context, land degradation resulting from soil sealing by urbanization, but also other threats such as soil erosion, desertification, salinization, contamination and acidification, is monitored by ongoing activities of the technical organism of the European Commission, the Directorate-General, Joint Research Centre (DG-JRC), which observes land use change at the European scale to make more efficient use of resources.

Despite further attention being given to soil related issues, at the moment soil is not subject to a commonly defined and systematic set of rules in the EU. Thus, existing EU policies in other areas are not sufficient to ensure an adequate level of soil protection. Furthermore, having been blocked at the Council's table in 2010, the proposal for a Soil Framework Directive was finally withdrawn in May 2014.

Despite the lack of success in the top-down approach, the right to recommend EU legislation via the European Citizens' Initiative (ECI) is always possible with bottom-up approaches such as petitions. One of the emergent proposals, 'People 4 Soil', is a free and open network of European NGOs, research institutes, farmers' associations and environmental groups that aims to relaunch the Soil Framework initiative. The petition was prepared during the International Year of Soils (2015), launched in 2016, and will last for 12 months.

## Soil sealing: drivers and impacts

Soil sealing is the most impacting effect of land take, because when the top-soil (which is the upper layer of soil) is covered by impermeable materials, all biological functions are compromised. This is why the equation 'soil sealed' is equal to 'soil lost' is true: even if practices of soil recovery are applied for desealing, neither the quantity nor the quality of biological functions can be restored. Hence the permanent covering of land by impermeable artificial material (e.g. asphalt and concrete) directly affects essential ecosystem services (e.g. food production, water absorption and the filtering and buffering capacities of soil) and biodiversity (European Commission, 2012).

We therefore here detail the drivers of soil sealing by urbanization in order to reveal how 'wide' the phenomenon is, and how economic, social, legislative and specific planning programmes combine as causes of the current trend of land consumption.

1   The need for housing, industry, business locations and infrastructure in response to the growth of population and the demand for a better quality of life and living standards. The EEA points out that urban expansion is more a reflection of changing lifestyles and consumption patterns rather than an increase of population (EEA, 2006).
2   The preference to live in a place far from the compact city, even if the commute between home and work is long. This model presents a spatial consumption of different resources (soil but also energy) and generates a long-term social cost in terms of services and health.

3   The dependency of some local authorities on incomes provided by urbanization feeds, which stimulates competition between municipalities for offering cheap land for development.
4   The idea that there is no need to worry about additional soil sealing because of the relative abundance of open space in rural areas.
5   The general lack of appreciation of soil in terms of environmental values (soil sealing is an irreversible process of transformation of the topsoil), social values (soil sealing reflects a private use of land pushed by real estate market dynamics) and economic values (soil sealing results in high public costs in the long term).

Even if drivers of soil sealing are diverse (e.g. economic, lifestyle, social and fiscal, mobility, planning legislation and so on) the impacts mostly relate to (the loss of) soil biodiversity, which implies a reduction in ecosystem service supply. Indeed, the action of covering topsoil with asphalt or concrete has a direct impact on soil-related ecosystem services. The most important effects:

- Soil sealing affects the normal flow of water drainage.
- Soil sealing has an impact on ground biodiversity and fragments ecosystems and the landscape.
- Soil sealing normally occurs in the most fertile areas, and impacts food security.
- Soil sealing has a direct impact on the carbon cycle.
- Soil sealing reduces evapotranspiration.
- Soil sealing reduces air quality.
- Soil sealing directly affects the quality of life in urban and rural areas.

Enforcing a political agreement on the problematic of soil sealing also requires that technical aspects in limiting soil sealing and land take in general be taken into account. Spatial planning has a specific role to play in this regard. At first, an integrated approach is required, which means mixing different planning theories that aim to apply an ecological approach to the salvage of green areas. An integrated approach requires also the full commitment of all relevant public authorities, in particular those levels of government that normally are directly responsible for the management of land use.

Local planning requires exploiting fully the possibilities offered by the Strategic Impact Assessment (SEA) Directive and, when relevant, the Environmental Impact Assessment (EIA) Directive, collecting detailed soil data and establishing suitable indicators, regular monitoring, critical assessments, as well as providing information, training and capacity-building for local decision makers (European Commission, 2012).

At all scales, the starting point has to be to assume that dealing with land take is a necessity. The process of urbanization affects all European Countries with varying intensity, and still affects states experiencing combined demographic/economic recession. The first requirement is to limit the phenomenon, acting at a political level but also using spatial planning measures.

# Reducing soil sealing needs a tiered approach

The efficient protection of soils from further sealing can only be achieved by following an integrated approach, requiring the full commitment of all governmental units (and not only those dealing with spatial planning and environment), by improving awareness and competence within all concerned stakeholders, by freezing counterproductive policies (i.e. funding of single family houses at urban fringes, commuter bonus, etc.), by establishing clear financial incentives, and by introducing binding legal requirements. In this context the European Commission in 2012 published Guidelines on best practice to limit, mitigate or compensate soil sealing. The guidelines demand a three-tiered approach, similar to the logic used in waste materials streams. The priority solution must be to limit soil sealing, as sealing is an almost irreversible process. Where it is not possible to avoid sealing, the second best option is to mitigate its impacts, reducing the worst effects where possible. The third option, a last resort, is to compensate for sealing soil in one location by soil-related remediation activity in another. Different options for translating these measures into practice are explored below.

## *Successful examples for limiting soil sealing*

There are two ways to limit soil sealing: by reducing land take, the rate at which natural areas are converted into developed areas; or by continuing to seal soil, but only on land that has been previously developed. To 'pave the way' for successful prevention of soil loss the following basic principles need to be implemented at the policy level:

- Establish the principle of sustainable development in spatial planning by following an integrated approach, requiring the full commitment of all governmental sectors (and not only spatial planning and environment). An example of best practice would be that the majority of EU member states establish the principle of sustainable development in their key spatial planning regulations, making reference to the economic use of soil resources and avoidance of unnecessary urban sprawl. However, without binding measures, regular monitoring and critical assessment soil functions this cannot be protected adequately.
- Define realistic land take targets at the national and the regional level. One of the best practices is placing quantitative limits on annual land take. Such limits exist only in six member states – Austria, Belgium (Flanders), Germany, Luxembourg, the Netherlands and the United Kingdom. In all cases the limits are indicative and are used as monitoring tools. In the United Kingdom and Germany the national targets are taken most seriously and their progress is regularly assessed. Only in the United Kingdom are development targets also defined at the regional level.

- Streamline existing funding policies accordingly by freezing subsidies that encourage land take and soil sealing (i.e. public subsidies for private housing on undeveloped land, subsidies for developments on green field sites, commuter bonuses, etc.). No best practices for this point have been identified.
- Develop specific regional approaches according to the actual land use pressures and, in particular,

1   Steer new developments to already developed land and provide financial incentives for the development of brownfield sites. One of the best practices is initial or supportive funding to encourage new infrastructure developments on brownfield sites, which exists in several member states and which is usually co-ordinated by designated brownfield organizations. Brownfield redevelopment projects are mostly realized in the form of public–private partnerships: (1) English Partnerships is probably the most experienced public land developer in the European Union and provides funding for social housing developments on derelict areas; (2) France disposes of a network of more than 20 public land development agencies, which among other activities develop brownfield land for social housing; (3) the land development agencies Czech Invest and Invest in Silesia are in charge of developing major industrial brownfields for new industrial investors; (4) in Flanders specific contracts (brownfield covenants) are negotiated between the government and private investors to promote brownfield redevelopment.

2   Improve the quality of life in large urban centres. Best practices to mention are several urban renewal programmes recently launched with the objective of attracting new residents and creating new jobs in central urban areas in decline. Best practice examples in this respect are (1) the urban renewal programmes of Porto and Lisbon and the neighbourhood renewal programme in Catalonia, all of which are supported by European Regional Development Funds, (2) the Västra hamnen project in Malmö, which is built on derelict harbour premises providing 1,000 new dwellings with the lowest possible environmental impact, (3) the Erdberger Mais development in Vienna, which is built on five inner urban brownfield areas, providing housing for 6,000 new inhabitants and 40,000 work places, (4) the Randstad programme in the Netherlands, which puts special emphasis on improving the attractiveness of inner urban areas in the metropolitan agglomeration of Amsterdam, Rotterdam and Den Haag.

3   Make small city centres more attractive in order to counteract dispersed settlement structures in rural regions with shrinking population. One of the best practices to mention is the Danish Spatial Planning Act, which puts clear restrictions on the construction of large shops and shopping centres on green fields outside the largest cities and promotes small retailers in small and medium-sized towns.

4    Impose development restrictions on top agricultural soils and valuable landscapes. Best practices are established where member states have promoted specific policies to avoid further land take and sealing on their best agricultural soils and most valuable landscapes, as is the case (1) in Spain where building activities within the first 500 metres from the sea are strictly controlled, (2) in France and the Netherlands where designated 'green and blue' landscapes are protected from infrastructure developments, (3) in the Czech Republic and Slovakia where the conversion of top agricultural soils requires a fee.

### Mitigating soil sealing (examples of technical measures)

Permeable surfaces can help to conserve soil functions and mitigate the effects of soil sealing to a certain extent. They contribute to the local water drainage capacity and can in some cases also fulfil biological or landscaping functions. Another advantage is their positive contribution to the micro-climate thereby trapping the heat and moderating temperatures in the area. Unsealed, green shaded surfaces have lower surface temperatures than sealed surfaces, the difference can amount to up to 20°C. In the case of storm water a parking area built with permeable surfaces discharges to the local sewage system by at least 50 per cent compared to a conventional asphalt surface. It can even be designed as an independent system without discharges to the local sewage system.

A broad range of materials and concepts is available for permeable surfaces. In addition to their clear ecological advantages most types of surfaces have lower lifespan costs compared to conventional impermeable surfaces. With regard to sustainability most permeable surfaces are made of materials that are locally available and reusable. Key barriers to implementation are currently the fact that site-specific know-how and building competence is required to construct them correctly. Furthermore, regular maintenance is needed to make sure that they function properly. Parking areas have the greatest potential for permeable surface application, in particular large parking areas in urban fringes. Most advanced in this respect is the United Kingdom, where permeable

*Figure 17.1* Overview of most common surfaces: (1) lawn, (2) gravel turf, (3) plastic grass grids, (4) concrete grass grids, (5) water bound macadam, (6) permeable pavers, (7) porous asphalt, (8) conventional asphalt

*Table 17.1* Comparison of benefits and limitations of most common permeable surfaces (in relation to asphalt)

| | Application range | | | | Benefits | | | | | Limitations | | | | | Unsealed surface (%) | Run-off coefficient | Cost: Asphalt = 100% |
|---|---|---|---|---|---|---|---|---|---|---|---|---|---|---|---|---|---|
| | Pedestrians | Parking, small vehicles | Parking, medium vehicles | Road traffic | Visual appearance | Vegetation possible | High drainage capacity | Regional materials | Improves micro-climate | High maintenance | Bad walking comfort | No disabled parking | Sludge accumulation | Dust formation | | | |
| Lawn, sandy soil | Y | | | | +++ | +++ | +++ | +++ | +++ | +++ | | +++ | +++ | +++ | 100% | <0.1 | < 2% |
| Gravel turf | Y | Y | Y | | ++ | ++ | +++ | +++ | +++ | + | + | + | | | 100% | 0.1–0.3 | 50–60% |
| Grass grids (plastic) | Y | Y | | | ++ | ++ | ++ | + | +++ | +++ | +++ | +++ | + | | 90% | 0.3–0.5 | 75% |
| Grass grids (concrete) | Y | Y | Y | Y | ++ | ++ | + | +++ | ++ | +++ | +++ | +++ | + | | 40% | 0.6–0.7 | 75–100% |
| Water bound surfaces | Y | Y | Y | | + | | + | +++ | + | | + | + | ++ | ++ | 50% | 0.5 | 50% |
| Permeable pavers | Y | Y | Y | | | | ++ | | | + | | | | | 20% | 0.5–0.6 | 100–125% |
| Porous asphalt | Y | Y | Y | Y | | | | | | | | | | | 0% | 0.5–0.7 | 100–125% |
| Asphalt | Y | Y | Y | Y | | | | | | | | | | | 0% | 1.0 | 100% |

Source: Prokop and Jobstmann (2011).

Note: * Indicative costs in relation to asphalt are provided; in 2010 average costs for conventional asphalt layers amounted to approximately 40 €/m² (without VAT), including construction costs. For each surface type, material costs and labour costs were considered.

surfaces are broadly used – even in big cities – and where research is continuously developed and many guidelines exist.

Figure 17.1 shows the most common surfaces for 'artificial' open areas. The surfaces are presented according to their permeability; i.e. the first picture shows a conventional lawn which can be considered 100 per cent unsealed, pictures 2 to 7 refer to various permeable surfaces, and the last shows asphalt, being 100 per cent sealed. Table 17.1 compares the benefits and limitations of most common permeable surfaces.

Parking areas have the greatest potential for permeable surface application. In Europe there are definitely more parking lots than cars. The number of cars is increasing from year to year and together with this trend also the number of parking lots; hence the application of reinforced grass systems with gravel or grass grids is ideal for use in large short-term parking areas such as in:

- Recreational sites: e.g. ski resorts, football stadiums, golf courses, touristic sites and trade fairs. Such surfaces improve the local drainage capacity and contribute positively to the landscape.
- Households: private driveways have great potential for the application of permeable surfaces. For this type of use almost all surface types are applicable.
- Supermarkets: the use of permeable concrete pavers in combination with drainage ditches is a long-lasting solution that allows heavy traffic. This type of surface is increasingly being applied in supermarket parking areas.

The use of such surfaces has some limitations: areas with sensitive groundwater resources or shallow groundwater (below 1 metre) are in general not suitable for surface drainage. Moreover the costs have to accounted because apart from natural stone pavements, it can be said that permeable surfaces do not bear higher costs than conventional asphalt and are not dependant on the crude oil price (unlike asphalt).

Moreover, gravel turf and concrete bricks are made of sustainable materials, which are readily available in most European regions. As these materials can easily be reused their life span is almost unlimited. Conventional asphalt on the contrary has to be recycled for re-application with more energy input.

The above-mentioned reasons explain why many planning authorities in Europe are currently revising their technical regulations towards surface sealing. Increased drainage capacity has many advantages, in particular in areas with flood risk or overloaded sewage systems. The fact that permeable surfaces can reduce or even avoid costs related to flood prevention, flood damage repair or enlargement of existing sewage systems is attractive for local planning authorities. For example, planning authorities in England, in the Alto Adige region (Italy) and in selected cities in Germany and Austria already restrict surface sealing for new building activities.

## Compensating soil sealing

The idea behind compensating for soil sealing is to make up for sealing in one place by restoring soil functions elsewhere in the same area. As a rule, compensation measures should be equivalent to the ecosystem functions lost.

Environmental impact assessments of large projects and for planning purposes can be used to identify the most appropriate compensation measure. Examples of compensation schemes include:

- Reuse of topsoil: topsoil can be removed from a construction site and used, for example, to upgrade agricultural sites, or to regenerate contaminated land and encourage seed germination, for example on a golf course, or to improve soil quality in gardens.
- Desealing (soil recovery): removing asphalt or concrete and replacing them with topsoil on subsoil can help renew the soil functions of a previously sealed site, as well as restoring the beauty of the landscape. Desealing is mainly used in urban regeneration projects, following the removal of derelict buildings to create green spaces, for example. Sadly, this option is not taken up often enough because the costs are perceived to be too high.
- Sealing fee: authorities can impose fees for land take and soil sealing. This could be used as a tool to limit soil sealing, but in practice fees are rarely high enough to discourage land take. Instead, the money collected is used to support soil-protection projects. Some countries in Europe use sealing fees to protect the best farmland.
- Eco-accounts and trading development certificates: in an eco-accounts system, the ecological cost of soil sealing is determined and developers have to ensure that compensation measures of equal value to sealing are carried out elsewhere. Official compensation agencies oversee the system.

## Conclusions

Despite a constant demand for urgent intervention and regulation that will tackle the incessant consumption of open space calculated at an aggregated scale, it seems that the problems of improvement of particular land-use development patterns have not yet been properly addressed. Even if analysis on land take is becoming much more significant, less successful cases of land take reduction are registered. The application of the Guidelines approach demand the greater advancement of research on land use management practices.

A simple contextualization of the analysis on land use trends gives simple but clear indications: traditional tools for land use/cover analysis are not adequate for the evaluation of impacts on ecosystem services and insufficient to steer local policies for land conservation. New approaches at the regional scale are required to introduce more detailed evaluation of the impact of land take on ecosystem services, with particular attention to the major effect of sealing on soil, air and water.

In order to implement soil sealing guidelines and activate a sustainable soil and land governance, a multidisciplinary approach is needed to bridge the gap between general, theoretical targets (e.g. land-take limitation) and the development of specific patterns of land-use management at the local scale.

The need to go beyond the simplistic approach of land use change analysis and to provide better information and more comprehensive data will enable policy and decision makers to activate the right prescriptions, limitations or regulations for land use management.

## References

Prokop, G. and Jobstmann, H. (2011) *Report on Best Practices for Limiting Soil Sealing and Mitigating Its Effects.* Technical Report – 2011 – 050, European Commission, Brussels. http://ec.europa.eu/environment/soil/pdf/sealing/Soil%20sealing%20-%20Final%20 Report.pdf, accessed 18 January 2016.

European Commission (2012) *Guidelines on Best Practice to Limit, Mitigate or Compensate Soil Sealing.* SWD(2012) 101 final/2, European Commission, Brussels.

EEA (2006) *Urban Sprawl in Europe: The Ignored Challenge.* EEA Report 10/2006. Luxembourg: Office for Official Publications of the European Communities.

# 18 Policy, strategy and technical solutions for land take limitations

*Stefano Salata*

## Introduction: from knowledge to policy

Approximately 75 per cent of Europe's population lives in urban environments and a quarter of the EU's land surface has been directly affected by urbanization. This type of land use change affects environmental resources, and thus the quality of human life is dependent on the capacity to govern the process of urbanization.

If sustainable urban development is focused on 'quality of life', it is necessary to assess and evaluate the effects of land take on soil-related ecosystems. But the development of strategies against land take require the quantitative/qualitative assessment of the environmental effects of urbanization (e.g. the impact on specific ecosystem service (ES) degradation).

The recent 'Scoping Study for DG ENV' (PRACSIS, 2014) defines soil resource as less attractive than other natural resources, and accordingly the creation of a common consciousness on soil related questions is far from being set. Moreover, soil scientists and ecologists attribute land take to planning weakness and then planners turn accusations to politics. Since no one bears responsibility, land take happens. The reasons are many:

- the urban development mode is still based on expansion (with low population density);
- the land take by urbanization is mostly concentrated on prime quality land;
- the urban development pattern promotes the private use of vehicles;
- the urban development model is less concentrated on a re-use approach;
- the lack of appreciation of soil as a finite, non-renewable resource.

Nevertheless, even if the causes are in the main addressed, the general impression is that the gap between analysis (quantification and cause–effect qualification of the land take phenomena) and regulation (improvement of particular land use development patterns) is still unfilled (Nuissl *et al.*, 2009). This problem occurs due to a deep epistemological issue: while 'land cover' refers to the ecological state and physical appearance of the land surface based on a classification system, the 'land use' refers to human purposes in relation to different things (e.g. the morphological characteristics of the soil, the proximity to a centre/service, the landscape value etc.) (Turner *et al.*, 1994; Dale and Kline, 2013).

Furthermore, academic positions are problematic: although many disciplines recognize land take as a central environmental issue, a large part of the research is still descriptive, rather than focused on supporting local policies for land use management. The knowledge of soil quality is too poor for planning disciplines, but even though an environmental phenomenon can be slightly shaded or undervalued, nowadays new economic paradigms emerge as crucial issues for further discussion: are the long-term social costs of urbanization higher than short-term incomes for operators over land use transformations? The economic quantification of ES demonstrates that the value of natural capital is higher than any single income of real estate operation. However, as long as it is not possible to demonstrate, and fully assess, the real economic side of land take, the objective to achieve quality of life and public health by land use control is far from being set.

Hence deeper analysis on the environmental effect of land take on ES provided by natural soils is required (Helian *et al.*, 2011), especially because such analysis requires integrative assessment across different disciplines (Breure *et al.*, 2012). The lack of a more systemic and holistic agreement on common considerations that land use/cover require a higher integration of knowledge between ecological, social and economic studies is weakening the possibility of achieving real 'sustainable development'.

Moreover, the construction of an analytical framework for territorial planning using ES as a real proxy for land use management can be pursued using different steps: (1) framing a key policy issue related to ES preservation or restoration, (2) identifying ES and users (e.g. the definition); (3) mapping and assessing status; (4) valuation; (5) assessing policy options including distributional impacts.

Among the different approaches, explained below, the use of ES as a central element for the re-definition of land use regulation seems to be the way to overcome national/regional quantitative policies of reduction with 'in-depth' qualitative support for decision-making processes. Innovations in using ES are mainly two: on one hand ES introduces qualitative elements of trade-off among alternative uses of soil (acting against the flat dichotomy between urban/rural uses), which allows secondary considerations; on the other hand ES helps not only in understanding how much soil will move into the urban category of land use (quantification) rather than define which kinds of soil are affected by urbanization, but also in assessing what are the environmental effects of such transformation (qualification). Finally, ES provides the possibility to associate both biophysical and economic values, allowing the economic quantification of losses or gains due to the uses of natural capital.

## How to limit land take

It is impossible to outline a unique approach to land take limitation. Cultural context, legislative frameworks, but also planning tools and territorial con-textualization require a mix of strategies rather than a single approach. While the standardization of a methodological assessment is necessary for defining a common knowledge system (e.g. definition, indicators), strategies for limita-tions are many, and advanced experiences show that a single approach is not strong enough to achieve great limitations targets – rather a mixed framework of policy, strategies and planning tools is required.

First, it is necessary to take into account that land take limitation is, at least, directly controlled by land use regulations: excluding informal or illegal uses, all other land use changes happen according to the local land regulation. Thus, it is at the local scale that the power of land take limitation, mitigation and com-pensation has to be enforced. If it is accepted that land use zoning is the tool that allows land use transformations, then strategies pursued consequently have to be adaptive and progressive, and communitarian as national/regional/local authorities have to work together using dispositive approaches (e.g. guidelines, or communitarian directives at the EU level), incentives (e.g. economics, since economic and fiscal measures are mainly addressed to nation states), limitations (e.g. thresholds on land take based on regional land use monitoring inventories) and regulations (e.g. definitions of Urban Growth Boundaries, greenbelts or land use prescriptions to stop land take on greenfields).

Above all, it has to be considered that land use regulation will only bring about effective results through long-term application, and the construction of a legislative context settled by an agreement between a communitarian approach and national/regional legal frameworks is still laborious. For such reasons, among others, a great deal of attention has to be dedicated to take, as soon as possible, the right decisions.

To date no single measure appears to have achieved great success in limit-ing land take. Data on urbanization trends demonstrate that the phenomenon is affecting, with varying intensity, all countries. Among the numerous sets of measures, the 'market-based' approach seems to be the most transversal way to reduce the amount of land turned to urban uses. Fiscal measures designed to address extra feeds for land take disincentives are gaining in efficacy, and this seems to be the only sovra-local 'strategy' that offers a real possibility of reducing land take.

When national fiscal measures are integrated with regional instruments for monitoring land use change and local regulations to mitigate or compen-sate land take based on ES assessment, the possibility of effective reduction increases. The regulation of land use through fiscal measures at the national level gives much greater power to local administration to act with additional regulative planning measures. Taxation is also necessary for capturing urban rent and provides the possibility of reducing speculation through the urban transformation of greenfields.

Some experiences are demonstrating how, over land use regulation by local planning, the introduction of additional fees on free land is able to decrease requests for transformation made by real estate operators. In Italy, the reintroduction of IMU (the Municipal Immobiliar Tax), which is far from being settled as an environmental tax for land take limitation, has changed behaviour in relation to speculative plans by land owners. If previously requests were commonly based on the extension of building rights on free land, nowadays the introduction of IMU has turned requests upside down: land owners ask to reconvert urban uses to agricultural ones where properties are located. Even if such requests reduce the possibility of building, and thus local plans reduce the amount of land designated to transformation, this measure allows them to avoid taxation.

The above-mentioned experience shows that market-based policy, based on fiscal measures such as land taxation, should achieve better results than other regulative prescriptions (e.g. land take thresholds).

The following sections detail experiences of taxation application as the main paradigm for land take regulation.

## For a theory of land taxation

Not all the costs of open, agricultural space are internalized in market transactions involving agricultural land. This may cause an excessive penetration of urban land into open spaces. In other words, the phenomenon of urban sprawl may occur (Korthal Altes, 2009). These few words essentially explain why land take happened with varying degrees of density or morphologies (dispersed, fragmented, ribbon, leapfrog) mainly in agricultural land.

By such definition, urban sprawl is the effect of an economic cause: the price of agricultural land is generally lower then urban land. If the difference between the two values is not equalized, urban rent still persists as a parasitic income for private operators who ask for free agricultural land to transform. If this fundamental point is not solved, even between different alternatives of localization for land use transformation, investment will be in greenfields rather than in the existent stock of urbanized soils.

Thus sprawl is not a consequence of a citizen's living preference, but the effect of the real estate market economy. The fact that urban areas are still expanding, even in the context of population decrease, is mainly dependent on the old theory of urban rent. Indeed sprawl affects all countries, and even if the requests for new housing are not coming from the upper classes but from new immigrants and young populations, the real estate offer is still too unbalanced to provide market-price homes in the suburbs, with high dependencies on private mobility.

One of the studies on urban rent done in the 1990s tried to introduce economic values on open spaces, opening the way to ES economic quantification (Costanza et al., 1997). At the base of the theory is the value of soil as a natural, finite resource that, playing crucial functions (both tangible and intangible), guarantees human life on earth (Daily, 1997). In 1997 the study, entitled 'The Value of the World's Ecosystem Services and Natural Capital', presented the 'cost' of such services, and the environmental economy became a discipline focused on assessing the balances of environmental damages done by human transformation.

At the end of the 1990s, the first systematic studies were launched on the environmental taxation of free land in order to avoid the decrease in the value of ES, and debates for and against using fiscal measures to dissuade land use changes on open fields arose.

If local planning prescriptions aimed to control urban growth were historically rooted in the planning discipline, for the first time the economic side of regulation was considered a key driver in land use control. Immediately a question arose: the creation of a hard limitation to the offer of urban land does not directly imply that transformation should be concentrated in the already urbanized land. The condition of a monopoly on the few urban areas suitable for transformation gives rise to an increase in the value of brownfields and thus a 'block' on marked real estate. Consequently, if taxation on greenfields is not accompanied by additional measures to stimulate transformation on brownfields, the risk is to create only a block on the real estate market rather than to limit the land take.

Despite all the technical questions that are slowing down the possibility of applying a common theory to land taxation against land take, it is important to recognize that, since the physical control of the city dominated the welfare state of the twentieth century, today the attitude is to consider the market approach, even for environmental good, as the only way to impact the real economy.

Taxing land take seems to be a necessity for an environmental policy directed at sustainable land use planning. It is quite evident that individual costs paid for land use transformation are not equal to the collective costs paid by society for greenfield urbanization (Nuissl et al., 2009). The effects on habitat quality (Price et al., 2006) soil and water buffering capacity (Haase and Nuissl, 2007), atmospheric particulate concentration (Yang and Lo, 2002), public health (Wells et al., 2007) and social segregation (Power, 2001) are quite evident, even if not systematically addressed together as a consequence of land take. Nonetheless an integrated approach is requested.

There are three main paradigms with regard to theories of land use taxation:

• the application of fiscal extra feeds on the urban transformation of greenfields (Korthal Altes, 2009);
• economic incentives for urban reuse (Ring, 2008);
• the introduction of a controlled market of land transformation (Nuissl et al., 2009).

A paradigmatic experience of taxation is given by the case of the Netherlands, where some prerequisites have facilitated the introduction of taxation as the base of a national strategy for land use transformation:

- the value of agricultural soil is high (because both agricultural rent and productivity are high);
- agricultural conservation has high costs;
- housing development has been mainly steered by national policies since the 1980s;
- even the planning system is regulated and it provides sufficient land for transformation (Faludi and Van der Valk, 1994).

In this context a framework for a methodology of land take taxation was devised, taking into consideration that:

- it is necessary to assess the welfare value of the free land for aesthetic purposes;
- it is required to capture the urban rent generated by land use plans changes;
- it is important to steer public/private economic resources on re-development towards existent urban areas, rather than towards greenfields;
- extra funds for public administrations from land taxation can be guaranteed.

Obviously, the proposal opened a debate on the economic evaluation of environmental goods such as the soil, and the debate turns back to the point stated above: it is crucial to estimate what are the economic values of soil services to achieve sustainability on land use. As long as economic values of ES are not considered during the decision-making processes, the target of sustainability will not be achieved.

As introduced, a great deal of research is dedicated to estimate the environmental effects of land take processes, especially using ES as a proxy (Breure et al., 2012; Jansson, 2013; Artmann, 2014; Li et al., 2014). From systematic studies on surface and covers, to the complete assessment of urban transformation effects in hydrologic systems, a huge amount of research is focused on the definition of 'what happened on topsoil, and under it, when a process of urbanization occurs' (Gardi et al., 2014).

In general, as an ES approach has emerged as the main paradigm to estimate quantitative and qualitative land transformation (Costanza et al., 1997; Daily, 1997), there is a lack of technical assessment to introduce indicators that hold different multidimensional aspects of soil transformation (e.g. the alteration of productive capacity – land capability, impermeabilization, biodiversity decrease, landscape and cultural values). Composite indicators on land take are far from being rooted in scientific literature (even if they are well defined) (Giovannini et al., 2008), and there is an impression that despite a great number of words written claiming an interdisciplinary approach on land management, no systematic results seem to be achieved. The demand for profound soil knowledge is high (Havlin et al., 2010), and the teaching of soil related disciplines is mainly housed in geology, geography, environmental science and agriculture programmes (Hopmans, 2007), but a major interaction of scientists from other disciplines is required in order to achieve a broad holistic role in society, and the context of 'fusion' between different backgrounds needs to be enforced (McBratney *et al.*, 2014).

Land Use Change (LUC) allows us to quantify the loss of ecosystem functions as an effect of change in cover or uses of land (Shuying *et al.*, 2011). Nowadays a weak assessment of indicators for specific ES functions demands a high account in research for ES identification and mapping, especially for local planning (Rutgers *et al.*, 2011; Dominati *et al.*, 2014).

For example, literature recognizes that the total ES value of each land use category can be obtained through multiplying the area of each land category by the value coefficient: ESV = Σ (Ai • VCi) – where ESV is the estimated ecosystem service value (Euro•a-1), Ai is the area (ha) and VCi is the value coefficient (Euro•ha-1•a-1) for land use category 'I' (Helian *et al.*, 2011). Such definition introduces the possibility of finding an economic overall evaluation of ES. Even oversimplified (Pimm, 1997; Toman, 1998), such possibility gives to public administration and planners an estimation of a variation of values for non-commodities (soil) through land use planning. Rather than absolute value, such methodology should be normally introduced as an economic computation of ES variation between present (net present value) and future (Bateman *et al.*, 2013; Baral *et al.*, 2014). Additional exploration of ES values for specific land use/cover categories is reported in the study 'Impact of Urbanization on Natural Ecosystem Service Values: A Comparative Study' (Shuying *et al.*, 2011).

The critical ways in which ecosystems support and enable human well-being are rarely captured in cost–benefit analysis for policy formulation and land use decision-making (Laurans *et al.*, 2013). Results showed that, although a conventional, market-dominated approach to decision-making chooses to maximize agricultural values, these monofunctional policies will reduce overall values (including those from other ES) from the landscape in many parts of the territory – notably in upland areas (where agricultural intensification results in substantial net emissions of GHG) and around major cities (where losses of greenbelt land lower recreation values). In comparison, an approach that considers all other ES for which robust economic values can be estimated yields net benefits in almost all areas, with the largest gains in areas of high population. Some analyses suggest that a targeted approach to land-use planning that recognizes both market goods and non-market ES would increase the net value of land to society by 20 per cent on average, with considerably higher increases arising in certain locations (Bateman *et al.*, 2013).

Even at the theoretical stage, the ES approach raises the possibility of estimating the net cost of an environmental service supporting the definition of a theory for land take taxation. Nevertheless, legislative and economic reasons seem to create obstacles: how can a theory of taxation influence the real-estate market?

The fear of lowering the few private resources dedicated to real estate dominates the position against the introduction of a land taxation system. Such a position is based on the fact that urban rent is, at least, the core of real estate investment. The taxation model cannot recover all the urban rent generated by land use change, otherwise the marginal incomes for operators would not be sufficient for transforming the land. Thus the risk, in such a case, is to stop all private real estate market operations (Korthal Altes, 2009).

Another kind of risk associated with the introduction of a purely fiscal approach to land use regulation based on taxation is the potential that it invites public administration to 'use' the tool to create extra income which would generate distortions: a rush to capture private resources, resulting in a huge number of transformations on greenfields. This would have the opposite effect to land take limitation.

Among others, the above-mentioned issues are sufficient to demonstrate that a land take taxation model is far from being regularly adopted and, moreover, that it is contextual and not a substitute for the traditional land use regulation system. If a method of 'low' taxation is introduced by the state, then local tools and instruments of land use regulation such as green-belts or Urban Growth Boundaries are enforced. Such integration between different instruments and strategies works better in a context where the strategic dispositions of land use transformations and prescriptions are fixed at a wider scale (sub-regional rather than at municipal level). It is widely recognized that green infrastructure design requires an intermediate scale of planning, detailed enough to be prescriptive for environmental issues, but synthetic enough to fix common rules for homogeneous territories (Crompton, 2007; EEA, 2014). Economic instruments of land use policy, therefore, will considerably mitigate the negative impacts of land development only when backed by traditional land use planning tools such as zoning, which allow for the spatial guidance of land development in the first place (Nuissl *et al.*, 2009).

Another option would be to offer buildable land only with a land use change permit. In this case building permits would be awarded to private operators through negotiation on the basis of a planning regulation system. Even with a market-based approach, this system does not introduce a system of taxation based on land use definition; rather it invites discussion of the single initiative of transformation and involves:

- fixing a 'price' for building permits based on the value of soil quality;
- creating market control of the offer of building rights.

In any case, at the base of the taxation theory, the debate for or against the possibility to attaching economic values to a purely environmental resource generates opposition. A branch of ecology states that it is impossible to fix the overall value of non-commodities, others claim there is a 'need to evaluate ecosystems economically', starting from the 'real' market value of some goods. For example, the economic evaluation of biodiversity should be derived from the market price of the 'reproduction' of specific land uses that provide such ES: the cost of planting a forest, or the cost of a public garden for urban green areas. This means using a market price of 'substitution' (How much does it cost to reproduce the goods?), using a biophysical environmental index as a proxy of distribution of the service.

## Local regulation

Tools focusing on the containment of land take mainly aim to define:

- a target of reduction for the amount of land development (quantitative thresholds);
- the improvement of specific land use patterns (qualitative control of settlement distribution).

While the first approach is rooted in environmental discipline and mainly based on the quantitative assessment of land take (Helming et al., 2011), the second is less covered by scientific studies even if it seems to be evident that limitation measures require full integration between quantitative and qualitative methods of assessment (Haberl and Wackernagel, 2004).

The process of land take generally implies a reduction of the ES delivered by soils on the basis of land use variation over different years. With the basilar knowledge of urban land use changes (quantitative), comes the evaluation (qualitative) of land take impact on ES (Shuying et al., 2011) that give support to local practices of land conservation.

There are two consolidated approaches for land take limitation over local land regulation:

- the introduction of green border areas for the containment of urban expansion (greenbelts);
- the definition of regulative borders by planning constraints between buildable and non-buildable areas (Urban Growth Boundaries – UGB).

The two approaches are traditionally used by local planning regulation to achieve a compact settlement system. Since the legal planning system has been theorized, the definition of city borders requires a project of green areas around the compact city. Garden cities first experimented with such an approach, which was exported over numerous European cases. It is a typical projectual measure aimed to design a green corridor between dense urban built-up areas and the countryside.

Nowadays international literature clearly talks about an explosion of the urban (Brenner, 2013; Ove Arup & Partners International Limited, 2014) in which the relationship between the central city and the surrounding regional space cannot be described anymore in terms of an 'inside' and 'outside', of a centre and a periphery, at least in a traditional way. Within this perspective the metropolis is gone. In its place, there seems to be a post-metropolis, a space without limits and with extremely diversified social and spatial models subject to continuous assembly and disassembly processes, leading to a progressive loss of meaning for terms such as city, countryside, suburbs. A fractal city, extremely heterogeneous, with constantly changing centres and peripheries. Furthermore, the typical regulative approach based on greenbelts seems to be obsolete.

New environmental issues have emerged and the approach to the city is largely dominated by new spatial paradigms: for example, urban vs rural is the matter of defining the efficiency of an urban fabric pattern, for example compact vs sprawl (Antrop, 2004; Millward, 2006). The land-take concept dominates the physical approach to city development and nowadays it is impossible to allow city expansion on new greenfields with the same intensity of previous years.

Some of the aims of local plans include sustainable economic strategies for cities; the preservation and development of favourable settlement structures; mixed land uses and social integration; higher development densities and the protection of open space; strengthened inner cities and local centres; the protection of urban heritage; sustainable urban infrastructure and urban and regional transportation systems (Hale and Sadler, 2012). A common key aim is to limit the rate of urbanization of previously undeveloped land to defined parameters (quantitative or qualitative) (Couch *et al.*, 2011).

Trying to recapture its capacity to regulate space and society, sustainable urban planning is today capturing the attention of politicians and opinion leaders. The academic debate is gradually returning to old categories (for example the opposition of urban vs rural), thus misinterpreting the original social nature of these categories rather than acknowledging their spatial value. Debate on 'post-metropolis' and 'regionalization' is pointing to how physical boundaries, administrative fragmentations and spatial policies can be fitted to the urban dimension which is based on a continuity and homogeneity of spaces and landscape, based on common lifestyles and flows of long distance mobility. After a period in which a de-regulative approach has been applied (Mazza, 1997, 2004), debate is reconsidering how the physical dimension of regulation is capable of governing territorial transformations.

Crucial to the further development of a policy of regeneration and re-use is to define borders between the compact and sprawled city rather than to define a perimeter where building rights are allowed or not. Nevertheless UGB are good tools to support monitoring strategies to control local planning decisions: when sovra-local authorities use UGB as a comparative border between the existent built-up system and the one planned by the land use scenario, it is possible to easily control if, and where, land take occurs.

One of the most representative uses of UGB is that applied by the Swiss Planning Policy, where local administrations are forced to present to sovra-local authorities a regular border of the land use plan settled for the built-up system definition. Every 10–15 years the border between dense and dispersed settlement is verified and GIS monitoring is constantly applied (Gennaio *et al.*, 2009).

This approach is based on the assumption that sovra-local monitoring systems of land use change are constructed with a bottom-up approach: the municipal level has to work with a high degree of precision for constructing a land use database, in accordance with sovra-local and national inventories. In this way local accountability should fit with regional or national tools of land use monitoring.

The success of the Swiss case is supported by analysis made between 1960–1970 and 1990–2000. Positive results where demonstrated by:

- the low number of buildings erected in non-dense zones;
- the rise of the density index in urban dense areas;
- the concentration of land use changes inside the existent stock of urban areas;
- the application of 'compact development' (almost 70 per cent of new buildings were built inside the UGB).

As noted above, a single measure cannot create a real reduction in the amount of land take; rather a mix of options need to be considered. For example, some regional authorities have tried to implement UGB using land use repertories with a low degree of precision and not shared with the land use maps adopted for local planning regulations. It has to be considered that the Swiss case represents an example where land use control is facilitated by the dimension of administrative boundaries rather than the geomorphological reasons that facilitate the 'compact' development of a city.

In any case, one of the factors limiting land take control is the fact that a large majority of soil indicators for land take assessment are consistent only as descriptive tools for soil scientists, but less consistent as tools to steer local policies for preserving soil degradation due to urbanization (Geneletti, 2013).

A national agenda of environmental policies would need to be supported by aggregated data concerning the levels of urbanization: all nations engaged in the discussion of an instrument that will limit the further growth of urban areas (Germany, Netherlands, UK, etc.) are supported by national databases of land cover/use. However, a theoretical model for land use management at the local scale, specifically created for limiting land take, is still lacking where advanced policies are designed (Dale and Kline, 2013; Calzolari et al., 2016).

For these reasons, among others, it is impossible to adopt a single approach to limit land take. Thus it is still impossible to define a common methodological framework for adopting policies against land take, even based on deep knowledge of cause–effect dynamics.

A great deal of research is dedicated to estimate the environmental effects of land take processes, especially using ES as a proxy (Breure et al., 2012; Jansson, 2013; Artmann, 2014; Li et al., 2014).

Although the most common application of ES mapping is done at the macro-scale using national inventories of land use rather than European ones (Corine Land Cover), the challenge is to propose an evaluation at the micro-scale (here intended as the urban scale). Thus the potential role played by open areas of vegetation with ecological characterization is to facilitate the planning choices made during the screening phase of local plan construction. The application of new operative methods of soil classification is also useful to provide relevant information to urban planners during the decision-making process (Dale and Kline, 2013).

Nowadays, collected data on the urbanization trend (land cover classification, rate of change, urbanization per capita) is being well analysed (Benini *et al.*, 2010; Bhatta *et al.*, 2010; Pileri and Salata, 2011; Munafò, 2013) and the proposed European guidelines for land-take reduction are supported by national databases of land cover/use.

The goal of reducing land take with an integrative approach between analysis and policies of local land regulation, for example, has to better consider the role of Strategic Environmental Assessment (SEA Directive, 200142/EC). SEA is aimed at monitoring the land take phenomenon, using environmental data, and assessing impacts of land use change due to urbanization (Treville, 2011). But SEA is not sufficiently qualified to perform a complete land take assessment when used only for quantitative purposes (Geneletti, 2011).

Integrated into the urban planning discipline, the knowledge of the soil functions in urban ecosystems – assumed by a scientifically recognized model with a soil evaluation method – can orient the urban planner to make decisions and choices for the rational use of land (Rametsteiner *et al.*, 2011; Clerici *et al.*, 2014). However, future research has to be dedicated to understanding the effects of land cover changes on ES, especially using both econometric and biophysical evaluation as a proxy for sustainable urban planning.

## References

Antrop, M. (2004) 'Landscape change and the urbanization process in Europe', *Landscape and Urban Planning*, 67, 9–26.

Artmann, M. (2014) 'Institutional efficiency of urban soil sealing management: from raising awareness to better implementation of sustainable development in Germany', *Landscape and urban Planning*, 131, 83–95.

Baral, H., Keenan, R. J., Sharma, S. K., Stork, N. E., Kasel, S. (2014) 'Economic evaluation of ecosystem goods and services under different landscape management scenarios', *Land Use Policy*, 39, 54–64.

Bateman, I. J., Harwood, A. R., Mace, G. M., Watson, R. T., Abson, D. J., Andrews, B., Termansen, M. (2013) 'Bringing ecosystem services into economic decision-making: land use in the United Kingdom', *Science*, 341, 45–50.

Benini, L., Bandini, V., Marazza, D., Contin, A. (2010) 'Assessment of land use changes through an indicator-based approach: a case study from the Lamone river basin in Northern Italy', *Ecological Indicators*, 10, 4–14.

Bhatta, B., Saraswati, S., Bandyopadhyay, D. (2010) 'Urban sprawl measurement from remote sensing data', *Applied Geography*, 30(4), 731–740.

Brenner, N. (2013) *Implosions. Explosions. Towards a Study of Planetary Urbanization*, Berlin, Jovis.

Breure, A. M., De Deyn, G. B., Dominati, E., Eglin, T., Hedlund, K., Van Orshoven, J., Posthuma, L. (2012) 'Ecosystem services: a useful concept for soil policy making! Current Opinion', *Environmental Sustainability*, 4, 578–585.

Calzolari, C., Ungaro, F., Filippi, N., Guermandi, M., Malucelli, F., Marchi, N., Tarocco, P. (2016) 'A methodological framework to assess the multiple contributions of soils to ecosystem services delivery at regional scale', *Geoderma*, 261, 190–203.

Clerici, N., Paracchini, M. L., Maes, J. (2014) 'Land-cover change dynamics and insights into ecosystem services in European stream riparian zones', *Ecohydrology & Hydrobiology*, 14, 107–120.

Costanza, R., d'Arge, R., de Groot, R., Farber, S., Grasso, M., Hannon, B. (1997) 'The value of the world's ecosystem services and natural capital', *Nature*, 387, 253–260.

Couch, C., Sykes, O., Borstinghaus, W. (2011) 'Thirty years of urban regeneration in Britain, Germany and France: the importance of context and path dependency', *Progress in Planning*, 75, 1–52.

Crompton, J. L. (2007) 'The role of the proximate principle in the emergence of urban parks in the United Kingdom and in the United States', *Leisure Studies*, 2(26), 213–234.

Daily, G. (1997) 'Introduction: what are ecosystem services?' in G. Daily, *Nature's Services: Societal Dependence on Natural Ecosystems*, Washington, DC, Island Press, 1–10.

Dale, V. H., Kline, K. L. (2013) 'Issues in using landscape indicators to assess land changes', *Ecological Indicator*, 28, 91–99.

Dominati, E., Mackay, A., Green, S., Patterson, M. (2014) 'A soil change-based methodology for the quantification and valuation of ecosystem services from agro-ecosystems: a case study of pastoral agriculture in New Zealand', *Ecological Economics*, 100, 119–129.

European Environment Agency (2014) 'Spatial analysis of green infrastructure in Europe'. www.eea.europa.eu/publications/spatial-analysis-of-green-infrastructure, accessed 15 October 2015.

Faludi, A., Van der Valk, A. (1994) *Rule and Order: Dutch Planning Doctrine in the Twentieth Century*, Dordrecht, Kluwer Academic Publishers.

Gardi, C., Panagos, P., Van Liedekerke, M. (2014) 'Land take and food security: assessment of land take on the agricultural production in Europe', *Journal of Environmental Planning and Management*, 58(5), 898–912.

Geneletti, D. (2011) 'Reasons and options for integrating ecosystem services in strategic environmental assessment of spatial planning', *International Journal of Biodiversity Science, Ecosystem Services et Management*, 7(3), 143–149.

Geneletti, D. (2013) 'Assessing the impact of alternative land-use zoning policies on future ecosystem services', *Environmental Impact Assessment Review*, 30, 25–35.

Gennaio, M., Hersperger, A., Buergi, M. (2009) 'Containing urban sprawl: evaluating effectiveness of urban growth boundaries set by the Swiss Land Use Plan', *Land Use Policy*, 26(2), 224–232.

Giovannini, E., Nardo, M., Saisana, M., Saltelli, A., Tarantula, A., Hoffman, A. (2008) 'Handbook on constructing composite indicators: methodology and user guide'. www.oecd.org/std/42495745.pdf, accessed 15 October 2015.

Haase, D., Nuissl, H. (2007) 'Does urban sprawl drive changes in the water balance and policy? the case of Leipzig (Germany)', *Landscape and Urban Planning*, 1(80), 1–13.

Haberl, H., Wackernagel, M. (2004) 'Land use and sustainability indicators: an introduction', *Land Use Policy*, 21, 193–198.

Hale, J. D., Sadler, J. (2012) 'Resilient ecological solutions for urban regeneration', *Engineering Sustainability*, 165 (ES1), 59–67.

Havlin, J., Balster, N., Chapman, S., Ferris, D., Thompson, T., Smith, T. (2010) 'Trends in soil science education and employment', *Soil Science Society of America*, 74(5), 1429–1432.

Helian, L., Shilong, W., Hang, L., Xiaodong, N. (2011) 'Changes in land use and eco-system service values in Jinan, China', *Energy Procedia*, 5, 1109–1115.

Helming, J., Diehl, K., Bach, H., Dilly, O., Konig, B., Kuhlman, T., Wiggering, H. (2011) 'Ex ante impact assessment of policies affecting land use, Part A: analytical framework'. www.ecologyandsociety.org/vol16/iss1/art27, accessed 10 September 2015.

Hopmans, J. (2007) 'A plea to reform soil science education', *Soil Science Society of America*, 71, 639–640.

Jansson, A. (2013) 'Reaching for a sustainable, resilient urban future using the lens of ecosystem services', *Ecological Economics*, 86, 285–291.

Korthal Altes, W. (2009) 'Taxing land for urban containment: reflections on a Dutch debate', *Land Use Policy*, 2(26), 233–241.

Laurans, Y., Rankovic, A., Billè, R., Pirard, R., Mermet, L. (2013) 'Use of ecosystem services economic evaluation for decision making: questioning a literature blinds-pot', *Journal of Environmental Management*, 119, 208–219.

Li, F., Wang, R., Hu, D., Ye, Y., Wenrui, Y., Hongxiao, L. (2014) 'Measurement methods and applications for beneficial and detrimental effects of ecological ser-vices', *Ecological Indicators*, 47, 102–111.

McBratney, A., Field, D. J., Koch, A. (2014) 'The dimension of soil security', *Geoderma*, 213, 203–313.

Mazza, L. (1997) 'Pubblico e privato nelle pratiche urbanistiche', in L. Mazza, *Trasformazioni del piano*, Milano, Franco Angeli, 105–126.

Mazza, L. (2004) *Progettare gli squilibri*, Milano, Franco Angeli.

Millward, H. (2006) 'Urban containment strategies: a case study appraisal of plans and policies in Japanese, British and Canadian cities', *Land Use Policy*, 24, 473–485.

Munafò, M. (2013) Il consumo di suolo in Italia. *Urbanistica Informazioni*, 41(247), 19–21.

Nuissl, H., Haase, D., Lazendorf, M., Wittmer, H. (2009) 'Environmental impact assessment of urban land use transitions: a context-sensitive approach', *Land Use Policy*, 26, 414–424.

Ove Arup & Partners International Limited (2014) 'City resilience framework'. www.rockefellerfoundation.org/app/uploads/City-Resilience-Framework1.pdf, accessed 9 September 2015.

Pileri, P., Salata, S. (2011) 'L'intensità del consumo di suolo. Lombardia, Emilia Romagna, Friuli Venezia Giulia e Sardegna', in *Rapporto 2010 CRCS*, Roma, INU Edizioni.

Pimm, S. (1997) 'The value of everything', *Nature*, 387, 231–232.

Power, A. (2001) 'Social exclusion and urban sprawl: is the rescue of cities possible?', *Regional Studies*, 8(25), 731–742.

PRACSIS (2014) 'International year of soil. Scoping study for DG ENV report'. ec.europa.eu/environment/soil/pdf/IYS%202015_%20Scoping%20Study.pdf, accessed 9 September 2015.

Price, S., Dorcas, M., Gallant, A., Klaver, R., Willson, J. (2006) 'Three decades of urbanization: estimating the impact of land-cover change on stream salamander populations', *Biological Conservation*, 4(133), 436–441.

Rametsteiner, E., Pulzl, H., Alkan-Olsson, J., Frederiksen, P. (2011) 'Sustainability indicator development: science or political negotiation?', *Ecological Indicators*, 11(1), 61–70.

Ring, I. (2008) 'Integrating local ecological services into intergovernmental fiscal transfer: the case of the ecological ICMS in Brazil', *Land Use Policy*, 4(25), 485–497.

Rutgers, M., van Wijnen, H. J., Schouten, A. J., Mulder, C., Kuiten, A. M., Brussaard, L., Breure, A. M. (2011) 'A method to assess ecosystem services developed from soil attributes with stakeholders and data of four arable farms', *Science of the Total Environment*, 415, 39–48.

Shuying, Z., Changshan, W., Hang, L., Xiadong, N. (2011). Impact of urbanization on natural ecosystem service values: a comparative study. *Environmental Monitoring and Assessment* 179, 575–588.

Toman, M. (1998) 'Special section: forum on valuation of ecosystem services: why not to calculate the value of the world's ecosystem services and natural capital', *Ecological Economics*, 25(1), 57–60.

Treville, A. (2011) Strategic Environmental Assessment as a tool for limiting land consumption. Special conference on Strategic Environmental Assessment, IAIA SEA. Prague, 1–8.

Turner, B. L., Meyer, W. B., Skole, D. L. (1994) 'Global land-use/land-cover change: towards an integrated study', *Ambio*, 23(1), 91–95.

Wells, N., Ashdown, S., Davies, E., Cowett, F., Yang, Y. (2007) 'Environment, design, and obesity: opportunities for interdisciplinary collaborative research. *Environment and Behavior*, 1(39), 6–33.

Yang, X., Lo, C. P. (2002) 'Using a time series of satellite imagery to detect land use and land cover changes in the Atlanta, Georgia metropolitan area', *International Journal of Remote Sensing*, 23(9), 1775–1798.

# 19 Soil sealing and land take as global soil threat

## The policy perspective

*Luca Montanarella*

Soil sealing and land take are a global threat to food security and social stability as well as to biodiversity and ecosystem services. The exponentially increasing consumption of the most fertile soils for urbanization and infrastructure is affecting the availability of cropland for feeding the global population. Expansion of infrastructure in pristine natural areas, like in the Amazon, is rapidly affecting biodiversity and ecosystems. The general perception is that urbanization and infrastructure are a sign of economic growth and increased well-being of the population. Unfortunately the facts and figures in front of us clearly demonstrate the contrary. The quality of life of the urbanized population has been hardly increasing and the number of undernourished is not significantly decreasing.

Taking stock of these facts, at the Conference for Sustainable Development in Rio de Janeiro in 2012 (Rio+20 Conference, as 20 years had passed since the first conference in 1992) the countries of the world adopted a new document (the 'Future We Want') and asked for defining necessary sustainable development goals in order to reverse the on-going unsustainable development trend. In 2015 the proposed Sustainable Development Goals (SDGs) were defined and adopted by the UN General Assembly. Soil and land are addressed specifically in three goals (Montanarella and Alva, 2015):

1 Goal 2. End hunger, achieve food security and improved nutrition and promote sustainable agriculture.
2 Goal 3: Ensure healthy lives and promote well-being at all ages.
3 Goal 15: Protect, restore and promote sustainable use of terrestrial ecosystems, sustainably manage forests, combat desertification, and halt and reverse land degradation and halt biodiversity loss.

These goals have, then, specific targets that address soils explicitly:

*Target 2.4* By 2030, ensure sustainable food production systems and implement resilient agricultural practices that increase productivity and production, that help maintain ecosystems, that strengthen capacity for adaptation to climate change, extreme weather, drought, flooding and other disasters and that progressively *improve land and soil quality*.

*Target 3.9* By 2030, substantially reduce the number of deaths and illnesses from hazardous chemicals and air, water and *soil pollution and contamination*.

*Target 15.3* By 2020, combat desertification, *restore degraded land and soil*, including land affected by desertification, drought and floods, and strive to achieve a *land-degradation-neutral world*.

Specific indicators are still in discussion in order to consistently monitor progress towards achieving these ambitious targets by 2030.

Particular attention by policy makers as well as by scientists and other stakeholders was triggered by the new concept of land degradation neutrality (Chasek *et al.*, 2015). Originally introduced by the United Nations Convention to Combat Desertification (UNCCD), if applied consistently also in countries not affected by desertification it could eventually also reverse the growing trend towards soil sealing and land take. Assuming that soil sealing is a form of land degradation (a concept still to be accepted by most policy makers), then achieving land degradation neutrality may also imply a reduction of land take and sealing of pristine natural areas and of productive fertile land as well as an increase in recycling and re-use of already sealed and urbanized areas. Indeed recycling of brownfield and abandoned industrial areas has been advocated as one of the possible good practices for achieving a substantial reduction of the consumption of fertile agricultural land for housing and infrastructure (European Commission, 2012). Unfortunately, neither at the EU level nor at the global level has a binding legal obligation emerged for national governments to limit the dramatically increasing consumption of fertile land. The newly adopted Sustainable Development Goals and the related target for achieving a land-degradation-neutral world by 2030 may help in initiating some positive developments at the national level.

As a first step there needs to be a consistent definition of land degradation neutrality to be adopted by all countries in the world. Recent debates within the UNCCD have resulted in a still on-going controversy about the precise definition of land degradation. From a scientific point of view the most current definition refers to the loss of ecosystem services that a degraded land can deliver. Indeed the Intergovernmental Science-Policy Platform on Biodiversity and Ecosystem Services (IPBES) has defined, for the purpose of its thematic land degradation and restoration assessment, land degradation as 'the many processes that drive the decline or loss in biodiversity, ecosystem functions or services, and includes the degradation of all terrestrial ecosystems' (IPBES, 2015). This definition implies that soil sealing and land take are to be recognized as major land degradation processes. This paves the way towards establishing appropriate policy measures in order to limit this degradation and to restore already degraded areas. Compensating the continuing trend of urbanization and soil sealing with equivalent areas being restored and un-sealed would allow the achievement of the goal of a land-degradation-neutral world.

Policies addressing soil sealing and land take need to limit the expansion of urbanization on pristine agricultural land. A good start has been recently made in Italy, with the proposal, still pending in the Italian parliament, for a law that would limit soil sealing in agricultural areas (Russo, 2013). The interesting approach developed by the Italian legislator is the limitation of soil sealing of agricultural land on the basis of the previous subsidies that the owner of the agricultural area has received in the framework of the Common Agricultural Policy (CAP). The principle that if land has been receiving public support from the taxpayers for a certain type of land use, specifically for agricultural production, the land owner cannot proceed, for a number of years after the last payment of subsidies, in any changes of land use. The original proposal for the period of restriction of land use change was of 10 years after the last payment, but recent debates in the parliament have already watered down the proposal to five years or less. Still the approach of restricting land use change if public funding has been made available for certain land uses can be the basis for future legislation at the national and also the EU level. For the moment only a few countries have implemented some national policies in order to limit land take and soil sealing (Prokop, 2011). Quantitative limits for annual land take exist in six EU member states: in Austria and Germany a limit of soil sealing is defined as hectares per day for a target year; in Belgium (Flanders), Luxembourg and the Netherlands there are limits based on inner urban development, for example 60 per cent of new developments within defined urban circles; in the United Kingdom (England) limits are based on brownfield redevelopment, for example housing on already developed land. Overall, policies for limiting soil sealing are rather scarce and are usually not very effective. Land take and soil sealing is continuing in Europe and worldwide at an increasing rate.

Effective limitation of soil sealing is actually only happening in protected areas, like the NATURA 2000 sites in the EU, national protected areas and national parks. Unfortunately the recent economic crisis in Europe and the need to stimulate economic growth and job creation has put the existing EU nature protection legislation under pressure. There is still a widespread opinion that protected areas are preventing economic growth by restricting economic activities like construction of houses, infrastructure and industrial installations. A very old-fashioned model of economic development is still considered the only way forward while alternative development models are a priori not taken into consideration. Dismantling the system of protected areas in Europe is proposed as the solution for reversing the negative economic trend of the EU area. Extensive literature exists proving the contrary (Schoukens, 2015), but nevertheless the dominating ideology is not taking these options into consideration.

Decoupling economic growth from soil sealing is the only way forward for a sustainable future for urbanization and infrastructure in the world. Developing alternative city models, incorporating to a large extent green infrastructure, urban gardening and more compact city designs can substantially contribute to better living and a smaller ecological footprint. Reverting from consumption patterns implying high soil sealing rates, like large commercial areas on city outskirts linked by extensive infrastructure for transport towards a more sustainable polycentric city model with short transport distances and more distributed commercial and productive areas could as well substantially improve the ratio between economic growth and soil sealing, with less sealing per GDP unit. This goes hand in hand with a change of the agricultural production model, moving away from a highly mechanized, energy intensive agricultural model towards a more distributed and labour intensive model attracting part of the urban population back to the rural environment. Smaller farms producing high added value food products mostly for local consumption are the alternative to the current system. Creating jobs in the agricultural sector is the way forward to reverse urban expansion. A large amount of scientific literature and theoretical work has already been done proving the advantages of such an alternative system (Schell, 2011; Kuyper and Struik, 2014; Loos *et al.*, 2014; Petersen and Snapp, 2015). Terms like sustainable intensification, agroecology, organic farming etc. have been debated in many conferences and scientific seminars. Translating these scientific findings into effective policy measures is still the missing step. There is the need for an efficient science–policy interface addressing sustainable soil management. The recent establishment by the Global Soil Partnership (Montanarella and Vargas, 2012) of the Intergovernmental Technical Panel on Soils (ITPS) is a good step forward. A panel of soil scientists nominated by governments and providing policy relevant scientific advice may initiate the necessary steps for a coherent soil protection policy at the national, regional and global scale. As a first deliverable, the ITPS has already revised the World Soil Charter and the new version has been adopted by FAO members (most countries in the world) (FAO, 2015). National governments have now a legal basis for initiating the process towards national legislation for sustainable soil management. The World Soil Charter recommends to national governments to 'incorporate the principles and practices of sustainable soil management into policy guidance and legislation at all levels of government, ideally leading to the development of a national soil policy'.

Not only action at global and national levels is needed, but also awareness and action at the local level is mandatory for actual implementation of sustainable soil management guidelines. Soil sealing is a direct consequence of urbanization and therefore spatial planning authorities have to play a key role in limiting soil sealing and land take. Spatial planning is usually a strictly local competence of municipalities and local administrations. It is at that level that effective measures could be taken, if sufficient political will is exercised by the local administrators. Involving the local population in the decision-making process for spatial planning is necessary, but will yield positive effects only if associated with extensive awareness raising and education campaigns.

Cultural ecosystem services, urban soils form an integral part of urban environmental education – bridging the gap between people and nature. The incorporation of urban soils into education and outreach programmes and linking urban soils to participatory urban restoration and gardening experiences are a key way to ground urban residents in their local ecology.

Ultimately, effective policies for limiting soil sealing need to address the underlying economic model of our society. As long as there will be a close coupling between economic growth and increased soil sealing there will be little hope to reverse the negative trend.

# References

Chasek, P., Safriel, U., Shikongo, S. and Fuhrman, V. F. (2015). Operationalizing zero net land degradation: the next stage in international efforts to combat desertification? *Journal of Arid Environments, 112,* 5–13. http://doi.org/10.1016/j.jaridenv.2014.05.020, accessed 23 January 2016.

European Commission (2012). Guidelines on best practice to limit, mitigate or compensate soil sealing. Commission Staff Working Document. http://doi.org/10.2779/75498, accessed 5 December 2015.

FAO (2015). Thirty-ninth Session Rome, 6–13 June 2015 Global Soil Partnership – World Soil Charter, April. www.fao.org/3/a-mn442e.pdf, accessed 23 January 2016.

IPBES (2015). Scoping for a thematic assessment of land degradation and restoration, January.

Kuyper, T. W. and Struik, P. C. (2014). Epilogue: global food security, rhetoric, and the sustainable intensification debate. *Current Opinion in Environmental Sustainability, 8,* 71–79. http://doi.org/10.1016/j.cosust.2014.09.004, accessed 7 February 2016.

Loos, J., Abson, D. J., Chappell, M. J., Hanspach, J., Mikulcak, F., Tichit, M. and Fischer, J. (2014). Putting meaning back into 'sustainable intensification'. *Frontiers in Ecology and the Environment, 12*(6), 356–361. http://doi.org/10.1890/130157, accessed 23 January 2016.

Montanarella, L. and Alva, I. L. (2015). Putting soils on the agenda: the three Rio Conventions and the post-2015 development agenda. *Current Opinion in Environmental Sustainability, 15,* 41–48. http://doi.org/10.1016/j.cosust.2015.07.008, accessed 23 January 2016.

Montanarella, L. and Vargas, R. (2012). Global governance of soil resources as a necessary condition for sustainable development. *Current Opinion in Environmental Sustainability, 4,* 559–564.

Petersen, B. and Snapp, S. (2015). What is sustainable intensification? Views from experts. *Land Use Policy, 46,* 1–10. http://doi.org/10.1016/j.landusepol.2015.02.002, accessed 23 January 2016.

Prokop, G. (2011). *Report on Best Practices for Limiting Soil Sealing and Mitigating Its Effects.* http://doi.org/10.2779/15146, accessed 7 February 2016.

Russo, L. (2013). Il consumo di suolo agricolo all'attenzione del legislatore. *Aestimum, 63* (December), 163–174.

Schell, E. E. (2011). Framing the megarhetorics of agricultural development: industrialized agriculture and sustainable agriculture. *Project Muse, 4,* 149–173. www.scopus.com/inward/record.url?eid=2-s2.0-84907667100&partnerID=tZOtx3y1, accessed 23 January 2016.

Schoukens, H. (2015). Habitat restoration on private lands in the United States and the EU: moving from contestation to collaboration? *Utrecht Law Review, 11*(1), 33–60. http://doi.org/10.1111/ele.12387/full, accessed 7 February 2016.

# 20 Conclusions

*Ciro Gardi*

The percentage of urban population, as well as the total population of the planet, will continue to grow in the future, and this will drive a further increase in urban areas worldwide. The intensity of this process will be differentiated, reaching a maximum in Africa and Asia, and the role of urban expansion, as a process of irreversible soil degradation, will be of crucial importance. It is essential to govern these processes, in order to limit, as much as possible, the areas affected and to mitigate and compensate the impact on ecosystem services.

Food production, in the absence of a second 'green revolution', could represent a critical issue in the future. Even if globally the demand for food could be satisfied, the agricultural areas lost in some parts of the globe will determine imbalances in the food supply chain. Areas that have traditionally been self-sufficient for food production will become importers, causing an off-set of agricultural area demand (indirect land use change).

Impacts of urban expansion will also be relevant for other ecosystem services. The effects on water regulation are already evident in some part of the world. In many cases the causes of flooding events are changes in the amount, intensity and distribution of precipitation, changes in land use and physical properties of soil (i.e. soil compaction), and also the increase of sealed surfaces.

The impact of urban expansion on climate, local and global, is another relevant issue. In addition to the Urban Heat Island effect, caused by the alteration of the radiative energy budget within urban areas, we have the indirect effect of the loss of important carbon sinks, which could contribute to climate change mitigation.

Other examples of the impacts of urban expansion on the capacity of soil to deliver ecosystem services have been widely discussed in Part II of the book.

Considering that urban areas are unavoidably going to grow in the future, the challenge is to define how much they should grow and which type of urban environment is desirable. There is of course a trade-off between population density in urban areas and quality of life. Modern architecture and urban planning (with very few exceptions) has exacerbated the trade-off: dense settlements are usually unliveable, while low-density residential areas are generally more attractive. This is where the challenge is, and it should involve not only architects and urban planners, but also ecologists, agronomists, economists and social scientists.

It is important, when we are envisioning the future, to have an honest retrospective view on the excellent lessons from the past. If we consider some shining examples, like the central Italian landscape with its beautiful small cities, it appears very clear that we have to learn from the past. These examples, like many other traditional landscapes in the world, often represent an optimal use of resources, ensuring a high quality of life that doesn't neglect the social dimension. Furthermore, we should consider the fast changes that our 'liquid society' is facing. Thanks to the digital revolution, we are moving towards a decentralised society. In contraposition to globalisation, there is a growing community of people who consider regionalism an added value. The need, or the possibility, to move everything from everywhere to everywhere will hopefully decrease in the future, halting the absurd alteration to the global biogeochemical cycle that humankind is imposing on the Earth.

# Index

# Taylor & Francis eBooks

---

## Helping you to choose the right eBooks for your Library

Add Routledge titles to your library's digital collection today. Taylor and Francis ebooks contains over 50,000 titles in the Humanities, Social Sciences, Behavioural Sciences, Built Environment and Law.

**Choose from a range of subject packages or create your own!**

**Benefits for you**

- » Free MARC records
- » COUNTER-compliant usage statistics
- » Flexible purchase and pricing options
- » All titles DRM-free.

**Benefits for your user**

- » Off-site, anytime access via Athens or referring URL
- » Print or copy pages or chapters
- » Full content search
- » Bookmark, highlight and annotate text
- » Access to thousands of pages of quality research at the click of a button.

| REQUEST YOUR **FREE** INSTITUTIONAL TRIAL TODAY | **Free Trials Available**<br>We offer free trials to qualifying academic, corporate and government customers. |
|---|---|

## eCollections – Choose from over 30 subject eCollections, including:

| | |
|---|---|
| Archaeology | Language Learning |
| Architecture | Law |
| Asian Studies | Literature |
| Business & Management | Media & Communication |
| Classical Studies | Middle East Studies |
| Construction | Music |
| Creative & Media Arts | Philosophy |
| Criminology & Criminal Justice | Planning |
| Economics | Politics |
| Education | Psychology & Mental Health |
| Energy | Religion |
| Engineering | Security |
| English Language & Linguistics | Social Work |
| Environment & Sustainability | Sociology |
| Geography | Sport |
| Health Studies | Theatre & Performance |
| History | Tourism, Hospitality & Events |

For more information, pricing enquiries or to order a free trial, please contact your local sales team:
www.tandfebooks.com/page/sales

**Routledge**
Taylor & Francis Group

The home of
Routledge books

**www.tandfebooks.com**

T - #0118 - 111024 - C332 - 234/156/15 - PB - 9780367172794 - Gloss Lamination